Satellite Rainfall Applications for Surface Hydrology

Mekonnen Gebremichael · Faisal Hossain
Editors

Satellite Rainfall Applications for Surface Hydrology

Editors
Dr. Mekonnen Gebremichael
University of Connecticut
Civil & Environmental Engineering Dept.
261 Glenbrook Road
Storrs CT 06269-2037
USA
mekonnen@engr.uconn.edu

Dr. Faisal Hossain
Tennessee Technological University
Civil & Environmental Engineering Dept.
Cookeville TN 38505-0001
USA
fhossain@tntech.edu

ISBN 978-90-481-2914-0   e-ISBN 978-90-481-2915-7
DOI 10.1007/978-90-481-2915-7
Springer Dordrecht Heidelberg London New York

Library of Congress Control Number: 2009937296

© Springer Science+Business Media B.V. 2010
No part of this work may be reproduced, stored in a retrieval system, or transmitted in any form or by any means, electronic, mechanical, photocopying, microfilming, recording or otherwise, without written permission from the Publisher, with the exception of any material supplied specifically for the purpose of being entered and executed on a computer system, for exclusive use by the purchaser of the work.

Cover: Background image is a satellite photo of the Ganges River Delta.

Printed on acid-free paper

Springer is part of Springer Science+Business Media (www.springer.com)

Preface

As individual topics, the terms "satellite rainfall" and "surface hydrology" have been much widely studied over the last few decades. Ever since rainfall products begun to be developed using space-borne infrared sensors in geostationary orbit in the seventies, satellite remote sensing of rainfall experienced tremendous progress. Microwave sensors on low earth orbits came along during the eighties to provide more accurate estimates of rainfall at the cost of limited sampling. As the contrasting but complementary properties of microwave and infrared sensors became apparent, merged rainfall products started to appear during the following decade. In 1997, the Tropical Rainfall Measuring Mission (TRMM) with the first space-borne active microwave precipitation radar (TRMM-PR), was launched. The success of TRMM in improving our understanding on Tropical and Sub-tropical rainfall distribution and precipitation structures consequently spurred a larger scale mission aimed at the study of global distribution of precipitation. Today, we now eagerly anticipate the Global Precipitation Measurement (GPM) mission, which envisions a global constellation of microwave sensors that will provide more accurate global rainfall products at high resolution from 2013 onwards.

It is therefore safe to claim three decades of research heritage on satellite remote sensing of rainfall. Similarly, the topic of "surface hydrology" requires no introduction for readers of environmental sciences and geosciences either. But what happens if we connect all the individual terms and name it – "satellite rainfall applications for surface hydrology"? A new topic is created. But little is known about this topic because satellite remote sensing of rainfall and surface hydrology have evolved rather independently of each other. Even though the potential for a space-borne source of rainfall data was always recognized for a variety of applications (such as flood forecasting in ungauged regions, transboundary water resources, global/regional drought and agricultural planning), the fields of satellite rainfall and surface hydrology have hardly intersected during their developmental stages during the last few decades. We are now faced with a myriad of questions ranging from common operational issues to detailed scientific inquiries. Some of these questions are: *There are so many satellite rainfall products currently available - which one does one use for a specific application to get the best results? What is the optimum scale of application of satellite rainfall data for a given surface application? What is the level of uncertainty in each satellite rainfall product and what is the implication for a given surface hydrologic prediction? Where do I acquire the data for research or for operational applications? How are these satellite rainfall products developed and how do they differ from one another?*

This book by Springer on "Satellite Rainfall Applications for Surface Hydrology" is a contribution to both scientific and practical questions regarding the use of satellite rainfall products in surface hydrology. Focusing on the evolution of the algorithm of the satellite precipitation products, the accuracy assessment of the products in different regions of the world, and application of the precipitation products for various Decision Support Systems, the book provides very useful and most up to date information to practitioners, researchers and graduate students who need to explore the latest on satellite precipitation products for various hydrological purposes.

The book is organized into three parts: (1) evolution of high resolution precipitation products, (2) evaluation of high resolution precipitation products, and (3) real time operations for decision support systems. Part one contains seven chapters that dwell on high resolution satellite rainfall products by presenting detailed overview of the algorithm behind seven of the major products available today that a hydrologist user would be interested in knowing. These products provide rainfall estimates at various scales of hydrologic interest (from 4 to 25 km, sub-daily to sub-hourly) using a myriad of estimation techniques. The purpose of this part is to enlighten the reader on the essential estimation features of each rainfall product. Hence, most of these chapters are outlined according to a common format, although deviations may exist from chapter to chapter. Each of these chapters outline the sensors and input datasets and the methodology used for calibration. These chapters also provide examples and comparisons across the globe and future directions in the algorithm development.

In Part two, seven chapters dwell on the evaluation of satellite rainfall products. Among the seven, two chapters deal with verification methodology and error structure. Three chapters address uncertainty of satellite rainfall products in various regions of the world through comparison with ground-based rainfall measurements. Two chapters characterize the error propagation of satellite rainfall products in hydrological applications. Finally, Part three provides examples of real-world applications of satellite rainfall products in operational hydrology and real-time decision support system. The aim of this part is to enable readers to understand the potential of real-time satellite rainfall products for societal applications ranging from agricultural/crop monitoring to flood and landslide detection in developing countries.

This Springer book is not meant to be confused with two other books on similar topics that have appeared in recent times – "Measuring Precipitation from Space" (editors Vincenzo Levizzani et al.) and "Precipitation: Advances in Measurement, Estimation and Prediction" (editor Silas Michaelides). Readers will find these two books a far more comprehensive and voluminous source on the topic of rainfall. Our Springer book focuses at the interface between satellite rainfall and surface hydrologic applications. It dwells on issues that are of concern at hydrologic scales of application. It is our hope that this book will therefore be of interest to graduate students and researchers who actively deal with satellite rainfall products for gaining a better understanding and prediction of surface hydrologic phenomenon.

As we began to work on compiling the book more than a year ago, we realized how difficult and impossible the task would be without the timely cooperation from our contributing authors, numerous reviewers and the dedicated Springer staff. We gratefully acknowledge their participation in this book effort. We would also like to acknowledge the NASA New Investigator Program and the Program for Evaluation of High Resolution Precipitation Products (PEHRPP) under the auspices of the International Precipitation Working Group (IPWG) of the World Meteorological Organization. These two programs provided us with a common platform and a mandate to bring the two communities spanning satellite rainfall estimation and surface hydrologic applications together. It is therefore a privilege for us that we finally managed to complete the book and experience it becoming available for the readers!

Storrs, Connecticut Mekonnen Gebremichael
Cookeville, Tennessee Faisal Hossain
April, 2009

Contents

Part I Evolution of High Resolution Precipitation Products

The TRMM Multi-Satellite Precipitation Analysis (TMPA)   3
George J. Huffman, Robert F. Adler, David T. Bolvin, and
Eric J. Nelkin

**CMORPH: A "Morphing" Approach for High Resolution
Precipitation Product Generation** 23
Robert J. Joyce, Pingping Xie, Yelena Yarosh, John E. Janowiak
and Phillip A. Arkin

**The Self-Calibrating Multivariate Precipitation Retrieval
(SCaMPR) for High-Resolution, Low-Latency Satellite-Based
Rainfall Estimates** 39
Robert J. Kuligowski

**Extreme Precipitation Estimation Using Satellite-Based
PERSIANN-CCS Algorithm** 49
Kuo-Lin Hsu, Ali Behrangi, Bisher Imam, and Soroosh Sorooshian

The Combined Passive Microwave-Infrared (PMIR) Algorithm 69
Chris Kidd and Catherine Muller

**The NRL-Blend High Resolution Precipitation Product and its
Application to Land Surface Hydrology** 85
Joseph T. Turk, Georgy V. Mostovoy, and Valentine Anantharaj

**Kalman Filtering Applications for Global Satellite Mapping of
Precipitation (GSMaP)** 105
Tomoo Ushio and Misako Kachi

Part II Evaluation of High Resolution Precipitation Products

Neighborhood Verification of High Resolution Precipitation Products . 127
Elizabeth E. Ebert

A Practical Guide to a Space-Time Stochastic Error Model for Simulation of High Resolution Satellite Rainfall Data 145
Faisal Hossain, Ling Tang, Emmanouil N. Anagnostou, and Efthymios I. Nikolopoulos

Regional Evaluation Through Independent Precipitation Measurements: USA 169
Mathew R.P. Sapiano, John E. Janowiak, Wei Shi, R. Wayne Higgins, and Viviane B.S. Silva

Comparison of CMORPH and TRMM-3B42 over Mountainous Regions of Africa and South America 193
Tufa Dinku, Stephen J. Connor, and Pietro Ceccato

Evaluation Through Independent Measurements: Complex Terrain and Humid Tropical Region in Ethiopia 205
Menberu M. Bitew and Mekonnen Gebremichael

Error Propagation of Satellite-Rainfall in Flood Prediction Applications over Complex Terrain: A Case Study in Northeastern Italy 215
Efthymios I. Nikolopoulos, Emmanouil N. Anagnostou, and Faisal Hossain

Probabilistic Assessment of the Satellite Rainfall Retrieval Error Translation to Hydrologic Response 229
Hamid Moradkhani and Tadesse T. Meskele

Part III Real Time Operations for Decision Support Systems

Applications of TRMM-Based Multi-Satellite Precipitation Estimation for Global Runoff Prediction: Prototyping a Global Flood Modeling System 245
Yang Hong, Robert F. Adler, George J. Huffman and Harold Pierce

Real-Time Hydrology Operations at USDA for Monitoring Global Soil Moisture and Auditing National Crop Yield Estimates ... 267
Curt A. Reynolds

Real-Time Decision Support Systems: The Famine Early Warning System Network 295
Chris Funk and James P. Verdin

Index 321

Contributors

Robert F. Adler Laboratory for Atmospheres, NASA/GSFC, Greenbelt, MD, USA; Earth System Science Interdisciplinary Center, University of Maryland, College Park, MD, USA, Robert.f.adler@nasa.gov

Emmanouil N. Anagnostou Department of Civil and Environmental Engineering, University of Connecticut, Storrs, CT 06269, USA, manos@engr.uconn.edu

Valentine Anantharaj GeoResources Institute, Mississippi State University, Starkville, MS 39762, USA, val@ERC.MsState.Edu

Phillip A. Arkin Cooperative Institute for Climate Studies, University of Maryland, College Park, MD, USA, parkin@essic.umd.edu

Ali Behrangi The Henry Samueli School of Engineering, Center for Hydrometeorology and Remote Sensing, UC Irvine, CA, USA, abehrangi@uci.edu

Menberu M. Bitew Civil and Environmental Engineering Department, University of Connecticut, Storrs, CT, USA, Menberu@engr.uconn.edu

David T. Bolvin Laboratory for Atmospheres, NASA/GSFC, Greenbelt, MD, USA; Science Systems and Applications, Inc., Lanham, MD, USA, david.t.bolvin@nasa.gov

Pietro Ceccato International Research Institute for Climate and Society (IRI), The Earth Institute at Columbia University, Palisades, NY, USA, pceccato@iri.columbia.edu

Stephen J. Connor International Research Institute for Climate and Society (IRI), The Earth Institute at Columbia University, Palisades, NY, USA, sjconnor@iri.columbia.edu

Tufa Dinku International Research Institute for Climate and Society (IRI), The Earth Institute at Columbia University, Palisades, NY 10964, USA, tufa@iri.columbia.edu

Elizabeth E. Ebert Centre for Australian Weather and Climate Research, Melbourne, Australia, E.ebert@bom.gov.au

Chris Funk Research Geographer, USGS EROS/UCSB Geography, Santa Barbara, CA 93106, USA, funk.cc@gmail.com

Mekonnen Gebremichael Civil and Environmental Engineering Department, University of Connecticut, Storrs, CT, USA, mekonnen@engr.uconn.edu

R. Wayne Higgins Climate Prediction Center/NOAA/NWS/NCEP, 5200 Auth Road, Camp Springs, MD, USA, wayne.higgins@noaa.gov

Yang Hong School of Civil Engineering and Environmental Sciences, University of Oklahoma, Norman, OK 73019, USA, yanghong@ou.edu

Faisal Hossain Civil and Environmental Engineering Department, Tennessee Technological University, Cookeville, TN 38505-0001, USA, fhossain@tntech.edu

Kuo-Lin Hsu Civil and Environmental Engineering, Center for Hydrometeorology and Remote Sensing (CHRS), UC IRvine, Irvine CA 92697-2175, USA, kuolinh@uci.edu

George J. Huffman Laboratory for Atmospheres, NASA/GSFC, Greenbelt, MD, USA; Science Systems and Applications, Inc., Lanham, MD 20771, USA, george.j.huffman@nasa.gov

Bisher Imam The Henry Samueli School of Engineering, Center for Hydrometeorology and Remote Sensing, UC Irvine, CA, USA, bimam@uci.edu

John E. Janowiak Cooperative Institute for Climate Studies, University of Maryland, College Park, MD, USA, jjanowia@umd.edu

Robert. J. Joyce Wyle Information Systems, McLean, VA, USA; Climate Prediction Center/NCP/NWS/NOAA, Camp Springs, MD, USA, Robert.joyce@noaa.gov

Misako Kachi Earth Observation Research Center (EORC), Japan Aerospace Exploration Agency (JAXA), 2-1-1 Sengen, Tsukuba, Ibaraki 305-8505, Japan, kachi.misako@jaxa.jp

Chris Kidd School of Geography, Earth and Environmental Sciences, University of Birmingham, Edgbaston, Birmingham, UK, c.kidd@bham.ac.uk

Robert J. Kuligowski NOAA/NESDIS Center for Satellite Applications and Research, 5200 Auth Road, Camp Springs, MD 20746-4304 USA, Bob.Kuligowski@noaa.gov

Tadesse T. Meskele Department of Civil and Environmental Engineering, Portland State University, Portland, OR 97201, USA, tadesse@pdx.edu

Hamid Moradkhani Department of Civil and Environmental Engineering, Portland State University, Portland, OR 97201, USA, hamidm@cecs.pdx.edu

Georgy V. Mostovoy GeoResources Institute, Mississippi State University, Starkville, MS 39762, USA, mostovoi@gri.msstate.edu

Contributors

Catherine Muller School of Geography, Earth and Environmental Sciences, University of Birmingham, Edgbaston, Birmingham, UK, clm64@leicester.ac.uk

Eric J. Nelkin Laboratory for Atmospheres, NASA/GSFC, Greenbelt, MD, USA; Science Systems and Applications, Inc., Lanham, MD, USA, eric.j.nelkin@nasa.gov

Efthymios I. Nikolopoulos Department of Civil and Environmental Engineering, University of Connecticut, Storrs, CT 06269, USA, ein06002@engr.uconn.edu

Harold Pierce Science Systems and Applications, Inc., Lanham, MD 20706, USA; NASA Goddard Space Flight Center, Greenbelt, MD 20771, USA, harold.f.pierce@nasa.gov

Curt A. Reynolds USDA's Foreign Agricultural Service (FAS), Office of Global Analysis (OGA), International Production Assessment Division (IPAD), Washington, DC 20250, USA curt.reynolds@fas.usda.gov

Mathew R.P. Sapiano Cooperative Institute of Climate Studies (CICS), Rm 3004, Earth System Science Interdisciplinary Center (ESSIC), University of Maryland, College Park, MD, 20742, USA, msapiano@essic.umd.edu

Wei Shi Climate Prediction Center/NOAA/NWS/NCEP, 5200 Auth Road, Camp Springs, MD, USA, Wei.Shi@noaa.gov

Viviane B.S. Silva Climate Prediction Center/NOAA/NWS/NCEP, 5200 Auth Road, Camp Springs, MD, USA, viviane.silva@noaa.gov

Soroosh Sorooshian The Henry Samueli School of Engineering, Center for Hydrometeorology and Remote Sensing, UC Irvine, CA, USA, soroosh@uci.edu

Ling Tang Department of Civil and Environmental Engineering, Tennessee Technological University, Cookeville, TN 38505-0001, USA, ltang21@tntech.edu

Joseph T. Turk Naval Research Laboratory, Marine Meteorology Division, Monterey, CA 93943, USA; Radar Science and Engineering/334, Jet Propulsion Laboratory, Pasadena, CA 91109-8099, USA, jturk@jpl.nasa.gov

Tomoo Ushio Osaka University, 2-1 Yamadaoka, Suita, Osaka 565-0871, Japan, ushio@comm.eng.osaka-u.ac.jp

James P. Verdin Research Geographer, USGS EROS/UCSB Geography, Santa Barbara, CA 93106, USA, verdin@usgs.gov

Pingping Xie Climate Prediction Center/NCP/NWS/NOAA, Camp Springs, MD, USA, pingping.xie@noaa.gov

Yelena Yarosh Wyle Information Systems, McLean, VA, USA; Climate Prediction Center/NCP/NWS/NOAA, Camp Springs, MD, USA, yelena.yarosh@noaa.gov

Part I
Evolution of High Resolution Precipitation Products

The TRMM Multi-Satellite Precipitation Analysis (TMPA)

George J. Huffman, Robert F. Adler, David T. Bolvin, and Eric J. Nelkin

Abstract The Tropical Rainfall Measuring Mission (TRMM) Multi-satellite Precipitation Analysis (TMPA) is intended to provide a "best" estimate of quasi-global precipitation from the wide variety of modern satellite-borne precipitation-related sensors. Estimates are provided at relatively fine scales (0.25° × 0.25°, 3-h) in both real and post-real time to accommodate a wide range of researchers. However, the errors inherent in the finest scale estimates are large. The most successful use of the TMPA data is when the analysis takes advantage of the fine-scale data to create time/space averages appropriate to the user's application. We review the conceptual basis for the TMPA, summarize the processing sequence, and focus on two new activities. First, a recent upgrade for the real-time version incorporates several additional satellite data sources and employs monthly climatological adjustments to approximate the bias characteristics of the research quality post-real-time product. Second, an upgrade for the research quality post-real-time TMPA from Versions 6 to 7 (in beta test at press time) is designed to provide a variety of improvements that increase the list of input data sets and correct several issues. Future enhancements for the TMPA will include improved error estimation, extension to higher latitudes, and a shift to a Lagrangian time interpolation scheme.

Keywords Precipitation · Satellite · Remote sensing · TRMM · GPM

1 Introduction

As elaborated elsewhere in this book, precipitation is a critical weather element for determining the habitability of different parts of the Earth, yet is difficult to measure adequately with surface-based instruments due to its small-scale variability in

G.J. Huffman (✉)
Laboratory for Atmospheres, NASA/GSFC, Code 613.1, Greenbelt, MD 20771, USA
e-mail: george.j.huffman@nasa.gov

space and time. Thus, satellite-borne sensors play a key role in estimating precipitation. The proliferation of precipitation-sensing satellites in the last 20 years (Fig. 1) has tremendously enhanced our ability to estimate precipitation over much of the globe, but the critical piece of the puzzle is deciding how to combine all of these individual estimates to form a single, best estimate. The desired result is a stable, long, quasi-global time series of precipitation estimates on a uniform time/space grid that has the finest scale that the data will reasonably support. Several factors work against these attributes. Starting at the finest granularity, each sensor, and associated algorithms, has strengths and weaknesses that can affect its accuracy, usually varying by region. A second factor is that the equatorial crossing times of the various low-Earth-orbit (leo) precipitation-sensing satellites is uncoordinated, although operational agencies typically strive to maintain one or two specific satellites in preferred orbits. This dependence on satellites of opportunity introduces larger gaps in temporal sampling than would be the case for a coordinated constellation. As well, many of the satellites drift (Fig. 1), giving interannual changes in the gaps, even for the same complement of satellites. Finally, the number and types of satellites change over time, implying that the input data cannot be considered homogeneous. Accordingly, schemes that seek to combine all of these inputs into a "best" dataset must be designed around, and examined for, these issues.

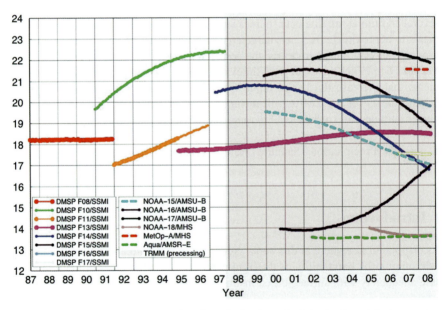

Fig. 1 Time history of Equator-crossing times (in local standard time) of precipitation-sensing microwave satellites/sensors through September 2008. All are ascending node, except for DMSP F08 and MetOp-A. The *thickest lines* denote the satellite used as a calibrator in the GPCP datasets. TRMM is denoted by shading because it precesses, covering all overpass times in the course of a 43-day period

The Tropical Rainfall Measuring Mission (TRMM) Multi-satellite Precipitation Analysis (TMPA) was designed with a heritage that includes the Adjusted Geosynchronous Operational Environmental Satellite (GOES) Precipitation Index (AGPI; Adler et al. 1994), the Global Precipitation Climatology Project (GPCP) monthly satellite-gauge (SG) combination (Huffman et al. 1997; Adler et al. 2003), and the GPCP One-Degree Daily (Huffman et al. 2001) combination estimates of precipitation. In common with these predecessor data sets, we identify a specific high-quality data set as the calibrator, and then work to make the remaining input data as consistent as possible with the calibrator data. In contrast to the predecessor data sets, the TMPA is designed to use "all" available data, meaning that we are accepting the potential inhomogeneities of a time-varying complement of inputs in return for potentially better combination results when more high-quality data are available. Table 1 summarizes key features of the TMPA data sets, including the input data sources. Another difference with the earlier data sets is that the TMPA is generated twice, first as a real-time (RT) product computed about 6–9 h after observation time, and then as a post-real-time research product computed about 15 days after the end of the month with additional data, including monthly surface rain gauge data.

The spatial resolution was chosen as 0.25° × 0.25° latitude/longitude to ensure that the grid box is somewhat larger than the typical footprint size for passive

Table 1 Summary of TMPA dataset characteristics. All inputs except the TRMM sensors (TMI and PR) and AMSR-E are from multiple satellites. The TMPA generally uses a subset of each sensor's period of record due to various procedural limitations

	Real-time product	Research product
Input sensor-algorithm datasets		TMI-and-PR–TCI[1]
	TMI–GPROF[1]	TMI–GPROF
	AMSR-E–GPROF	AMSR-E–GPROF
	SSM/I–GPROF	SSM/I–GPROF
	AMSU–NESDIS	AMSU–NESDIS
	MHS–NESDIS	
	geo-IR–VAR	geo-IR–VAR
		leo-IR–VAR
		Monthly raingauge–GPCC[2]
	Monthly climatological TMI-TCI quantile-quantile and TCI-3B43V.6 ratio calibrations[2]	
Spatial scale, extent	0.25°×0.25°, 50°N–50°S	0.25°×0.25°, 50°N–50°S
Temporal scale, extent	3-h, 1 October 2008-present[3]	3-h, 1 January 1998-present

[1] Microwave calibration standard.
[2] Final dataset calibrator(s).
[3] Estimates lacking the climatological calibration start 7 February 2005.

microwave (hereafter "microwave") precipitation estimates, which are the coarsest estimates in common use. The spatial domain was set to 50°N–50°S because all of the microwave and infrared (IR) estimates we are using tend to lose skill at higher latitudes. The temporal resolution was chosen as 3 h because (1) it allows us to resolve the diurnal cycle, (2) it matches the mandated interval for full-disk images from the international constellation of geosynchronous (geo) satellites, and (3) it provides a reasonable compromise between spatial coverage and temporal frequency for gridding the asynoptic microwave estimates from leo satellites. The time spans covered by the TMPA data sets are currently determined by the start of TRMM for the research product and the start of recent-version processing for the RT product, respectively.

The following sections briefly address the instruments and input datasets that are used in the TMPA (Section 1.2), the methodology used to combine them (Section 1.3), and their current status (Section 1.4). Then we display some comparisons and examples (Section 1.5) and end by discussing future plans (Section 1.6).

2 Instruments and Input Datasets

The TMPA depends on input from two different types of satellite sensors, namely microwave and IR. First, precipitation-related microwave data are being collected by a variety of leo satellites (Fig. 1), including the TRMM Microwave Imager (TMI) on TRMM, Special Sensor Microwave/Imager (SSM/I) and Special Sensor Microwave Imager/Sounder (SSMIS) on Defense Meteorological Satellite Program (DMSP) satellites, Advanced Microwave Scanning Radiometer for the Earth Observing System (AMSR-E) on Aqua, the Advanced Microwave Sounding Unit (AMSU) on the National Oceanic and Atmospheric Administration (NOAA) satellite series, and the Microwave Humidity Sounders (MHS) on later NOAA-series satellites and the European Operational Meteorological (MetOp) satellite. All of these data have a direct physical connection to the hydrometeor profiles above the surface, but each individual satellite provides a very sparse sampling of the time-space occurrence of precipitation. Even when composited into 3-h datasets, the current "full" microwave coverage averages about 80% of the Earth's surface in the latitude band 50°N–S and amounted to about 40% at the beginning of the TMPA record in 1998 with three satellites. Not all of the data shown in Fig. 1 can be used in the TMPA. For example, a signal contamination problem on the F15 DMSP that began in August 2006 suspended its use, while various new sensors are in the process of being incorporated into the products, including the SSMIS (DMSP F16 and F17) and MHS (NOAA 18 and MetOp).

Each pixel-level microwave observation from TMI, AMSR-E, SSM/I, and SSMIS is converted to a precipitation estimate with sensor-specific versions of the Goddard Profiling Algorithm (GPROF; Kummerow et al. 1996, Olson et al. 1999) for subsequent use in the TMPA. This takes place at the Precipitation Measurement

Missions' (PMM) Precipitation Processing System (PPS), formerly known as the TRMM Science Data and Information System (TSDIS). GPROF is a physically-based algorithm that applies a Bayesian least-squares fit scheme to reconstruct the observed radiances for each pixel by selecting the "best" combination of thousands of pre-computed microwave channel upwelling radiances based on TRMM precipitation radar (PR) data. As part of the processing the microwave data are screened for contamination by surface effects.

Pixel-level microwave radiances from AMSU-B and MHS are converted to precipitation estimates at the National Environmental Satellite Data and Information Service (NESDIS) using operational versions of the Zhao and Weng (2002) and Weng et al. (2003) algorithm. Ice Water Path (IWP) is computed from the 89- and 150-Ghz channels, with a surface screening that employs ancillary data. Precipitation rate is then computed based on the IWP and precipitation rate relations derived from cloud model data based on the NCAR/PSU Mesoscale Model Version 5 (MM5).

The AMSU-B algorithm detects solid hydrometeors, but not liquid. The multi-channel conical-scan passive microwave sensors (TMI, AMSR, SSM/I) similarly sense only solid hydrometeors over land, so the AMSU-B estimates are roughly comparable for land areas. However, over ocean the conical scanners also sense liquid hydrometeors, providing additional sensitivity, including to warm rain contributions from clouds that largely or totally lack the ice phase. As a result, the AMSU-B estimates over ocean are relatively less capable in detecting precipitation over ocean. An upgrade in 2007 added an emission component to increase the areal coverage of rainfall over oceans through the use of a liquid water estimation using AMSU-A 23.8 and 31 GHz (Vila et al. 2007). Additionally, an improved coastline rainrate module was added that computes a proxy IWP using the 183 GHz bands (Kongoli et al. 2007). (Despite the over-land focus of this book, some background on "coast" and "ocean" will be given for completeness.)

The second major data source for the TMPA is the geo-IR data, which provide excellent time-space coverage, in contrast to the microwave data. However, all IR-based precipitation estimates share the limitation that the IR brightness temperatures (T_b) primarily represent cloud-top temperature, and implicitly cloud-top height. Arkin and Meisner (1987) showed that IR estimates are poorly correlated to precipitation at fine time/space scales, but relatively well-correlated at scales larger than about 1 day and 2.5° × 2.5° of lat./long. The Climate Prediction Center (CPC) of the National Weather Service/NOAA merges geo-IR data from the five main international geo satellites into half-hourly 4 × 4-km-equivalent lat./long. grids for the domain 60°N–60°S (hereafter the "CPC merged IR"; Janowiak et al. 2001). This dataset contains IR T_b's corrected for zenith-angle viewing effects and inter-satellite calibration differences. At present, the research TMPA estimates generated prior to the start of the CPC merged IR data set in early 2000 are computed using a GPCP data set (also produced at CPC) that contains 24-class histograms of geo-IR T_b data on a 3-h, 1° × 1° lat./long. grid covering the latitude band 40°N–S (hereafter the "GPCP IR histograms"; Huffman et al. 2001). This data set also includes GOES

Precipitation Index (GPI; Arkin and Meisner 1987) estimates computed from leo-IR data recorded by the NOAA satellite series, averaged to the 1° × 1° grid. The TMPA fills gaps in the geo-IR coverage with these data, most notably before June 1998 in the Indian Ocean sector.

Finally, the research TMPA employs three additional data sources: the TRMM Combined Instrument (TCI) estimate, which combines data from both TMI and the PR (TRMM product 2B31; Haddad et al. 1997a, b); the Global Precipitation Climatology Centre (GPCC) monthly rain gauge analysis (Rudolf 1993); and the Climate Assessment and Monitoring System (CAMS) monthly rain gauge analysis developed by CPC (Xie and Arkin 1996).

3 General Methodology

The research-quality TMPA is computed in four stages; (1) the microwave precipitation estimates are inter-calibrated and combined, (2) IR precipitation estimates are created using the calibrated microwave precipitation, (3) the microwave and IR estimates are combined, and (4) rain gauge data are integrated. The real-time TMPA lacks the fourth step and has a few simplifications, as outlined in Section 1.3.3. Each TMPA precipitation field is expressed as the precipitation rate effective at the nominal observation time because most gridboxes contain data from one snapshot of satellite data. Figure 2 is a high-level summary of the following sections.

3.1 Combined Microwave Estimates

Each microwave precipitation data set is averaged to the 0.25° spatial grid over the time range ±90 min from the nominal 3-h observation times (00Z, 03Z, ..., 21Z). Probability matching to a "best" estimate using coincident matchups is used to adjust each sensor with a quantile-quantile relationship, similar to Miller (1972) and Krajewski and Smith (1991). Although we wish to adopt the TCI as the calibrating data source, the coincidence of TCI with any of the sensors other than TMI is sparse, so we establish a TCI–TMI calibration, then apply that to TMI. The remaining sensor data are all calibrated to TMI, and then adjusted to the TCI using the TCI–TMI calibration. The TCI–TMI relationship is computed on a 1° × 1° grid for each month with the coincident data aggregated on overlapping 3° × 3° windows. The TCI–TMI calibration interval for the research product is a calendar month, and the resulting adjustments are applied to data for the same calendar month. This choice is intended to keep the dependent and independent data sets for the calibrations as close as possible in time.

Preliminary work showed that the TMI calibrations of the other sensors' estimates are adequately represented by climatologically-based coefficients representing large areas. In the case of the TMI–SSM/I calibration, separate calibrations are used for five oceanic latitude bands (40–30°N, 30–10°N, 10°N–10°S, 10–30°S,

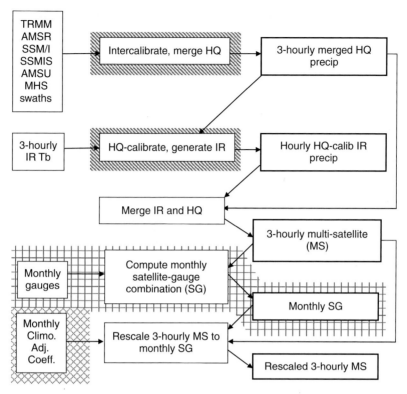

Fig. 2 Block diagram for the TMPA. Slanted hatched background indicates calibration steps that are different in the RT and research products. Square- and cross-hatched backgrounds indicate final RT and research calibration steps, respectively

30–40°S) and a single land-area calibration for each of four three-month seasons. The TMI–AMSR-E and TMI–AMSU/MHS calibrations are given one climatological adjustment for land and another for ocean. The AMSU/MHS calibration has two additional issues. First, the NESDIS algorithm changed on 31 July 2003 and 31 May 2007, so separate sets of calibrations are provided for the data periods. Second, in all periods the AMSU/MHS fractional occurrence of precipitation in the subtropical highs is notably deficient. After extensive preliminary testing, the authors judged it best to develop the ocean calibration as a single region, recognizing that the resulting fields would have a somewhat low bias. Huffman et al. (2007) show that the low bias is somewhat larger than expected, but this does not affect the over-land hydrology applications on which this book focuses. In all cases the calibrations in the 40°–50° latitude belts in both hemispheres are simply taken to be the calibrations that apply just Equator-ward of 40°.

Once the microwave estimates are calibrated for each satellite and quality-controlled, each gridbox is filled using the "best" available data to produce the

High Quality (HQ) microwave combination field. The TCI alone is used, if available. If not, when there are one or more overpasses available from TCI-adjusted TMI, AMSR-E, SSM/I, and SSMIS in the 3-h window for a given grid box, all of these data are used (averaging as necessary). The histogram of precipitation rate is somewhat sensitive to the number of overpasses averaged together, so it would be more consistent to take the single "best" overpass in the data window period. Finally, the TCI-adjusted AMSU/MHS estimates are only used if none of the other microwave estimates are available for the grid box, due to the detectability deficiency in the AMSU/MHS estimates over ocean discussed above. Detectability is equally problematic over land for AMSU/MHS and conical-scan sensors, so this rule is unnecessarily restrictive, but likely not a serious problem.

3.2 Microwave-Calibrated IR Estimates

The IR data are not provided at the 0.25° resolution, so some pre-processing is required. In the early period of the research product (1 January 1998 to 7 February 2000), each grid box's histogram in the 1°×1° 3-h GPCP IR histogram dataset is zenith-angle corrected, averaged to a single T_b value for the grid box, and plane-fit interpolated to the 0.25° grid. For the period from 7 February 2000 onwards, the CPC Merged IR is averaged to 0.25° resolution and combined into hourly files as ±30 min from the nominal time. Time-space matched HQ precipitation rates and IR T_b's are accumulated for a month into histograms of 3-h 0.25°×0.25° values on a 1°×1° grid, aggregated to overlapping 3°×3° windows, and then used to convert IR T_b's to precipitation rates. As in the TCI–TMI calibration for the HQ, the calibration period is the calendar month. Quality control is again applied to the HQ here to control artifacts.

The IR precipitation estimate is a simple "colder clouds precipitate more" approach, with the coldest 0.25°×0.25°-average T_b assigned the greatest observed HQ precipitation rate, and so on, with zero precipitation assigned for all T_b's warmer than a spatially varying threshold value determined by the fractional coverage of precipitation in the microwave data. We refer to this approach as the variable rainrate (VAR) algorithm. Calibration coefficients in grid boxes that lack coincident data throughout the month, usually due to cold-land dropouts or quality control, are computed using smooth-filled histograms of coincident data from surrounding grid boxes. Strict probability matching tends to show unphysical fluctuations at the highest precipitation rates, so we somewhat subjectively choose to replace the coldest 0.17% of the T_b histogram by a fourth-order polynomial fit to a climatology of coldest-0.17%–precipitation rate points around the globe. In each grid box a constant is added to the climatological curve to make it piecewise continuous with the grid box's T_b-precipitation rate curve at the 0.17% T_b. The HQ–IR calibration coefficients computed for a month are applied to each 3-h IR data set during that month.

3.3 Merged Microwave and IR Estimates

It is somewhat challenging to combine the HQ and IR precipitation estimates at individual times because the quantities being sensed tend to have different fine-scale patterns. Accordingly, we simply use the more physically-based HQ estimates "as is" where available, and fill the remaining grid boxes with HQ-calibrated IR estimates. This scheme provides the "best" local estimate, but the time series of precipitation estimates has heterogeneous statistics, including data boundaries in space and time.

3.4 Rescaling to Monthly Data

The final step in creating the research product is to introduce monthly rain gauge data. Huffman et al. (1997), among others, have demonstrated the advantages of including rain gauge data in combination data sets at the monthly scale, but we were skeptical of including sub-monthly data due to issues of data coverage and timeliness. Rather, we adopt the approach we took in the GPCP One-Degree Daily combination data set, which is to scale the short-period estimates to sum to a monthly estimate that includes monthly gauge data (Huffman et al. 2001). All available 3-h merged HQ-IR estimates are summed over the calendar month to create a monthly multi-satellite (MS) product. The MS and gauge are combined using inverse-error-variance weighting as in Huffman et al. (1997) to create a post-real-time monthly satellite-gauge combination (SG), which is posted by PPS as a separate TRMM product (3B43). Then for each gridbox the (monthly) SG/MS ratio is computed, then applied to scale each 3-h field in the month, producing the Version 6 3B42 product. The final fields have the detail of the satellite data, but have nearly neutral monthly bias compared to gauges (i.e., over land).

The output of the 3-h algorithm is best viewed as movie loops, examples of which are posted at http://trmm.gsfc.nasa.gov under the button labeled "Realtime 3 h & 7 Day Rainfall".

3.5 RT Algorithm Adjustments

The RT and research product systems are designed to be as similar as possible to ensure consistency between the resulting data sets. One important difference is that the research product's calibrator, the TCI, is not available in real time. In its absence we use the TMI estimates as the initial RT calibrator. A second important difference is that a real-time system cannot reach into the future, so the microwave-IR calibration "month" is taken as the five trailing and one current (partial) pentads, or 5-day calendar intervals, of accumulated coincident data. As in the research product, the inter-calibration of individual microwave estimates to the TMI is handled with climatological coefficients. The HQ-IR calibration is recomputed for each 3-h period to capture rapid changes in the calibration for rare heavy rain events. Third, the monthly gauge adjustment step carried out for the research product is not possible for the RT.

Starting 17 February 2009, we implemented a procedure to address the second and third differences listed above. It is labeled "Version 6", but this should not be confused with Version 6 of the official TRMM products 3B42 and 3B43. Preliminary testing showed that computing these adjustments on a climatological basis provoked fewer artifacts than attempting to use data from trailing months. Therefore, we first determine a matched histogram calibration of TMI to the TCI, computed for 10 years of coincident data to establish the climatology for each calendar month. Second, a climatological monthly calibration of TCI to the 3B43 research product is computed as a simple ratio, again on a $1° \times 1°$ spatial grid and using 10 years of data. Finally, the TMI–TCI and TCI–3B43 calibrations are successively applied to the preliminary 3-h RT multi-satellite product to create the final 3B42RT.

4 Current Status on Algorithm Development

The research product system is currently running as the Version 6 algorithm for TRMM product 3B42, although that product provides only the final gauge-adjusted merged microwave-IR field. The Version 6 TRMM 3B43 product provides the post-real-time monthly SG described above. Version 6 data are available for January 1998 to the (delayed) present at http://lake.nascom.nasa.gov/data/dataset/TRMM/. Users should be aware that beta testing is underway at press time for Version 7, as discussed in Section 1.6.

The RT system has been running routinely on a best-effort basis in the PPS (originally TSDIS) since late January 2002, and the last major upgrade occurred at 00 UTC 17 February 2009, at which point an archive of new Version 6 RT estimates starting 00 UTC 1 October 2008 was released. For simplicity, a fixed latency (currently 9 h after nominal observation time) triggers the processing. The combined microwave, microwave-calibrated IR, and merged microwave-IR estimates, which are labeled 3B40RT, 3B41RT, and 3B42RT, respectively, are available from ftp://trmmopen.gsfc.nasa.gov or http://precip.gsfc.nasa.gov. All RT estimates created before 7 February 2005 are considered obsolete because they have rather different processing, and therefore should not be used. As part of the release of 17 February 2009 the format for 3B42RT was augmented so that the "new" climatologically-calibrated precipitation estimate is provided in the first ("precip") field, but an additional field is appended to each file providing the "old" uncalibrated precipitation estimate. This configuration permits users to continue using their previously established analysis routines by accessing the additional precipitation field at the end of each file.

Both data sets (and other precipitation data) are also accessible in the interactive Web-based TRMM On-line Visualizations and Analysis System (TOVAS) at http://lake.nascom.nasa.gov/Giovanni/tovas. The TOVAS site can be particularly helpful for new users, since it allows them to quickly create graphics from any of the TMPA data sets. These results can be used as the standard to validate the data access/navigation/scaling carried out in the users' own application code.

5 Comparisons and Examples

Both versions of the TMPA have been produced for a sufficiently long time that researchers have had the chance to develop and start reporting various applications and validations that employ one or both versions. As well, the TMPA is examined in other chapters in this book, including Chapters 10–13. This section summarizes some of the previous results and provides some test results illustrating the new climatological calibration that was recently instituted in the real-time product.

5.1 Prior Results

Basic validation statistics have been reported for a number of locations, including primarily ocean locations (Huffman et al. 2007; Sapiano and Arkin 2009) and primarily land areas (Ebert et al. 2007 and Tian et al. 2007, among others). The latter set are most interesting to hydrologists. The Ebert et al. (2007) study introduces notable on-going systematic daily continental-scale validation of many different quasi-operational precipitation estimates for Australia, the continental United States, western Europe, parts of South America, and other sites, organized through the International Precipitation Working Group (IPWG) of the Coordinating Group for Meteorological Satellites. The various Web sites for these regions are accessible through http://www.bom.gov.au/bmrc/SatRainVal/validation-intercomparison.html, and each provides a variety of detailed and summary statistics.

The first major result arises from the fact that the histogram of precipitation rates in the microwave input data is generally more accurate over ocean than over land as discussed above. As such, the land estimates are best in convective regimes, where the icy hydrometeors that cause scattering are well-correlated to surface rainfall. It is also the case that the more-approximate IR estimates are better correlated to short-interval precipitation in convective conditions. Conversely, the stratiform clouds that tend to dominate in cool-season and frontal conditions lead to significant mis-estimation using IR algorithms. The behavior for numerical models tend to be the opposite, estimating precipitation more accurately when the model convective parameterizations are not a major factor. As a result, observational estimates, including the TMPA, tend to out-perform models in warm/convective conditions and vice-versa in cool-season stratiform conditions (Ebert et al. 2007).

A second major result is that fine-scale precipitation estimates tend to have high uncertainty, while averaging in space and time improves the error characteristics. As discussed in Huffman et al. (2007), the fine-scale uncertainty arises from a number of issues, including algorithmic uncertainty and variations in the observational characteristics of the various input sensors. Figure 4 in Huffman et al. (2007) exemplifies the variations among near-simultaneous estimates from different sensors. Hossain and Huffman (2008), among others, show the systematic improvement in uncertainty that occurs across a number of metrics when increasingly more time/space averaging is applied. The issue, of course, is that many users' applications require the full resolution. Nonetheless, the implicit averaging that results from, say, computing hydrologic drainage basin flows, can allow the relatively uncertain estimates to be useful (Nijssen and Lettenmaier 2004; Hong et al. 2007).

A third major result from previous studies is that the use of monthly gauge analysis data in the post-real-time research TMPA is beneficial, as demonstrated by comparison to the real-time TMPA, which lacks gauge input. This is true at monthly and longer time scales, as one might expect by construction, so that the climatology for the research TMPA is close to that of the undercatch-corrected GPCC analysis in most land regions. However, monthly gauge adjustment also brings improvement on relatively short intervals (Ebert et al. 2007). This result is the basis for the shift of the real-time TMPA to using calibration to the post-real-time research TMPA.

A few studies have examined how the histogram of precipitation values compares for the two TMPA products (and other combination algorithms). Typically, the TMPA shows somewhat too many high-precipitation-rate values and lacks precipitation events at the low rates (Fisher and Wolff 2008; Tian et al. 2007). The result is that rain areas tend to be too small in size and have conditional rates that are too high. These results are consistent with the finding in Jiang et al. (2008) that the fraction of precipitation produced by tropical cyclones is as much as 30–50% higher in the TMPA than in comparable radar and raingauge data. That is, regions that possess a concentration of high precipitation rates will likely have an excess in the TMPA due to the characteristic bias in the TMPA histograms.

More qualitatively, it is clear that the performance of the TMPA and other combination products is critically dependent on the quality of the input data sets. All of the combination schemes attempt to limit the impact of defects and disparities in the input precipitation data, but the options are limited. For example, the TMPA process starts by auditing the input microwave estimates for possible artifacts based on "ambiguous pixel" flags contained in the GPROF datasets. All microwave datasets are then intercalibrated to a single reference, which is the GPROF-TMI for the real-time product and the TCI (Haddad et al. 1997a, b) for the non-real-time research product. Likewise, the IR calibration is computed from the combined microwave product. These actions should minimize shifts in bias as various satellites contribute intermittently during the day. Issues that cannot be addressed with current tools include: orographic enhancement and warm rain processes in general over land, where only the solid-hydrometeor-based scattering signal is useful; lack of sensitivity to light or very small-scale precipitation; and lack of retrieval skill in frozen surface areas. The older versions of GPROF used up to now in the TMPA (and other combinations) display artifacts in some coastal regions, including around inland water bodies (Tian and Peters-Lidard 2007), but it is possible that the new

GPROF2008 will correct this issue (Kummerow, 2008, "personal communication"). There are deficiencies in the current gauge analysis in regions of complex terrain, usually underestimates, but the new analysis available from the GPCC (Schneider et al. 2008) should improve this situation. It will continue to be the case that some underdeveloped areas, such as central Africa, are highly deficient in gauge observations, leading to more uncertainty in those regions for the post-real-time research TMPA.

Additional studies of TMPA performance may be found listed in the document posted at ftp://precip.gsfc.nasa.gov/pub/trmmdocs/TMPA_citations.pdf.

5.2 Climatological Calibration of the RT

As noted above, the real-time TMPA was upgraded in early 2009 to include a climatological calibration to the post-real-time research TMPA product. Here, we summarize the test data that were used to validate improved performance. Months representative of each season, namely January, April, July, and October in 2007, were computed and then compared to the original uncalibrated RT product. As well, new calibrated RT estimates were computed for October-December 2008 and similarly examined. The months in 2007 are not fully independent of the calibration coefficients, since 2007 is one of the ten years used in the calibration, but we believe this should not be a critical factor.

The monthly accumulations of all matched 3-h estimates for July 2007 from each scheme (Fig. 3) have a very similar visual appearance, but there are important differences (Fig. 4 (top, middle)). The Version 6 product is taken as the standard in this discussion, since the reason for the calibration is to make the RT as consistent with Version 6 as possible. The excess precipitation displayed by the original RT (Fig. 4 (top)) in Africa, the U.S., and Mesoamerica for this particular month is consistent with typical warm-season results, as are the low values in northeastern Equatorial South America, the Western Ghats in India, and the monsoonal maximum in Bangladesh and surrounding areas. We examine the success of the calibration by defining the improvement (I_{cal}) as

$$I_{cal} = |3B42RT(cal) - 3B42V6| - |3B42RT(uncal) - 3B42V6|$$

where each term represents the monthly average of the product named. The calibrated RT is closer to Version 6 in regions where $I_{cal}<0$, while the uncalibrated RT is closer for $I_{cal}>0$. This metric tends to emphasize regions with high precipitation, since a larger dynamic range is possible, but this seems appropriate from a global water and energy balance perspective.

The I_{cal} for July 2007 (Fig. 4(bottom)) demonstrates that most of the regions with biases likely dominated by regimes noted above see improvement with the new

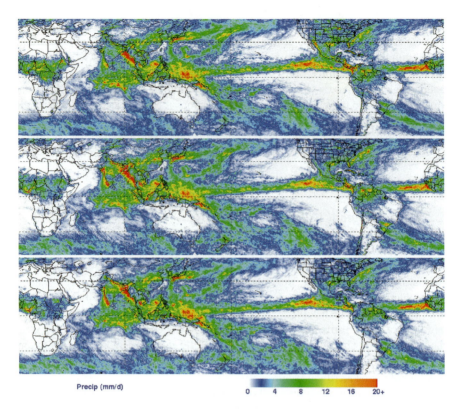

Fig. 3 Average precipitation rate in mm/d for July 2007 for the (*top*) uncalibrated RT, (*middle*) calibrated RT, and (*bottom*) V6 TMPA precipitation products

calibration scheme (green and blue colors). By comparison with the individual difference images in Fig. 4 (top, middle), it is clear that the improvement is frequently in the sense of simply reducing the bias, rather than fully correcting it. The climatological relationships are not always effective for an individual month. Most notably along the coast of Myanmar, but also in the Sahel, western coastal India, southern Japan, and southern Brazil, the calibration drives the result further from correspondence to 3B42V6 for this particular month. We should expect such fluctuations to occur when the selection of regimes experienced in individual months do not match up with the climatological distribution of regimes by month, but future work will include an analysis of such cases for possible design issues with the calibration scheme.

The two-dimensional histograms displayed in Fig. 5 give a better depiction of how the calibration works over the whole domain, broken into land and water areas, since gauge influence is only at work over land. In both regions the bulk of the calibrated points (bottom diagrams) are clustered more closely about the 1:1 line than for the uncalibrated (top). Note that the thin scatter of high values tends to be

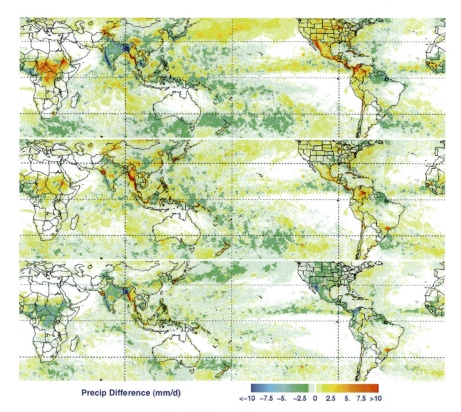

Fig. 4 Monthly average difference between (*top*) uncalibrated RT and V6, and (*middle*) calibrated RT and V6. (*bottom*) Improvement metric defined in Eq. (1), I_{cal}, which is the absolute value of the *top image* subtracted from the absolute value of the *middle image*. In all three panels the units are mm/d, the data are for July 2007, and the fields referenced are displayed in Fig. 3

shifted toward higher RT values when calibration is applied, reducing the negative bias that appears to characterize the high end before calibration in this month.

Bias and root-mean-square (RMS) differences are summarized in Table 2 for all, ocean, and land areas in the latitude band 50°N–50°S for the seven test months. Changes in the bias are small and only favor the calibration scheme in about half of the months for each of the regions. These changes are not considered important, since the overall bias is a small residue of regions of opposing sign. That is, improvements predominantly in regions of one sign could easily drive the average result. The RMS is a more sensitive measure because it quantifies the degree to which the RT is close to the V6. Near-unanimous improvement in both land and ocean with the calibration confirms our qualitative impression from Figs. 4 and 5 that the calibration is working as intended. The lack of skill over land for January 2007 and December 2008 perhaps indicates that the results are sensitive to the treatment of artifacts in winter land conditions, which dominate in those months due to the preponderance of land in the Northern Hemisphere.

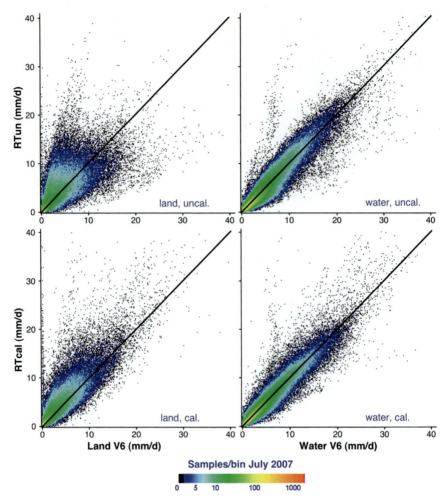

Fig. 5 Two-dimensional histograms of monthly precipitation for TMPA-RT without (*top*) and with (*bottom*) climatological calibration against V.6 TMPA for land (*left*) and ocean (*right*) for July 2007, corresponding to the maps in Fig. 3

6 Future Plans/Conclusions

The TMPA is intended to provide a "best" estimate of quasi-global precipitation from the wide variety of modern satellite-borne precipitation-related sensors. Estimates are provided at relatively fine scales (0.25° × 0.25°, 3-h) in both real and post-real time to accommodate a wide range of research applications. However, the errors inherent in the finest scale estimates are large. The most successful use of the TMPA data takes advantage of the fine-scale data to create averages appropriate to the user's application.

Table 2 Bias and root-mean-square (RMS) difference statistics comparing monthly accumulations of uncalibrated and calibrated 3B42RT estimates, taking monthly accumulations of Version 6 3B42 as the standard. Results are displayed for all, ocean, and land regions in the latitude band 50°N–50°S for selected months in units of mm/d

		Bias (mm/d)		RMS (mm/d)	
		RT(uncal)–V6	RT(cal)–V6	RT(uncal)–V6	RT(cal)–V6
Jan'07	Land	0.203	0.511	2.09	2.41
	Ocean	0.020	0.016	1.04	0.88
	All	0.069	0.147	1.40	1.45
Apr'07	Land	0.603	0.467	2.16	1.86
	Ocean	0.046	0.073	1.00	0.90
	All	0.194	0.178	1.41	1.23
Jul'07	Land	0.399	0.535	2.31	1.90
	Ocean	−0.079	0.053	1.04	0.98
	All	0.048	0.181	1.49	1.29
Oct'07	Land	.650	0.439	2.17	1.79
	Ocean	0.057	0.100	0.93	0.92
	All	0.214	0.190	1.37	1.21
Oct'08	Land	0.162	−0.012	1.60	1.56
	Ocean	−0.012	−0.042	0.90	0.80
	All	0.034	−0.034	1.13	1.05
Nov'08	Land	−0.364	−0.315	1.78	1.66
	Ocean	−0.007	0.000	0.99	0.87
	All	−0.101	−0.083	1.25	1.13
Dec'08	Land	−0.194	−0.605	1.66	1.86
	Ocean	0.042	0.035	0.94	0.86
	All	−0.020	−0.134	1.17	1.21

At press time an upgrade of the research quality post-real-time TMPA from Version 6 to Version 7 was in beta test, providing a variety of improvements that modernize the input data sets and correct several issues. Specifically, the latest GPROF code is being introduced for SSM/I (including a correction for the channel interference on F15), SSMIS (thus bringing in the F16 and F17 records), AMSR-E, and TMI; improved AMSU estimates are used; and MHS estimates from NOAA-18 and MetOp-A are included. The GPCC's improved raingauge analyses are included for both retrospective and initial (i.e., new-data) processing. Finally, we have substantially augmented the fields available in the output product data files. In 3B42, we are adding fields of merged microwave precipitation, microwave-calibrated IR precipitation, and microwave overpass time, in addition to providing more detail as to the particular sensor on which the final precipitation estimate is based. In 3B43 we are including a data field for the relative weighting that the gauges receive in each grid box.

The immediate task at hand is to complete the current beta test of the Version 7 TMPA system, reprocess the TRMM archive, and commence Version 7 computations on new observations. Once Version 7 is established, presumably in early 2010,

Version 6 will be considered obsolete. As well, at that point the climatological calibrations in the RT will be updated to Version 7, rendering the Version 6-calibrated RT obsolete. Status messages are posted routinely in the "Information" hot link buttons on http://precip.gsfc.nasa.gov for the respective products.

Looking to the future, we are studying how best to extend the TMPA to higher latitudes, for example by incorporating fully global precipitation estimates based on Television Infrared Observation Satellite (TIROS) Operational Vertical Sounder (TOVS), Advanced TOVS (ATOVS), and Advanced Infrared Sounder (AIRS) data. It is also a matter of research to work toward a Lagrangian time interpolation scheme, conceptually along the lines of the CPC Morphing algorithm (CMORPH; Joyce et al. 2004) and the Global Satellite Map of Precipitation (GSMaP; Kubota et al. 2007). Finally, it is still the case that the study of precipitation in general needs a succinct statistical description of how errors in fine-scale precipitation estimates should be aggregated through scales up to global/monthly (Hossain and Huffman 2008).

On the instrumentation side there is a concerted effort to provide complete 3-h microwave data. Most of this effort is focused on the National Aeronautics and Space Administration's Global Precipitation Measurement (GPM) project. Besides simply increasing the frequency of coverage, it is planned for GPM to provide a TRMM-like "core" satellite to calibrate all of the microwave estimates on an ongoing basis over the latitude band 65°N–S. We expect the geo-IR–based estimates to have a long-term role in filling the inevitable gaps in microwave coverage, as well as in enabling sub-3-h precipitation estimates at fine spatial scales.

References

Adler RF, Huffman GJ, Chang A, Ferraro R, Xie P, Janowiak JE, Rudolf B, Schneider U, Curtis S, Bolvin DT, Gruber A, Susskind J, Arkin PA, Nelkin EJ (2003) The Version 2 Global Precipitation Climatology Project (GPCP) monthly precipitation analysis (1979-present). *J. Hydrometeor.* **4**:1147–1167

Adler RF, Huffman GJ, Keehn PR (1994) Global tropical rain estimates from microwave-adjusted geosynchronous IR data. *Remote Sensing Rev.* **11**:125–152

Arkin PA, Meisner BN (1987) The relationship between large-scale convective rainfall and cold cloud over the Western Hemisphere during 1982–1984. *Mon. Wea. Rev.* **115**:51–74

Ebert EE, Janowiak JE, Kidd C (2007) Comparison of near real time precipitation estimates from satellite observations and numerical models. *Bull. Amer. Meteor. Soc.* **88**:47–64

Fisher BL, Wolff DB (2008) Validating microwave-based satellite rain Rate retrievals over TRMM Ground Validation sites. Amer. Geophys. Union (AGU) Fall Meeting, 15–19 December 2008, San Francisco, CA

Haddad ZS, Smith EA, Kummerow CD, Iguchi T, Farrar MR, Durden SL, Alves M, Olson WS (1997a) The TRMM "Day-1" radar/radiometer combined rain-profiling algorithm. *J. Meteor. Soc. Japan* **75**:799–809

Haddad ZS, Short DA, Durden SL, Im E, Hensley S, Grable MB, Black RA (1997b) A new parameterization of the rain drop size distribution. *IEEE Trans. Geosci. Rem. Sens.* **35**:532–539

Hong Y, Adler R, Hossain F, Curtis S (2007) Global runoff simulation using satellite rainfall estimation and SCS-CN method. *Water Resources Res* 43(W08502) doi:10.1029/2006WR005739

Hossain F, Huffman GJ (2008) Investigating error metrics for satellite rainfall data at hydrologically relevant scales. *J. Hydrometeor.* **9**:563–575

Huffman GJ, Adler RF, Arkin P, Chang A, Ferraro R, Gruber A, Janowiak J, McNab A, Rudolf B, Schneider U (1997) The Global Precipitation Climatology Project (GPCP) combined precipitation data set. *Bull. Amer. Meteor. Soc.* **78**:5–20

Huffman GJ, Adler RF, Bolvin DT, Gu G, Nelkin EJ, Bowman KP, Hong Y, Stocker EF, Wolff DB (2007) The TRMM Multi-satellite Precipitation Analysis: Quasi-global, multi-year, combined-sensor precipitation estimates at fine scale. *J. Hydrometeor.* **8**:38–55

Huffman GJ, Adler RF, Morrissey M, Bolvin DT, Curtis S, Joyce R, McGavock B, Susskind J (2001) Global precipitation at one-degree daily resolution from multi-satellite observations. *J. Hydrometeor.* **2**:36–50

Janowiak JE, Joyce RJ, Yarosh Y (2001) A real-time global half-hourly pixel-resolution IR dataset and its applications. *Bull. Amer. Meteor. Soc.* **82**:205–217

Jiang H, Halverson JB, Simpson J (2008) On the differences in storm rainfall from Hurricanes Isidore and Lili. Part I: Satellite observations and rain potential. *Wea. Forecasting* **23**:44–61

Joyce RJ, Janowiak JE, Arkin PA, Xie P (2004) CMORPH: A method that produces global precipitation estimates from passive microwave and infrared data at high spatial and temporal resolution. *J. Hydrometeor.* **5**:487–503

Kongoli, C, Pellegrino P, Ferraro R, 2007: The utilization of the AMSU high frequency measurements for improved coastal rain retrievals. *Geophys. Res. Lett.,* **34**:L17809, doi:10.1029/2007GL029940.

Krajewski WF, Smith JA (1991) On the estimation of climatological Z–R relationships. *J. Appl. Meteor.* **30**:1436–1445

Kubota T, Shige S, Hashizume H, Aonashi K, Takahashi N, Seto S, Hirose M, Takayabu YN, Nakagawa K, Iwanami K, Ushio T, Kachi M, Okamoto K (2007) Global precipitation map using satellite-borne microwave radiometers by the GSMaP Project: Production and validation. *IEEE Trans. Geosci. Remote Sens.* **45**:2259–2275

Kummerow, C., W.S. Olson, L. Giglio, 1996: A simplified scheme for obtaining precipitation and vertical hydrometeor profiles from passive microwave sensors. *IEEE Trans. Geosci. Remote Sens.* **34**:1213–1232

Miller JR (1972) A climatological Z-R relationship for convective storms in the northern Great Plains. *15th Conf. on Radar Meteor.* 153–154

Nijssen B, Lettenmaier DP (2004) Effect of precipitation sampling error on simulated hydrological fluxes and states: Anticipating the Global Precipitation Measurement satellites. *J. Geophys. Res.* 109 D02103

Olson WS, Kummerow CD, Hong Y, Tao W-K (1999) Atmospheric latent heating distributions in the Tropics derived from satellite passive microwave radiometer measurements. *J. Appl. Meteor.* **38**:633–664

Rudolf B (1993) Management and analysis of precipitation data on a routine basis. *Proc. Internat. Symp. on Precip. and Evap.* (Eds. B. Sevruk, M. Lapin), 20–24 Sept. 1993, Slovak Hydrometeor. Inst., Bratislava, Slovak Rep. 1:69–76

Sapiano MRP, Arkin PA (2009) An inter-comparison and validation of high resolution satellite precipitation estimates with three-hourly gauge data. *J. Hydrometeor.* **10**:149–166

Schneider U, Fuchs T, Meyer-Christoffer A, Rudolf B (2008) Global precipitation analysis products. File GPCC_intro_products_2008.pdf at http://gpcc.dwd.de/. Accessed 8 December 2008

Tian Y, Peters-Lidard CD (2007) Systematic anomalies over inland water bodies in satellite-based precipitation estimates. *Geophys. Res. Lett.* **34**:L14403

Tian Y, Peters-Lidard CD, Choudhury BJ, Garcia M (2007) Multitemporal analysis of TRMM-based satellite precipitation products for land data assimilation applications. *J. Hydrometeor.* **8**:1165–1183

Vila, D, Ferraro R, Joyce R, 2007: Evaluation and improvement of AMSU precipitation retrievals. *J. Geophys. Res.,* **112**: D20119, doi:10.1029/2007JD008617

Weng F, Zhao L, Ferraro R, Poe G, Li X, Grody N (2003) Advanced microwave sounding unit cloud and precipitation algorithms. *Radio Sci.* **38**:8068–8079

Xie P, Arkin PA (1996) Gauge-based monthly analysis of global land precipitation from 1971 to 1994. *J. Geophys. Res.* **101**:19023–19034

Zhao L, Weng F (2002) Retrieval of ice cloud parameters using the advanced microwave sounding unit. *J. Appl. Meteor.* **41**:384–395

CMORPH: A "Morphing" Approach for High Resolution Precipitation Product Generation

Robert J. Joyce, Pingping Xie, Yelena Yarosh, John E. Janowiak and Phillip A. Arkin

Abstract The CMORPH technique was developed to synergize the most desirable aspects of passive microwave (high quality) and infrared (spatial and temporal resolution) data. CMORPH is a global (in longitude; 60°N–60°S) high-resolution (∼0.10° latitude/longitude, 1/2-hourly) precipitation analysis technique that uses motion vectors derived from half-hourly geostationary satellite IR imagery to propagate precipitation estimates derived from passive microwave data. Multi-hour precipitation totals derived via the CMORPH methodology are an improvement over both simple averaging of all available microwave-derived precipitation estimates and over other merging techniques that blend microwave and infrared information but which derive estimates of precipitation directly from infrared data when passive microwave data are not available.

Keywords CMORPH · Precipitation · Remote sensing

1 Introduction

The period from 1998 – present has witnessed an unprecedented number of satellites with passive microwave (PMW) instruments. These instruments provide the most accurate passive spaceborne estimates of precipitation to date. But because those sensors are housed on low-earth orbit spacecraft there are significant spatial and temporal sampling issues with data provided by them. Conversely, infrared (IR) data are spatially complete and are available every 15–60 min over the entire globe equatorward of about 60° latitude, but estimates of rainfall from IR data are not as accurate as those derived from PMW data. This situation has spurred the development of methods to combine these highly disparate data.

J.E. Janowiak (✉)
ESSIC/Cooperative Institute for Climate Studies, University of Maryland, College Park, MD, USA
e-mail: jjanowia@umd.edu

The combination of precipitation estimates that are derived separately from PMW and IR data has been done in a variety of ways. Turk et al. (2000) were the first to do this on a global basis by developing a rain rate frequency matching procedure on the precipitation estimates that are generated from IR and PMW data that updates continually. This process provides a calibration between the IR brightness temperatures and PMW-derived rainfall rates that provides a means to produce rainfall estimates directly from the IR brightness temperature information that have been "trained" to mimic the PMW-derived rainfall estimates. The TMPA method (Huffman et al. 2007) uses PMW data to calibrate the IR-derived estimates and creates analyses that contain PMW-derived rainfall estimates when and where PMW data are available and the calibrated IR estimates where PMW data are unavailable. The PERSIANN technique (Hsu et al. 1997) uses a neural network approach to derive relationships between IR and PMW data which are applied to the IR data to generate rainfall estimates. In each of these methods, some or all of the rainfall estimates in the resulting analyses are generated directly from IR data.

CMORPH (Joyce et al. 2004) uses a different approach in which IR data are used only to derive a cloud motion field that is subsequently used to propagate raining pixels; thus, only rainfall estimates that have been derived from PMW data are used in the procedure. A similar technique has been developed recently (GSMaP; Kubota et al. 2007) in which IR data are used to derive a motion field except that a Kalman filter approach is used in the blending methodology; however, morphing is not performed in that technique as of this writing.

2 Description of the CMORPH Data and Methodology

2.1 Infrared Data

IR brightness temperature information from all available geostationary satellites that have been collected, quality-controlled, and merged into global data sets (Janowiak et al. 2001) are used by the CMORPH technique. Every 30 min, new full earth-disc IR images are obtained from the European geostationary meteorological satellites ("Meteosat") spacecraft (sub-satellite points 0° and 63°E), but only every three hours from the Geostationary Operational Environmental Satellite spacecraft (GOES; sub-satellite points 75°W and 135°W) although northern and southern hemispheric images are available from the GOES spacecraft during the intervening 30 min intervals. Full earth disk images are available from Japan's Multi-Function Transport Satellite (MTSAT; sub-satellite point 140°E) at hourly intervals. The MTSAT spacecraft replaced the Geostationary Meteorological Satellite (GMS) spacecraft series (same sub-satellite point for both) in 2005.

The IR data management procedure constructs datasets for each satellite IR image in which the data are interpolated to a rectilinear grid at 0.03635° of latitude and longitude resolution ($\sim$ 4 km at the equator). Then the data are subjected to a parallax adjustment which corrects for the mis-navigation of high cloud (Vicente et al. 2002), and corrects for erroneously cold limb effects at the edges of the scans due to atmospheric attenuation effects (Joyce et al. 2001). After these processes have been completed, datasets for the individual satellite domains are merged to form global fields (60°N–60°S) for each half-hourly period.

2.2 Passive Microwave Data

The PMW-derived precipitation estimates that are presently used in CMORPH are generated from observations obtained from the NOAA polar orbiting operational meteorological satellites, the U.S. Defense Meteorological Satellite Program (DMSP) satellites, and from NASA's Tropical Rainfall Measuring Mission (TRMM; Simpson et al. 1988) and Aqua satellites. The PMW instruments aboard these satellites are the Advanced Microwave Sounding Unit (AMSU-B), the Special Sensor Microwave Imager (SSM/I), the TRMM Microwave Imager (TMI), and the Advanced Microwave Scanning Radiometer – Earth Observing System (AMSR-E), respectively. The characteristics of these sensors are summarized in Table 1.

Table 1 Characteristics of PMW sensors and associated satellites used in CMORPH

Sensor	Spatial resolution (footprint at ~85 GHz) (km)	Observing time (ascending orbit)	Frequencies (GHz)	Altitude (Km)
TMI	4.6 × 6.9	Precessing orbit	10, 19, 21, 37, 85	403[a]
SSM/I (F-13)	13 × 15	18:11	19, 22, 37, 85	830
SSM/I (F-14)	13 × 15	20:30	19, 22, 37, 85	830
SSM/I (F-15)	13 × 15	21:32	19, 22, 37, 85	830
AMSR-E (Aqua)	6 × 4	13:30	7, 11, 19, 24, 37, 89	705
AMSU-B (N-15)	15 × 15[b]	18:58	89, 150, 183	830
AMSU-B (N-16)	15 × 15[b]	13:57	89, 150, 183	830
AMSU-B (N-17)	15 × 15[b]	22:08	89, 150, 183	830

[a] Altitude raised from 360 km to 403 km over the August 7–24, 2001 period to save fuel and thus extend the mission life.
[b] footprint for nadir view, limb footprint extends to 45 km diameter.

The TMI instrument is a nine-channel radiometer that operates at five frequencies, four of which are quite similar to the frequencies of the SSM/I instrument (Table 1). Although the TRMM spacecraft orbit limits the geographic coverage to 38°N–38°S latitude, TMI offers higher spatial resolution than SSM/I due to the lower orbit of the TRMM spacecraft. Surface rainfall derived from the TMI instrument is a product of NASA's TRMM Science Data and Information System (TSDIS) 2A12 algorithm (Kummerow et al. 1996). This algorithm uses a look-up table that relates the vertical profiles of liquid and ice to surface rain rates to produce rainfall rates both over land and ocean surfaces.

The SSM/I sensors aboard the DMSP platforms are operational on the F-13, F-14 and F-15 satellites at the time of this writing. CMORPH ingests precipitation estimates that are generated from an SSM/I rainfall algorithm (Ferraro 1997) which utilizes the 85 GHz vertically polarized channel to relate the scattering of upwelling radiation by precipitation sized ice particles within the rain layer and in the tops of convective clouds to surface precipitation rate that is run operationally at NOAA's National Environmental Satellite and Data Information Service (NESDIS). A precipitation rate is derived empirically from the relationship between retrieved ice amount in the rain layer and in the tops of convective clouds to actual surface rainfall. This scattering technique is applicable over land and ocean. In addition, the NOAA/NESDIS algorithm provides rainfall estimates over ocean surfaces via methodology that is based upon the absorption of the upwelling radiation by rain and cloud water ("emission" technique) at 19 and 37 GHz. Rainfall estimates derived from such thermal emission channels are more directly related to precipitation than scattering techniques (and thus more accurate) since they directly sense the thermal radiation that is emitted from liquid hydrometeors. Error attributes of the algorithm are described by Li et al. (1998) and McCollum et al. (2002).

The AMSU-B instrument is currently operational aboard the NOAA-15, NOAA-16 and NOAA-17 polar orbiting satellites. The AMSU-B sensor has five window channels and its cross track swath width (approximately 2200 km) contains 90 fields of view (FOV) per scan. The NOAA/NESDIS AMSU-B rainfall algorithm (Ferraro et al. 2000; Zhao and Weng 2002; Weng et al. 2003) performs a physical retrieval of ice water path (IWP) and particle size from the 89- and 150 GHz channels. Then a conversion from IWP to rain rate is made based on cloud simulations from the Fifth-Generation NCAR/Penn State Mesoscale Model (MM5) and on comparisons with in situ data. The 183 GHz channel combined with surface temperature is used to screen out desert and the 23, 31, and 89 GHz channels are used to screen out snow as described in Zhao and Weng (2002).

2.3 Rainfall Mapping

The globally merged IR data analyses are available at $\frac{1}{2}$ h intervals; thus, a 30 min time resolution was selected for CMORPH. The 0.0727° latitude and longitude (8 km at the equator) grid resolution used in CMORPH was determined by considering the spatial resolution of the various input data sources: 4 km (GOES IR), 3–5 km (Meteosat IR), and the coarser resolution of the AMSU-B and SSM/I (~13 km) derived precipitation estimates, combined with the notion that the grid must be fine enough to represent the propagation of rainfall systems in half-hourly time increments. Because the PMW-derived rainfall estimates are coarser than the 8 km grid scale, the estimates are first mapped to the nearest grid point on global (60°N–60°S) rectilinear grids at 0.0727° of latitude and longitude resolution, separately for each half hour and for each satellite. Then satellite rainfall maps are combined by sensor type so that when the processes described above are completed, remotely sensed rainfall estimates for half-hourly periods for each sensor type (TMI, SSM/I, AMSU-B, AMSR-E) are saved to separate files. The half-hourly global IR data that are used to propagate the PMW precipitation estimates are averaged to approximately 0.727° latitude/longitude resolution (8 km at the equator), to match the grids that contain the PMW estimated rainfall.

A precedence of sensor type was established to determine which estimate to use when a PMW-derived estimate from more than one sensor type is available at the same location for a given half hourly period. Several estimates may be available for a given time and location because TRMM under-flies all other low orbiters used in this study. The order of precedence was established based on spatial resolution and the availability of both emission-based and scattering-based estimates over the oceans. The resulting order of precedence in regions of overlap is to use estimates from TMI first, then from AMSR-E if no estimate from TMI is available, followed by SSMI and finally AMSU-B. Each pixel in the half hourly analyses is tagged with a satellite identifier representing the orbiter used to produce the estimate and this information is stored as part of the CMORPH archive suite.

2.4 Intercalibration of the Various PMW-Derived Precipitation Estimates

It is very important that the PMW-derived precipitation estimates are calibrated to some reference because the various PMW sensors are not uniform and have some significant differences among them. For example, the TMI and SSMI instruments have channels that are sensitive to emission (37 GHz) and scattering (85 GHz) while the AMSU-B sensor has only scattering channels. Therefore, large differences in precipitation frequency and amount are apparent in comparisons of precipitation estimates derived from AMSU-B and SSMI or TMI data over the oceans (Fig. 3 in Joyce et al. 2004). The channel differences are only part of the reason for the differences seen in precipitation rate. In addition, AMSU-B is a cross-track scanner thus the field of view is a function of scan angle while SSMI and TMI are conical-scanning sensors and thus the field of view and pixel size is constant for those instruments. Furthermore, the AMSU (Vila et al. 2007) and SSMI GPROF (Kummerow et al. 2001; McCollum and Ferraro 2003; Wiheit et al. 2003) rainfall estimation algorithms have a 35 mm/h rain rate maximum which are much less than the 50 mm/h of the TMI and AMSR-E land and ocean algorithms.

The AMSR-E and TMI estimates were chosen as the normalization standard because of the finer spatial resolution and emission detection capability (over oceans) of those sensors. For both the AMSU and SSMI sensors, revised rain rate scales are determined dynamically by frequency matching (from the heaviest to lightest rain rates) with TMI and AMSR-E precipitation estimates that are temporally and spatially coincident with the precipitation estimates over the most recent ten-day period. The scaled rainfall rates are calculated separately for land and ocean in 10° latitude bands from 60°S to 60°N by matching estimates in overlapping 30° latitude domains. If the rainfall frequency of the "adjusted" rainfall is less than the frequency of the TMI/AMSR-E reference, then the cumulative total of the adjusted rainfall will be too low relative to the TMI-AMSR-E reference rainfall. In this case a "retrofit" of the matched PDF is forced such that the cumulative adjusted rainfall matches the reference rainfall.

2.5 CMORPH Methodology

The CMORPH methodology will be described very briefly here so readers who wish for more detail are referred to Joyce et al. (2004). In essence, two concurrent processes are initiated each half hour in preparation for constructing the precipitation analyses. Passive microwave rainfall estimates generated from SSMI, AMSR-E, AMSU, and TMI are collected (provided by NOAA/NESDIS), assembled and intercalibrated (Section 2.3) for each half-hour period, while cloud motion vectors are computed from successive ½ h images of global geosynchronous IR data (Janowiak et al. 2001).

In order to determine cloud speed and direction, an iterative spatial lag correlation process is applied to full resolution IR data. In essence, successive IR images are shifted spatially with respect to each other in all possible directions, and the shift that yields the highest correlation forms the basis of the motion field. However, because the spatial distribution of cloud-top temperatures and precipitation are not the same, the precipitation motion vectors are refined via a statistical model that relates cloud motion vectors to radar rainfall advection.

The rainfall propagation process begins by propagating fields of 8-km, half-hourly instantaneous PMW-derived precipitation estimates (t+0 h) forward in time using the IR derived precipitation motion vectors described in the previous paragraph (Fig. 1). All pixels with rainfall derived from PMW data (including those with

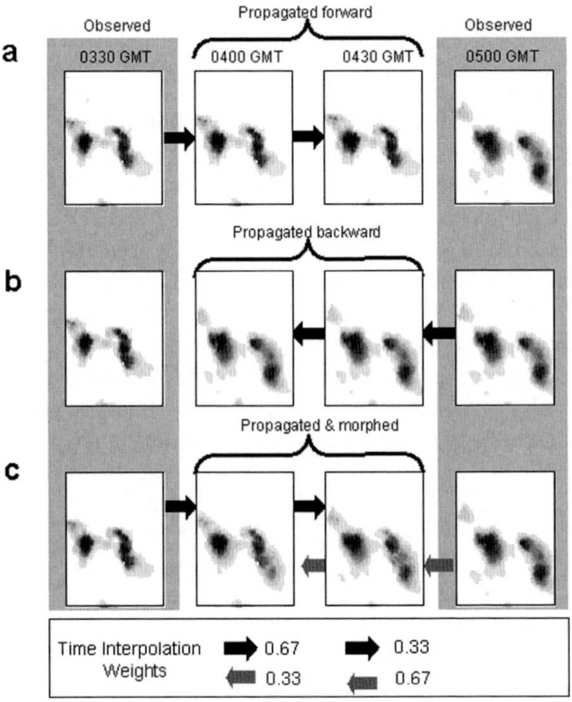

Fig. 1 Example of the CMORPH process. The 0330 and 0500 UTC images are actual rainfall estimates from passive microwave; the 0400 and 0430 images are (**a**) propagated forward in time; (**b**) propagated backward in time; (**c**) propagated and morphed. The time interpolation weights are the weights assigned (in this example) for the corresponding images and propagation directions (From Joyce et al. 2004: their Fig. 8)

zero precipitation) that reside within each 2.5° latitude/longitude region (centered within the 5°×5° grid box of IR data) are propagated in the same direction and distance to produce the analysis for the next half-hour (t+0.5 h). If a rainfall feature is on the border between two of the 2.5° latitude/longitude regions, the rainfall field is propagated evenly if the vector pairs from both regions match exactly. If two pixels from different regions are propagated to the same pixel location by convergence, an average of the two values is computed. If a data gap in the rainfall field occurs due to a divergence of features, a bilinear interpolation of the rainfall features across the gap is computed. Finally, if a PMW-derived precipitation estimate from a new scan at "t+0.5 h" is available at a particular pixel location, then that estimate overwrites the "older" propagated estimate. This entire process is repeated each half hour.

In addition to propagating rainfall estimates forward in time, a completely separate process is invoked in which the rainfall analyses are propagated backward in time using the same propagation vectors used in the forward propagation. However, this "backward propagation" is applied to the first PMW-derived rainfall field that is available after the time of the current ("t+0") estimates. A simple propagation of the features themselves will not change the character of those features but will merely translate them to new positions. However, changes in the intensity and shape of the rainfall features are accomplished by inversely weighting both forward- and backward-propagated rainfall by the respective temporal distance from the initial and updated PMW-derived rainfall fields which results in "morphing".

Several variations of CMORPH have been developed for various reasons since the technique was developed initially. "GMORPH" is exactly like CMORPH except that 700 hPa winds from the NCEP global data assimilation system (GDAS) are used for the motion field instead of the inferred motion that is derived from the displacement of features in successive IR images. The reason for developing this method is because of concerns that the use of IR to determine the motion of the rainfall can result in erroneous motion of tropical systems since IR data are retrieved from cloud-tops and the motion at the tops of clouds may not correspond to the motion of the rain at the surface. Another variant, dubbed "CMORPH-IR" was developed because the performance of CMORPH is directly related to the frequency of updates with new PMW information (Fig. 2.).

Therefore, more accurate rainfall estimates will be obtained by using IR-based rainfall estimates when the time between overpasses of a PMW sensor is "sufficiently long". Hence, an algorithm was developed to estimate rainfall directly from IR data. The algorithm ("IRFREQ") was modeled after Turk and Miller (2005) which is a histogram matching approach. Then, a decision model was instituted for the blending process to objectively decide when to use IR-based rainfall estimates (IRFREQ) instead of morphed estimates (see Joyce et al. 2004 for details). IRFREQ estimates are substituted for morphed estimates approximately 20% of the time.

Finally, because of the relatively long time between real time and when CMORPH analyses are available (i.e. "latency", which is about 13 h), a "quick look" product dubbed "QMORPH" was developed that has a latency of 3–4 h. QMORPH is not a morph because it merely propagates PMW-derived rainfall estimates forward in time, thus the features do not change in shape or intensity until new PMW

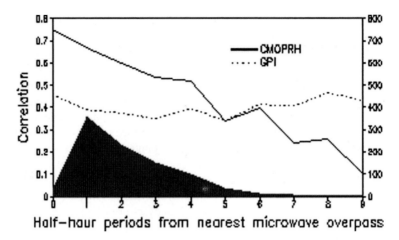

Fig. 2 The correlation between radar and CMORPH (*solid line*) as a function of $\frac{1}{2}$-h periods from nearest future or past microwave satellite overpass (x-axis) for the period April–May, 2003. For reference, the dotted line represents the correlation between radar and GPI over the same locations and times that were used in the radar-CMORPH correlation calculations. The shaded plot at bottom of the figure shows the number of pairs in the correlations; the scale for that plot is on the right (y-axis label in 1000's) (Figure from Joyce et al. 2004, their Fig. 13)

data become available. The distinction from CMORPH is that QMORPH "waits" for the availability of the "future" PMW rainfall data so that rainfall estimates can be morphed between the older and most recent PMW-derived rainfall estimates. This wait time is part of the reason for the long latency in CMORPH data.

3 Applications

One of the desirable features of CMORPH is the fine spatial and temporal character of the dataset which permits these analyses to be used for multiple purposes ranging from mesoscale to global in the spatial domain and diurnal to interannual in the temporal domain. Many precipitation analyses are available for exploring global/multi-seasonal variability but CMORPH is one of the few that permits the study of phenomena on mesoscale (spatial) and diurnal (temporal) over a global (60°N–60°S) domain. To demonstrate the utility of CMORPH for studies at fine time/space scales, we present a few examples.

In Fig. 3 we show the evolution of a strong squall line over the southeastern U.S. as depicted by hourly CMORPH and radar data. The temporal continuity of the rainfall pattern between the radar and CMORPH data is remarkably good over the sequence of hourly plots when considering that the PMW-derived rainfall that CMORPH relies on heavily is updated about every two–three hours. There are differences in intensity, particularly early in the sequence, although the relative intensities of CMORPH and radar agree reasonably well over the period. The utility of CMORPH over regions with dense rain gauge networks and radar coverage is

CMORPH Satellite Product

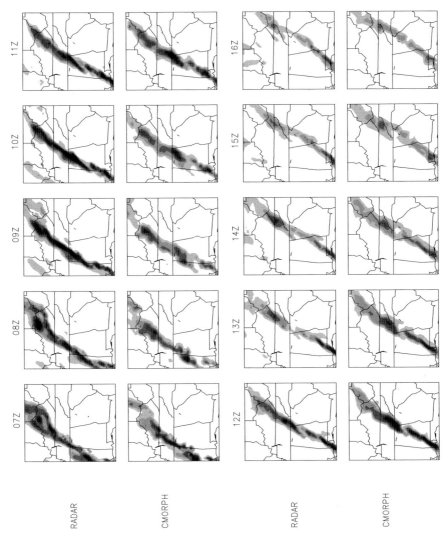

Fig. 3 Progression of a squall line over the southeastern United States on February 6, 2008 over successive hours from 0700 to 1600 UTC. *Top two rows* are for the 0700 to 1100 UTC period, with the *top row* depicting radar and the *bottom* depicting CMORPH. The *bottom tow rows* are for the 1200 to 1600 UTC period. The gray shades range from 1 mm day^{-1} (*lightest shading*) to 25 mm day^{-1} (*darkest shading*)

limited. However, by demonstrating that CMORPH provide radar-like information, it is reasonable to conclude that CMORPH can be of considerable value in regions with little or no direct observational data (particularly over the oceans), and can greatly facilitate research on systems that possess similar spatial and temporal characteristics as those presented in this example.

The diurnal cycle of precipitation is another application that demonstrates the utility of CMORPH. For example, a diagnosis of the performance of model precipitation forecasts in resolving the diurnal cycle of precipitation can be conducted by taking advantage of the spatial and temporal resolution of CMORPH. To this end we present some results from Janowiak et al. (2007) who used a version of CMORPH (referred to as "RMORPH") that was adjusted by rain gauge data (to reduce bias) over several regions of the U.S. to assess the ability of two models to resolve the diurnal cycle of precipitation during the convective season. In Fig. 4, the mean diurnal cycle of precipitation over the period June 12–August 15, 2004 is shown over four different regions as indicated in the titles of the four panels. The most prominent result of the comparison among these estimates is the large and early onset of peak convection in the southeast U. S. in the global model forecasts (GFS) that contrast with RMORPH and the regional model (labeled "Eta" but present-day vernacular is "NAM"). And although the

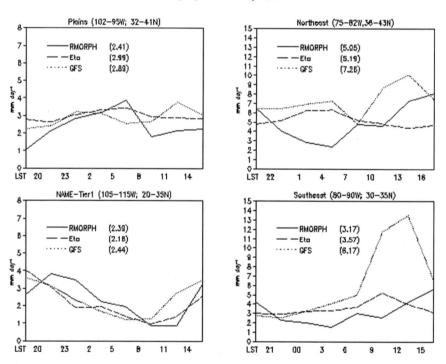

Fig. 4 The mean diurnal cycle of precipitation over several regions of the U.S. for raingauge adjusted CMORPH ("RMORPH"; Janowiak et al. 2007), and short-range precipitation forecasts from the NCEP regional North American Model ("Eta"), and the NCEP global model ("GFS"). Note that time of day (x-axis) is in local time (specific to each region) and that the x-axis labels correspond to the beginning of the 3 h accumulation periods (Figure from Janowiak et al. 2007; their Fig. 6)

Eta model exhibits an amplitude similar RMORPH, it peaks even earlier in the day than the global model. In addition, over the "Plains" states, where a nocturnal to early morning (depending upon location) precipitation maximum has been well-documented, RMORPH depicts a rainfall peak near sunrise while the global model peaks near noon. The regional model has a peak that coincides with RMORPH but has much less amplitude in the diurnal cycle.

4 CMORPH Improvements

4.1 Backward Extension and Reprocessing

CMORPH processing began on December 6, 2002. Numerous improvements have been developed and implemented to the processing system since production began which have lead to discontinuities in the time history of CMORPH. Several years after routine processing began, it was discovered that oceanic rainfall rates from the AMSU-B sounders were unrealistically low. Thus, a "retrofit" TMI-AMSR-E calibration was developed and implemented to eliminate the unrealistically low magnitude of oceanic AMSU-B rainfall estimates. Consequently, the total oceanic rainfall has risen in the operational CMORPH since the May 2005 implementation of this calibration change. Furthermore, an AMSU-B algorithm update in August 2005 removed a major coastline rainfall deficiency. Later, the AMSU algorithm was again vastly improved (Vila et al. 2007) with the inclusion of Cloud Liquid Water (CLW) derived from AMSU-A and a Convective Index from the 183 GHz channel. The resulting discontinuities in the CMORPH period of record have prompted a reprocessing effort for the entire CMORPH time history.

4.2 Backward Extension of the CMORPH Period of Record

We believe that CMORPH can be extended back to November 1998 which is a time when three passive microwave sensors were in orbit (TMI and two SSMI) and is as far back as the archive of globally merged IR data (Janowiak et al. 2001) are available. However, because the accuracy of CMORPH is highly dependent on the time between passive microwave sensor overpasses (Fig. 2), rainfall from geostationary satellite IR (IRFREQ) must also be incorporated over this extension period via a combination of those estimates with the relatively scarce PMW estimates by using an advanced Kalman filter methodology (Section 4.3). Upon completion of this effort, which is ongoing, CMORPH will have a time history that will exceed a decade at the time of this writing.

4.3 Kalman Filter

The initial CMORPH scheme is a simple one that uses the inverse of the distance (in time from a PMW overpass) as the weighting factor to derive "morphed" precipitation estimates for time periods between PMW overpasses. At the present time, CMORPH ingests precipitation estimates from multiple sources and creates analyses in a hierarchical framework after intercalibrating them against predetermined standards, thus decisions are made in a sequential, step-wise fashion. Recent work by Kubota et al. (2007) has strongly suggested that a more comprehensive Kalman filter approach applied to the data merging process can result in higher accuracy. We have developed a Kalman filter approach that is currently being tested which is designed to combine the various PMW-derived precipitation estimates based on the relative accuracy of these CMORPH inputs. The main inputs into this new Kalman scheme are weights that reflect the accuracy of the satellite estimates that were determined from comparisons over the U. S. of radar with PMW-derived rainfall from various sensors and satellites (with different observing times).

4.4 Bias Reduction

Satellite precipitation estimates contain bias, particularly over semi-arid continental regions during the warm season (Rosenfeld and Mintz 1988) and one way to reduce bias is by integrating rain gauge data. Fortunately, rain gauge data over the globe are increasingly available in near-real time so as to be useful for integration in near real-time global precipitation analyses despite variations in the daily collection times round the globe: 0000 UTC – 0000 UTC over Australia, 0600 UTC – 0600 UTC over Africa, 1200 UTC – 1200 UTC over the Americas, and 1800 UTC – 1800 UTC over most of Europe. An Optimum Interpolation approach (OI; Gandin 1963) to merge the satellite estimates and rain gauge reports with the goal of providing global precipitation estimates at 0.25° latitude/longitude spatial and daily temporal resolution is being developed. This procedure will be considerably different and more complex than the RMORPH methodology (Janowiak et al. 2007). RMORPH estimates are simple disaggregation of daily rain gauge values that are partitioned via the use of half-hourly CMORPH data. Besides employing the use of OI, the new methodology will incorporate in situ precipitation information over the oceans thus the resulting analyses will provide bias-adjusted oceanic precipitation estimates as well as over land.

5 CMORPH Data Availability and Performance

The primary repository for CMORPH data is the Climate Prediction Center (CPC), which is a division of the National Centers for Environmental Prediction (NCEP) and is part of NOAA's National Weather Service. Subsets of the CMORPH product suite, for several temporal and spatial resolutions, are available at http://www.cpc.ncep.noaa.gov/http://www.cpc.ncep.noaa.gov/products/janowiak/cmorph.shtml

Due to the voluminous nature of the data and to the finite resources at NCEP, the entire time history of CMORPH at the highest spatial (~8 km) and temporal (30 min) resolution is not available at the time of this writing. However, work is in progress to make these data available via the World Wide Web.

Time series of the spatial correlation between the CPC daily rain gauge analysis (Higgins et al. 1996) with CMORPH and several other merging methods are shown in Fig. 5. The upper-left panel ("Merged PMW") shows the spatial correlation with a simple mosaic of all available PMW-derived precipitation estimates (dotted line) that are ingested into the CMORPH process. Note that CMORPH (solid line) outperforms that product consistently over the time period. Similarly, CMORPH generally performs better than the techniques in the remaining panels in terms of spatial correlation. The mean bias of CMORPH (Fig. 5) is similar in magnitude to the other methods, except for PERSIANN (Hsu et al. 1997) which has very low bias. The remaining methods for which statistics are displayed in Fig. 5

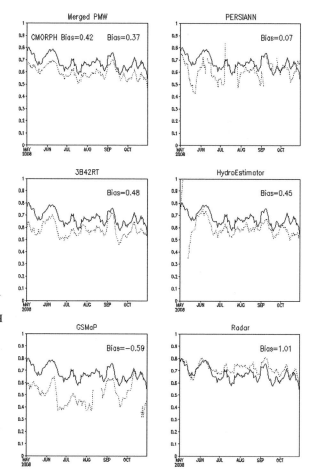

Fig. 5 Time series of spatial correlation between various merged precipitation estimates and rain gauge data over the U.S. For each panel, the *solid line* is for CMORPH and the *dotted line* is for the technique listed in the title above the panel. The mean bias for each technique is printed in the *upper-right* corner of each panel, and the CMORPH mean bias (0.42 mm day^{-1}) is printed in the *upper-left* corner of the panel labeled "Merged PMW"

are "3B42RT" which is the real-time version of the TMPA technique (Huffman et al. 2007) without bias correction, the "Hydroestimator" (Scofield and Kuligowski 2003), "GSMAP" (Kubota et al. 2007), and estimates from the "Stage II" U.S. radar mosaics (Lin and Mitchell 2005).

References

Ferraro, R. R., 1997: SSM/I derived global rainfall estimates for climatological applications. *J. Geophys. Res.*, 102, 16715–16735.

Ferraro, R. R., F. Weng, N. C. Grody, and L. Zhao, 2000: Precipitation characteristics over land from the NOAA-15 AMSU sensor. *Geophys. Res. Lett.*, 27, 269–2672.

Gandin, L. S., 1963: Objective analysis of meteorological fields. Leningrad, 1963; translated by Israel Program for Scientific Translations, Jerusalem, 1965.

Higgins, R. W., J. E. Janowiak, and Y. Yao, 1996: A gridded hourly precipitation data base for the United States (1963–1993). NOAA/NCEP/Climate Prediction Center Atlas No. 1, U. S. Dept. of Commerce, (available via world-wide-web http://www.cpc.ncep.noaa.gov/products/outreach/research_papers/ncep_cpc_atlas/1/)

Hsu, K., X. Gao, S. Sorooshian, and H. V. Gupta, 1997: Precipitation estimation from remotely sensed information using artificial neural networks. *Journal of Applied Meteorology*, 36, pp.1176–1190.

Huffman, G. J., R. F. Adler, D. T. Bolvin, D. Gu, E. J. Nelkin, Y. Hong, D. B. Wolf, K. P. Bowman, and E. F. Stocker, 2007: The TRMM Multisatellite Precipitation Analysis (TMPA): Quasi-global, multiyear, combined-sensor precipitation estimates at fine scales. *J. Hydromet.*, 8, 38–55.

Janowiak, J. E., R. J. Joyce, and Y. Yarosh, 2001: A real-time global half-hourly pixel-resolution IR dataset and its applications. *Bull. Amer. Meteor. Soc.*, 82, 205–217.

Janowiak, J. E., V. D. Dagostaro, V. E. Kousky, and R. J. Joyce, 2007: An examination of precipitation in observations and model forecasts during NAME with emphasis on the diurnal cycle. *J. Climate*, 20, 1680–1692.

Joyce, R., J. Janowiak, and H. Huffman, 2001: Latitudinally and seasonally dependent zenith-angle corrections for geostationary satellite IR brightness temperatures. *J. Appl. Meteor.*, 40, 689–703.

Joyce, R., J. E. Janowiak, P. A. Arkin, and P. Xie, 2004: CMORPH: a method that produces global precipitation estimates from passive microwave and infrared data at high spatial and temporal resolution. *J. Hydrometeor.*, 5(3), 487–503.

Kubota, T., S. Shige, H. Hashizume, K. Aonashi, N. Takahashi, S. Seto, M. Hirose, Y. N. Takayabu, K. Nakagawa, K. Iwanami, T. Ushio, M. Kachi, and K. Okamoto, 2007: Global precipitation map using satelliteborne microwave radiometers by the GSMaP Project: Production and validation. *IEEE Trans. Geosci. Remote Sens.*, 45, 2259–2275.

Kummerow, C., W. S. Olson, and L. Giglio, 1996: A simplified scheme for obtaining precipitation and vertical hydrometeor profiles from passive microwave sensors. *IEEE Trans. Geosci. Remote Sensing*, 34, 1213–1232.

Kummerow, C., Y. Hong, W. Olson, S. Yang, R. Adler, J. McCollum, R. Ferraro, G. Petty, and T. Wilheit, 2001: The evolution of the Goddard Profiling Algorithm (GPROF) for rainfall estimation from passive microwave sensors. *J. Appl. Meteor.*, 40, 1801–1820.

Li, Q., R. Ferraro, and N. C. Grody, 1998: Detailed analysis of the error associated with the rainfall retrieved by the NOAA/NESDIS SSM/I Rainfall Algorithm: Part I. Tropical oceanic rainfall. *J. Geophys. Res.*, 103, 11419–11427.

Lin, Y. and K. E. Mitchell, 2005: The NCEP Stage II/IV hourly precipitation analyses: development and applications. Preprints, 19th Conf. on Hydrology, American Meteorological Society, San Diego, CA, 9–13 January 2005, Paper 1.2.

McCollum, J. R., W. F. Krajewski, R. R. Ferraro, and M. B. Ba, 2002: Evaluation of biases of satellite rainfall estimation algorithms over the continental United States. *J. Appl. Meteor.*, **41**, 1065–1080.

McCollum, J. R. and R. R. Ferraro, 2003: The next generation of NOAA/NESDIS SSM/I, TMI and AMSR-E microwave land rainfall algorithms. *J. Geophys. Res.*, **108**, 8382–8404.

Rosenfeld, D. and Y. Mintz, 1988: Evaporation of rain falling from convective clouds as derived from radar measurements. *J. Appl. Meteor.*, **27**, 209–215.

Scofield, R. A. and R. J. Kuligowski, 2003: Status and outlook of operational satellite precipitation algorithms for extreme-precipitation events. *Mon. Wea. Rev.*, **18**, 1037–1051.

Simpson, J. R., R. F. Adler, and G. R. North, 1988: A proposed Tropical Rainfall Measurement Mission (TRMM) satellite. *Bull. Amer. Meteor. Soc.*, **69**, 278–295.

Turk, F. J., G. Rohaly, J. Hawkins, E. A. Smith, F. S. Marzano, A. Mugnai, and V. Levizzani, 2000: Meteorological applications of precipitation estimation from combined SSM/I, TRMM and geostationary satellite data. In: Microwave Radiometry and Remote Sensing of the Earth's Surface and Atmosphere, P. Pampaloni and S. Paloscia Eds., VSP Int. Sci. Publisher, Utrecht (The Netherlands), 353–363.

Vicente, G. A., J. C. Davenport, and R. Scofield, 2002: The role of orographic and parallax correction on real time high resolution satellite rainfall rate distribution. *Int. J. Remote Sens.*, **10**, 221–230.

Vila, D., R. Ferraro, and R. Joyce, 2007: Evaluation and improvement of AMSU precipitation retrievals. *J. Geophys. Res.*, **112**, D20119, doi:10.1029/2007JD008617.

Weng, F., L. Zhao, R. R. Ferraro, G. Poe, X. Li, and N. C. Grody, 2003: Advanced microwave sounding unit cloud and precipitation algorithms. *Radio Sci.*, **38**(4), 8068, doi:10.1029/2002RS002679.

Wiheit, T., C. Kummerow, and R. Ferraro, 2003: Rainfall algorithms for AMSR-E. *IEEE Trans. Geosci. And Rem. Sens.*, **41**, 204–214.

Zhao, L. and F. Weng, 2002: Retrieval of ice cloud properties using the Advanced Microwave Sounding Unit (AMSU). *J. Appl. Meteor.*, **41**, 384–395.

The Self-Calibrating Multivariate Precipitation Retrieval (SCaMPR) for High-Resolution, Low-Latency Satellite-Based Rainfall Estimates

Robert J. Kuligowski

Abstract The Self-Calibrating Multivariate Precipitation Retrieval (SCaMPR) is an algorithm for retrieving rainfall rates using visible (VIS)/infrared (IR) and microwave-frequency data from Earth-orbiting satellites. Rainfall rates derived from microwave-frequency data are used as a calibration target for an algorithm framework that both selects the optimal VIS/IR predictors and determines their optimal calibration coefficients in real time. This algorithm is highly flexible and its short data latency makes it well-suited for rapidly-changing heavy rainfall situations that trigger flash flooding.

Keywords Satellite · Rainfall · Infrared · Microwave · Flash flood

1 Introduction

The impacts of flood-related disasters are exacerbated in many parts of the world by the unavailability of critical rainfall information in real time. Rainfall information from satellites in geostationary orbit offer high spatial and temporal resolution with minimal data latency, which is well-suited for rapidly-developing heavy rainfall events for instances where more direct measurements of rainfall from gauges or ground-based radars are not available. While these satellite-based estimates are generally not as accurate as gauges or radar, they can provide a level of accuracy in many situations that makes them highly useful for real-time forecasting of heavy rainfall and resulting flood hazards.

Early estimates of rainfall rate from satellites were based on cloud-top brightness temperatures from the infared (IR) window region of the spectrum (e.g., Scofield 1987; Adler and Negri 1988) and sometimes also from visible (VIS) albedo (e.g., Griffith et al. 1978). These approaches are generally based on the assumption

---

R.J. Kuligowski (✉)
NOAA/NESDIS Center for Satellite Applications and Research, Camp Springs, MD, 20746-4304, USA
e-mail: Bob.Kuligowski@noaa.gov

that cloud-top temperature (and, by extension, cloud-top height) is a reliable predictor of rainfall rate, which works relatively well for convective rainfall but quite poorly for stratiform rainfall (e.g., Ebert et al. 2007). However, they also suffered from weaknesses in discriminating cold, non-raining cirrus clouds from cumulonimbus clouds, which often resulted in significant overestimation of the extent of heavy rainfall (e.g., Rozumalski 2000).

Beginning with the launch of the Special Sensor Microwave/Imager (SSM/I) in 1987, microwave (MW)-frequency information became routinely available for estimating rainfall rates. The advantage of these frequencies is that the radiative signal is sensitive to the properties of clouds throughout their depth and not just their uppermost portions. The disadvantage is that at present these frequencies can be monitored only from low-Earth orbit, so data at a particular location are generally available only twice per day for each satellite. This greatly limits the utility of microwave-based estimates of rainfall for rapidly-evolving situations. Also, MW instruments tend to have much coarser spatial resolution than IR instruments, making it difficult to discern the small-scale details of heavy rainfall systems.

Numerous authors have attempted to address this issue by combining geostationary IR and microwave data in order to optimize both accuracy and availability. An overview of all of the available techniques is far beyond the scope of this chapter (though some of them are addressed in detail elsewhere in this book), but it includes histogram matching (e.g., Turk et al. 2003; Huffman et al. 2007), artificial neural networks (e.g., Sorooshian et al. 2000), and using the IR as an interpolation template for MW-based rain rates (e.g., Joyce et al. 2004).

A combination technique that is aimed toward short-term flash flood forecasting applications is the Self-Calibrating Multivariate Precipitation Retrieval (SCaMPR), first introduced in Kuligowski (2002). It uses microwave rain rates as the calibration source for a framework that uses data from a selection of VIS/IR bands (or other data) as predictor information. The algorithm uses discriminant analysis to select and calibrate predictors for identifying rainfall areas, and uses stepwise forward regression to select and calibrate predictors for estimating rainfall rate. The resulting calibration coefficients are then applied to subsequent GOES imagery at the full spatial and temporal resolution of the GOES imager, allowing estimates with a temporal refresh and data latency of the GOES data—a temporal refresh of 15 min at most times over the CONUS and surrounding areas, and a latency of just minutes between the time of the GOES image and the availability of the SCaMPR rainfall rate field. The calibration is updated as new MW rainfall data become available. The algorithm has been run in real time over the CONUS and nearby regions since November 2004.

Section 2 of this chapter describes the data inputs to SCaMPR, while Section 3 summarizes the general architecture of the algorithm. Changes to the algorithm that have been made since Kuligowski (2002) are outlined in Section 4, while Section 5 will show some examples and validation from the current version of SCaMPR. Future plans for the algorithm will be summarized in Section 6.

2 Instruments and Input Datasets

The real-time version of SCaMPR that began running in November 2004 uses IR (bands 3–6) brightness temperature data from GOES-West (located at 135°W) and from GOES-East (located at 75°W). Additional predictors are derived from these inputs as described in Section 3. The new version of SCaMPR also uses VIS (0.6 μm) data from GOES; additional details are provided in Section 4.

Microwave rainfall rates from the SSM/I (Ferraro 1997) and Advanced Microwave Sounding Unit (AMSU; Vila et al. 2007) are used in the real-time version that began running in November 2004. Rainfall rates from the Tropical Rainfall Measuring Mission (TRMM) Microwave Imager (TMI; Kummerow et al. 2001) and Precipitation Radar (PR; Iguchi et al. 2000) have been added to the SCaMPR calibration data set, as described in Section 4.

3 General Methodology

SCaMPR is described in detail in Kuligowski (2002), but a brief overview is provided here. The SCaMPR algorithm can be broken down into a calibration component and an application component. The former uses previous IR/VIS data and MW-derived rainfall rates to select predictors and derive associated coefficients for identifying raining clouds and deriving rainfall rates. The latter component applies these coefficients to current IR/VIS data to produce corresponding rainfall rate fields with the spatial coverage and refresh rate of the IR/VIS data.

The calibration portion begins with the matching of IR/VIS data and MW-derived rain rates. GOES pixels within 15 min of a MW pixel of interest are spatially aggregated to the MW pixel resolution (assumed to be roughly 15 km) and placed in a file of matched GOES and MW pixels that corresponds to the 15 × 15° regions (they overlap, so there may be as many as nine) containing that pixel. When new pixels are obtained, the oldest ones are cycled out of the data set such that there are at least 500 pixels with MW rain rates exceeding 2.5 mm/h—this was done to ensure that the training data sample contained a sufficient number of significant rainfall events for a statistically reliable calibration. Separate calibration files are maintained for each calibration region.

Whenever new data are added to the calibration file, the calibration is updated. This involves selecting and optimizing predictors for rain/no rain separation and for rain rate estimation. The specific predictors that are available are GOES bands 3 (6.5–6.7 μm), 4 (10.7 μm), and 5 (12.0 μm – GOES-11 and earlier) or 6 (13.3 μm – GOES-12 and after). In addition, all brightness temperature difference (BTD) combinations are considered as predictors, along with cloud texture parameters that were adopted from Adler and Negri (1988) by Ba and Gruber (2001):

$$G_t = T_{avg} - T_{min} \tag{1}$$

$$S = 0.568(T_{min} - 217\,K) \tag{2}$$

where T_{min} is the minimum brightness temperature within the surrounding 5 × 5-pixel area and T_{avg} is the mean of the six nearest pixels (four along the scan line and two across the scan). G_t has a high value in highly textured clouds and S is used as a scaled threshold value of G_t in Adler and Negri (1988).

For the rain/no rain separation phase, the pixels in the calibration data set are divided into those with zero and non-zero MW rain rates. Discriminant analysis (basically a variant of linear regression with target values of 0 for non-raining pixels and 1 for raining pixels) is used to select and calibrate rain-no rain predictors in order to optimize the Heidke Skill Score:

$$HSS = \frac{2(c_1 c_4 - c_2 c_3)}{(c_1 + c_2)(c_2 + c_4) + (c_3 + c_4)(c_1 + c_3)} \qquad (3)$$

where c_1 is the number of dry/dry pixels (dry estimate, dry observation), c_2 is the number of wet/dry pixels, c_3 is the number of dry/wet pixels, and c_4 is the number of wet/wet pixels. The discriminant analysis is run stepwise until two predictors are chosen; additional predictors have not been shown to improve performance.

Only the raining pixels are used in the rain rate calibration phase. Because the relationship between rainfall rate and brightness temperature is clearly nonlinear (e.g., Fig. 1a of Vicente et al. 1998), linear regression will not adequately capture these relationships. To deal with this, a second set of rain rate predictors is produced by regressing each predictor against the rain rate in log-log space, creating a power-law regression that is equivalent to $y = ax^b$, where y is the target value, x the predictor, and a and b are the regression coefficients. Note that the exponential fit ($y = ae^{bx}$) used in Vicente et al. (1998) was tested in early work, but the power-law function produced better results in this work. Both sets of predictors are then used in the regression pool for selecting and calibrating the rain rate parameters, and the stepwise forward regression is stopped after two predictors are selected since tests have shown no benefit in adding a third rain rate predictor.

Once the calibration is completed, the selected predictors and coefficients are used to derive estimates of rainfall rate for subsequent IR imagery at the full resolution of the imagery. This means that the SCaMPR rain rate estimates have the same spatial resolution (4 km), temporal refresh (15 min over the CONUS and surrounding regions, less frequent elsewhere) and data latency (minutes) of the GOES data.

It should be noted that this is a finer resolution than the resolution at which the calibration was performed, which does introduce some error since it is known that for a nonlinear function f(x):

$$\overline{f(x)} \neq f(\overline{x}) \qquad (4)$$

The magnitude of this error has not been evaluated quantitatively; however, the algorithm appears to produce satisfactory results compared to other rain rate algorithms in spite of the difficulties of this assumption.

4 Current Status on Algorithm Development

The version of SCaMPR that has been running in real time at NESDIS since November 2004 is somewhat different from the version presented in Kuligowski (2002); the modifications are described in this section. As mentioned in Section 2, the AMSU is used in addition to the SSM/I as a calibration data source, and as mentioned in Section 3, the length of the training period now depends on the number of raining pixels available instead of being a fixed length of time.

The original version of SCaMPR in Kuligowski (2002) noted that even with the nonlinearly transformed predictors, the regression scheme did not adequately fit high rain rates. In Kuligowski (2002) this was addressed by padding the training data with additional copies of the highest rain rates in order to improve the representation of heavier precipitation events in the algorithm calibration. Subsequent investigation revealed that the problem was actually due to the lack of an intercept in the power-law regression, which led to incorrect curve fits since one end of the curve was effectively anchored at the origin. Adding an intercept (i.e., $y = ax^b + c$) led to much better curve fits and eliminated the need for data padding.

Also, whereas SCaMPR in Kuligowski (2002) uses rain rate fields from other algorithms as predictors, this is not done in the current real-time version of SCaMPR; it is a stand-alone algorithm. Finally, since the real-time version of SCaMPR covers a much larger area (20–60°N, 135–60°W), the algorithm was modified to produce separate calibrations for $15 \times 15°$ boxes with 5° of overlap on each side, as alluded to in Section 3. The rainfall rate at a given pixel is thus a weighted average of the values from the 9 different $15 \times 15°$ boxes which include that pixel, with the weighting being determined by the inverse squared distance to each box center.

Additional modifications to SCaMPR were implemented into the real-time version of the algorithm in 2009. The first is an improvement in the calibration data by using rainfall rate estimates from TRMM. The TMI and PR provide an additional source of microwave rainfall rates south of 38°N, and in a variant of the method used by Joyce et al. (2004), the TMI rainfall rates are used to bias-adjust the SSM/I and AMSU rainfall rates. This is done by separately matching the cumulative distribution functions (CDF's) of the SSM/I and AMSU against corresponding TMI data (after aggregating the TMI data to the footprint size of the SSM/I and AMSU) to create look-up tables (LUT's) for adjusting the SSM/I and AMSU rain rates. The previous 30 days of matched data are used, and the LUT's are updated daily.

Another modification was based on work that was done when applying SCaMPR to Spinning Enhanced Visible/InfraRed Imager (SEVIRI) data when evaluating its utility for GOES-R applications (see Section 6 for additional details). It was found during this work that the brightness temperature difference between the IR window and water vapor (WV) absorption bands could be used to distinguish convective from stratiform precipitation (e.g., Tjemkes et al. 1997). Specifically, the relationship between band 4 (10.7 μm) brightness temperature and rainfall rate was analyzed as a function of the difference between GOES bands 3 (6.5–6.7 μm) and 4, and it was determined (Fig. 1) that a significant transition took place below −2 K for

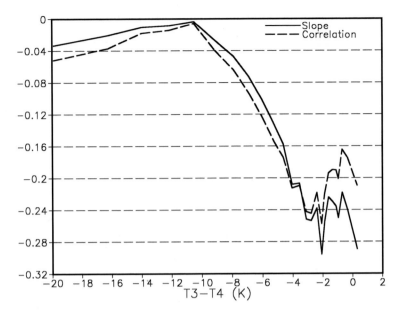

Fig. 1 Plot of the slope of the GOES T4 (10.7- μm)-rain rate relationship as a function of the difference between GOES T3 (6.5 μm) and T4 for GOES-12. Note the sharp transition that takes place at differences below –3 K

GOES-11 and –3 K for GOES-12. Consequently, separate calibrations of SCaMPR were performed for these two regimes.

In addition, the SEVIRI work revealed that the optimal training regions for SCaMPR were much larger than those used in the real-time version; four latitudinal bands of 30° in width (60–30°S, 30°S-EQ, EQ-30°N, 30–60°N) gave better results than larger or smaller regions. As a result, the real-time version of SCaMPR was modified to include only four spatial regions separated at 30°N and 105°W (the latter made necessary by the interface between GOES-West and GOES-East coverage). Subsequent work confirmed that this was the optimal configuration for GOES. To prevent discontinuities across the region boundaries, the calibration regions actually extend 5° beyond these boundaries and points in the 10°-wide overlap zone are assigned a distance-weighted value of the rain rates for all relevant regions.

The issue of the optimal number of training data points was also revisited, and it was found that for regions of this size, 2,000 pixels with rain rates exceeding 0.25 mm/h were required when T3–T4 was strongly negative, but only 400 such pixels were required when T3–T4 was weakly negative or positive. It is believed that this is because higher rain rates are much more common in the latter regime than in the former, and so less data is required to obtain a representative sample of significant rainfall events.

Finally, the SEVIRI work also revealed the usefulness of visible-band data (normalized by dividing by the cosine of the solar zenith angle), but only for stratiform clouds (the VIS data did not have any discernible benefit for deep convective clouds). The same conclusions were drawn for the GOES work, and so two separate calibrations are used: daytime-only, and a 24-h calibration that does not use visible data and is applied only during the night time hours.

5 Comparisons and Examples

Figure 2 is an example of SCaMPR output for 2015 UTC 30 August 2007 that illustrates the expanded coverage area. It should be noted that because the current-generation GOES requires half an hour to scan the full disk and do so only every 3 h (and then half an hour apart), instantaneous SCaMPR rain rates (and those from any other technique that uses GOES data) are never available simultaneously for the entire coverage area; however, for this coverage area any particular region seldom experiences a period of longer than 30 min without a rain rate estimate.

In Fig. 3, the original and new versions of SCaMPR are compared with the Stage IV radar/rain gauge field and the Hydro-Estimator (Scofield and Kuligowski 2003), which is the current operational satellite rainfall estimate at the National Oceanic and Atmospheric Administration (NOAA) National Environmental Satellite, Data, and Information Service (NESDIS). Although all three satellite rainfall estimates tend to spread out the heavier rainfall, both versions of SCaMPR avoid the significant overestimation of the heaviest rainfall over Iowa and Nebraska. The new version of SCaMPR is too light relative to the original in this particular case, however. In terms of overall performance, the changes to SCaMPR have produced slight

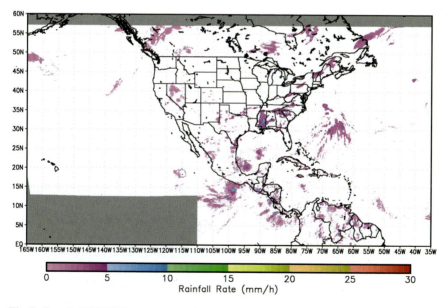

Fig. 2 Sample SCaMPR instantaneous rain rate field for 2015 UTC 30 August 2007

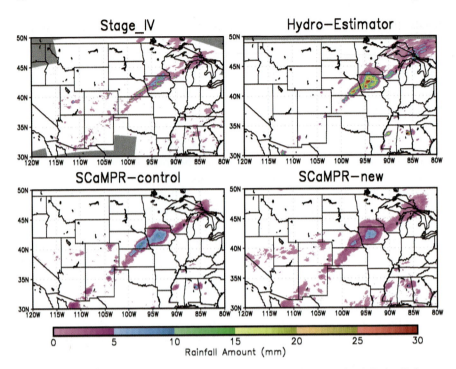

Fig. 3 Comparison of 1-h rainfall totals Stage IV (radar/gauge), operational Hydro-Estimator, original (control) version of SCaMPR, and new version of SCaMPR at 0200 UTC 29 August 2007

overall improvement in the algorithm in terms of both reducing an overall dry bias of SCaMPR (which is admittedly not readily apparent in the example in Fig. 3) and slightly improving the correlation between the estimates and observed values.

6 Future Plans and Conclusions

As mentioned briefly in Chapter 4, SCaMPR has been selected for use in the GOES-R operational ground system. GOES-R is scheduled for launch in 2015 (as of this writing) and represents the next generation of NOAA's geostationary satellites, and features a 16-band Advanced Baseline Imager (ABI) and a Geostationary Lightning Mapper (GLM). The GOES-R Algorithm Working Group (AWG) Hydrology Algorithm Team tested several candidate algorithms for potential operational use for GOES-R, and SCaMPR was chosen to move forward with operational implementation. SCaMPR will thus be used to produce estimates of rainfall rate for the full GOES disk (up to 60° latitude or 70° local zenith angle (LZA), whichever is less) every 5 min at the full 2-km resolution of the ABI. The SCaMPR rainfall rates will also be extrapolated forward in time to produce nowcasts of the probability of measurable rainfall and the predicted rainfall accumulation during the subsequent 3 h.

Additional modifications to SCaMPR may be implemented prior to operational deployment, perhaps including the use of GLM data for precipitation classification and/or rain rate retrieval. Previous work by the author showed a positive impact from using lightning data for both classifying precipitation systems and as a rain rate predictor, and the GLM data could prove to have even greater impact since they contain total lightning data whereas the previous work used only cloud-to-ground lightning data.

Acknowledgments This work was supported at various stages by the NESDIS GOES I-M Product Improvement Plan (GIMPAP), the GOES-R Program Office, and the NASA Research Opportunities in Space and Earth Sciences (ROSES). Ruiyue Chen, Wei Guo, Jung-Sun Im, and Yaping Li have all made significant contributions to the development of SCaMPR. The figures in this chapter were produced using the Grid Analysis and Display System (GrADS). The contents of this chapter are solely the opinions of the author and do not constitute a statement of policy, decision, or position on behalf of NOAA or the U. S. Government.

References

Adler, R. F. and A. J. Negri, 1988: A satellite infrared technique to estimate tropical convective and stratiform rainfall. *J. Appl. Meteor.*, **27**, 30–51.

Ba, M. and A. Gruber, 2001: GOES Multispectral Rainfall Algorithm (GMSRA). *J. Appl. Meteor.*, **29**, 1120–1135.

Ebert, E. E., J. E. Janowiak, and C. Kidd, 2007: Comparison of near-real-time precipitation estimates from satellite observations and numerical models. *Bull. Amer. Meteor. Soc.*, **88**, 47–64.

Ferraro, R. R., 1997: Special sensor microwave imager derived global rainfall estimates for climatological applications. *J. Geophys. Res.*, **102**, 16715–16735.

Griffith, C. G., W. L. Woodley, P. G. Grub, D. W. Martin, J. Stout, and D. N. Sikdar, 1978: Rain estimation from geosynchronous satellite imagery – Visible and infrared studies. *Mon. Wea. Rev.*, **106**, 1153–1171.

Huffman, G. J., R. F. Adler, D. T. Bolvin, G. -J. Gu, E. J. Nelkin, K. P. Bowman, Y. Hong, E. F. Stocker, and D. B. Wolff, 2007: The TRMM Multisatellite Precipitation Analysis (TMPA): Quasi-global, multiyear, combined-sensor precipitation estimates at fine scales. *J. Hydrometeor.*, **8**, 38–55.

Iguchi, T., T. Kozu, R. Meneghini, J. Awaka, and K. Okamoto, 2000: Rainprofiling algorithm for the TRMM precipitation radar. *J. Appl. Meteor.*, **39**, 2038–2052.

Joyce, R. J., J. J. Janowiak, P. A. Arkin, and P. Xie, 2004: CMORPH: A method that produces global precipitation estimates from passive microwave and infrared data at high spatial and temporal resolution. *J. Hydrometeor.*, **5**, 487–503.

Kuligowski, R. J., 2002: A self-calibrating real-time GOES rainfall algorithm for short-term rainfall estimation. *J. Hydrometeor.*, **3**, 112–130.

Kummerow, C., Y. Hong, W. S. Olson, S. Yang, R. F. Adler, J. McCollum, R. Ferraro, G. Petty, D. B. Shin, and T. T. Wilheit, 2001: The evolution of the Goddard Profiling Algorithm (GPROF) for rainfall estimation from passive microwave sensors. *J. Appl. Meteor.*, 40, 1801–1820.

Rozumalski, R. A., 2000: A quantitative assessment of the NESIDS auto-estimator. *Wea. Forecasting*, **15**, 397–415.

Scofield, R. A., 1987: The NESDIS operational convective precipitation estimation technique. *Mon. Wea. Rev.*, **115**, 1773–1792.

Scofield, R. A. and R. J. Kuligowski, 2003: Status and outlook of operational satellite precipitation algorithms for extreme-precipitation events. *Wea. Forecasting*, **18**, 1037–1051.

Sorooshian, S., K. Hsu, X. Gao, H. V. Gupta, B. Imam, and D. Braithwaite, 2000: Evaluation of PERSIANN system satellite-based estimates of tropical rainfall. *Bull. Amer. Meteor. Soc.*, **81**, 2035–2046.

Tjemkes, S. A., L. van de Berg, and J. Schmetz, 1997: Warm water vapor pixels over high clouds as observed by Meteosat. *Beitr. Phys. Atmos.*, **70**, 15–21.

Turk, F. J., E. E. Ebert, H. J. Oh, B. -J. Sohn, V. Levizzani, E. A. Smith, and R. Ferraro, 2003: Validation of an Operational Global Precipitation Analysis at Short Time Scales. Preprints, *3rd Conf. on Artificial Intelligence*. Long Beach, CA, Amer. Meteor. Soc, CD-ROM, JP1.2.

Vicente, G. A., R. Scofield, and W. P. Menzel, 1998: The operational GOES infared rainfall estimation technique. *Bull. Amer. Meteor. Soc.*, **79**, 1883–1898.

Vila, D., R. Ferraro, and R. Joyce, 2007: Evaluation and improvement of AMSU precipitation retrievals. *J. Geophys. Res.*, **112**, D20119.

Extreme Precipitation Estimation Using Satellite-Based PERSIANN-CCS Algorithm

Kuo-Lin Hsu, Ali Behrangi, Bisher Imam, and Soroosh Sorooshian

Abstract The need for frequent observations of precipitation is critical to many hydrological applications. The recently developed high resolution satellite-based precipitation algorithms that generate precipitation estimates at sub-daily scale provide a great potential for such purpose. This chapter describes the concept of developing high resolution Precipitation Estimation from Remotely Sensed Information using Artificial Neural Networks-Cloud Classification System (PERSIANN-CCS). Evaluation of PERSIANN-CCS precipitation is demonstrated through the extreme precipitation events from two hurricanes: Ernesto in 2006 and Katrina in 2005. Finally, the global near real-time precipitation data service through the UNESCO G-WADI data server is introduced. The query functions for viewing and accessing the data are included in the chapter.

Keywords Extreme precipitation · Image segmentation · Hurricane Katrina · Probability matching method · Self-organizing feature map

1 Introduction

Flood events caused by extreme precipitation are considered as the most frequent and widespread natural disaster in human history. It is estimated hundred millions of people are affected by flood events each year. Hurricane Katrina, for example, impaired the South Eastern US States, including Mississippi, Louisiana, Alabama, and Florida. The total damage is estimated at more than $80 billion and it is confirmed there were more than 1300 deaths (NWS NOAA, 2006). While accurate precipitation monitoring is a key element for improving flood-forecasting, precipitation observations from traditional means (e.g. rain gauges) are limited to point

K-L. Hsu (✉)
Civil and Environmental Engineering, The Henry Samueli School of Engineering,
Center for Hydrometeorology and Remote Sensing, UC Irvine, CA 92697-2175, USA
e-mail: kuolinh@uci.edu

M. Gebremichael, F. Hossain (eds.), *Satellite Rainfall Applications for Surface Hydrology*, DOI 10.1007/978-90-481-2915-7_4,
© Springer Science+Business Media B.V. 2010

measurements. Recent improvements in the ability of satellite remote sensing techniques provide a unique opportunity for precipitation observation for hydrologic applications of remote ungauged basins.

Satellite-based precipitation retrieval algorithms use information ranging from visible (VIS) to infrared (IR) spectral bands of Geostationary Earth Orbiting (GEO) satellites and microwave (MW) spectral bands from Low Earth Orbiting (LEO) satellites. GEO satellites give frequent observations every 15–30 min, but the information provided is indirect to the surface rainfall. Some improvements were reported using cloud classification approaches using texture measures and cloud-patch identification as well as combining information from multi-spectral imagery (Ba and Gruber, 2001; Behrangi et al. 2009a, b; Bellerby et al., 2000; Bellerby, 2004; Capacci and Conway, 2005; Hong et al., 2004; Turk and Miller, 2005). On the other hand, MW sensors on LEO satellites provide more direct sensing of rain clouds. Its low sampling frequency, however, limits the effectiveness of the rainfall retrievals. Integration of multiple LEO satellite information is considered to be effective in improving rainfall retrieval at short-time scales. Improvements in precipitation retrieval were also reported by locally adjusting GEO-IR retrievals using near-real-time LEO MW-based rainfall estimation (Bellerby et al., 2000; Bellerby, 2004; Hsu et al., 1997; Huffman et al., 2007; Kidd et al., 2003; Marzano et al., 2004; Nicholson et al., 2003a, b; Sorooshian et al., 2000; Todd et al., 2001; Turk and Miller, 2005; Xu et al., 1999). More recent developments include morphing of MW rainfall according to cloud advection from GEO-IR imagery and were found to be effective in improving rainfall retrievals (Bellerby et al., 2009; Hsu et al., 2009; Joyce et al., 2004; Ushio et al., 2009).

In this chapter, the Precipitation Estimation from Remotely Sensed Information using Artificial Neural Networks – Cloud Classification System (PERSIANN – CCS) is presented (Hong et al., 2004; Hsu et al., 2007). Instead of extracting local texture information in PERSIANN (Hsu et al., 1997, 1999; Sorooshian et al., 2000), PERSIANN-CCS extracts information from the whole cloud patch and provides multiple infrared brightness temperature versus rainfall rate (T_b–R) relationships for different cloud classification types. The chapter will be described as follows: In Section 2, the development of PERSIANN-CCS and its application to the rainfall retrieval using geostationary satellite IR imagery is introduced. In Section 3 and 4, evaluation of PERSIANN-CCS rainfall retrieval on extreme events and near real-time visualization of global precipitation estimation through the UNESCO G-WADI (Water And Development Information for Arid Lands-A global Network) map server are discussed. Finally, conclusions and future directions are discussed in Section 5.

2 Methodology

Rainfall estimation from the PERSIANN-CCS consists of four major steps: (1) IR cloud image segmentation, (2) feature extraction from IR cloud patches, (3) patch feature classification, and (4) rainfall estimation. The classification and rainfall

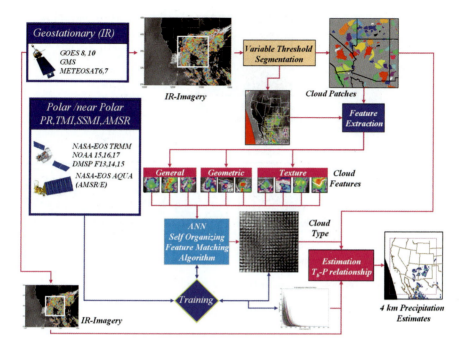

Fig. 1 The cloud image segmentation, feature extraction, classification, and rainfall estimation of PERSIANN-CCS algorithm

estimation using PERSIANN-CSS is demonstrated in Fig. 1. The description of the CCS system components is outlined below:

2.1 Cloud Image Segmentation

Studies show that different cloud systems consist of distinct thermal features and distributions of cloud precipitable water. In a satellite-based cloud image, a number of cloud systems may co-exist and be inter-connected in space and time. Although cloud systems may be distinguished from our visual observation, it is useful to develop an automatic method for classifying clouds. This is facilitated due to the growing computational capability of modern computers. Overall, an effective image segmentation procedure is one key step towards a better cloud image classification.

In this study a watershed-based (WA) segmentation approach (Vincent and Soille, 1991; Dobrin et al., 1994) is used for cloud patch segmentation. This algorithm starts with locating the local brightness temperature minimums in the IR cloud image. Each local minimum is considered as a seed point to form a cloud system (see "+" marker in Fig. 2a). Next, a threshold temperature is assigned and continued to increase; image pixels under the threshold connecting to those local minimum

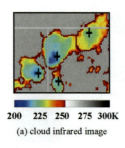

(a) cloud infrared image

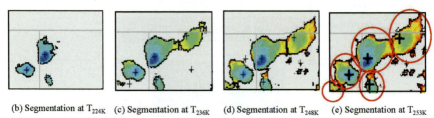

(b) Segmentation at T_{224K} (c) Segmentation at T_{236K} (d) Segmentation at T_{248K} (e) Segmentation at T_{253K}

Fig. 2 Cloud segmentation using a watershed-based algorithm: (**a**) the GOES IR imagery below 253 K and local temperature minimum points; (**b**)–(**e**) pixels are merged to several cloud patch groups slowly by increasing IR temperature thresholds from 210 to 253 K

seeds are merged to the same patch group. As temperature continues to rise (see Fig. 2b–c), image basins with pixels surrounding the local minimums are formed. When two basins are connected to each other, a border is formed to separate them (see Fig. 2d–e). The process stops when the assumed no-rain (or clear sky) threshold of IR temperature is reached. In our case, the no-rain IR threshold temperature is assigned as 253 K.

2.2 Input Feature Extraction

After the image segmentation, the cloud image is separated into a number of distinct patches, having various sizes and shapes. Transformation of the image under the cloud patch coverage to fewer features can be useful to reduce the redundant information in the patches, which as a result, may give rise to a more effective classification of cloud images. This transformation is called feature extraction. In our case, from the appearance of cloud systems, convective clouds are those puffy clouds with sharp temperature gradient near the coldest cloud top, while stratiform clouds are layer clouds with mild temperature gradients. Warm and cold clouds are also separable based on their cloud top temperature features. Features are selected from cloud patches with characteristics of (1) coldness, (2) geometry, and (3) texture (see Table 1). The features, except for Tb_{min} and TOPG, are extracted at three temperature thresholds (220, 235, and 253 K).

Table 1 The input features extracted from cloud patches

Category	Features
Coldness Features	Minimum temperature of a cloud patch: Tb_{min}
	Mean temperature of a cloud Patch: Tb_{mean}
Geometric Features	Cloud patch area: $AREA$
	Cloud patch shape index: SI
Texture Features	Standard deviation of cloud patch: STD
	Mean value of local standard deviation of cloud patch: $MSTD(5x5)$
	Standard deviation of local standard deviation: $STD_{std}(5x5)$
	Temperature gradient near Tb_{min}: $TOPG$

2.3 Image Classification

After the representative features are extracted from the cloud patches, the next stage is to the classify cloud patch features into a number of categories arranged in a two dimensional classification map according to the similarity between cloud patch features. This is accomplished by implementing an unsupervised clustering algorithm named Self-Organizing Feature Map (SOFM).

A SOFM is a two layer data classification network (See Fig. 3). The first layer connecting to the input features is called the input layer, while the second layer, with units arranged in two dimensional arrays, represents the classification layer. The array size in the 2nd layer is relevant to the cloud system (or patch) classification groups. In our case, an array of 20×20 units is assigned to the SOFM 2nd layer, which implies that 400 cloud groups were assigned. After SOFM training, similar cloud patch features are assigned to the nearby cloud groups. The mean classification feature of the cluster j is presented as the connecting parameters

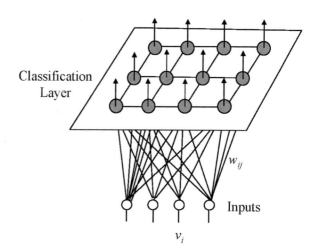

Fig. 3 A two layer self-organizing feature map

between the input features and the output unit j of the 2nd layer. The cloud patches with input features close to the parameters of unit j are assigned to the cluster j. Description of SOFM can be found from Kohonen (1995) and Hsu et al. (1999). A brief discussion of training SOFM is listed below:

Step 1: Initialize the weights (or parameters) w_{ij} as uniform random number in [0 1].

Step 2: Collect a set of cloud patches and extract their features. These features are normalized and are presented as: $\vec{V}(p) = [\vec{V}_{220}(p), \vec{V}_{235}(p), \vec{V}_{253}(p)]$, $p=1..N$, where N is the number of cloud patch samples used in the training; $\vec{V}_{220}(p)$, $\vec{V}_{235}(p)$, $\vec{V}_{253}(p)$ are features of patch p extracted from the temperature threshold of 220, 235, and 253 K, respectively.

Step 3: Select a cloud patch sample m having normalized features as $\vec{V}(m)$, $m \in [1,N]$. Find the best unit in the 2nd layer that has minimum distance between normalized input features and connect parameters ($\vec{w}_j$):

$$j^* = \arg\min_{j} \|\vec{w}_j - \vec{V}(m)\| \quad (1)$$

Step 4: Update the connection parameters for the neighborhood (radius r) units of j^* with learning rate α:

$$\vec{w}_j = \vec{w}_j + \alpha \left(\vec{V}(m) - \vec{w}_j\right) \quad (2)$$

Step 5: Terminate if the $\vec{w}_j$ is converged, or reduce r and α and go to step 3.

2.4 Mapping Patch to Pixel Rainfall

After the IR cloud image is processed through the patch segmentation and classification, the following stage is to assign rainfall rates to the pixels under the patch coverage. This requires a set of training data including GEO-IR image and rainfall measurement from radar or LEO-MW sensors. Over the continental United States (CONUS), ground-based radar rainfall can be used for this purpose. The process includes assigning IR cloud patches to the SOFM 2nd classification units and assigning surface rainfall rates, observed from radar or LEO-MW sensors, to the classified SOFM output units.

Taking rainfall estimation of the j-th SOFM output unit for example, a large number of cloud patches were assigned to this unit. The corresponding IR brightness temperature (Tb) and rainfall rate (R) in the cloud patch at pixel level are presented as:

$$\vec{X}^j(k) = [Tb^j(k), R^j(k)], k = 1..n_1. \quad (3)$$

Where n_1 is the total number of pixels assigned to the SOFM output unit j; $R^j(k)$ is the rainfall rate assigned to the SOFM output unit j.

PERSIANN Satellite Product

The probability matching method (PMM; Atlas et al, 1990; Vincent et al., 1998) is then used to define the relationship of Tb and R at pixel scale, with the assumption that the lower the Tb the higher the rainfall rate. The cumulative distribution functions of the data pairs of Tb-R of the j-th SOFM classification unit can be presented as:

$$\int_0^{R^*} P(R)dR = \int_{Tb_{min}}^{Tb_{max}} P(Tb)dTb - \int_{Tb_{min}}^{T_b^*} P(Tb)dTb = 1 - \int_{Tb_{min}}^{T_b^*} P(Tb)dTb \quad (4)$$

Where the $P(.)$ is the probability distribution function and the estimated value of Tb and R is in the range $[0\ R_{max}]$ and $[Tb_{min}\ Tb_{max}]$, respectively. To generalize the Tb and R relationship for future events, a power-law function of Tb and R is suggested (Vicente et al., 1998):

$$R^j(k) = f(\vec{\theta}^j, Tb^j(k)) = \theta_0^j + \theta_1^j \cdot \exp[\theta_2^j (Tb^j(k) + \theta_3^j)^{\theta_4^j}] \quad (5)$$

Where $\theta_k^j, k = 0, 1, \ldots, 4$ are parameters and j is the j-th *SOFM* output unit. In our case, 400 (i.e. 20 × 20) cloud patch groups were assigned and each group consists of a power law function of 5 parameters. Parameters for each cloud patch group are defined calibration data.

Figure 4(a) shows the 400 cloud patch groups, while each group is assigned a unique Tb-R function; for those groups being close to each other, the Tb-R curves are similar. Here we have selected a few cloud groups and their Tb-R functions, as they are marked as G0~G6 in Fig. 4a, b. For the group in region G0, the cloud Tb is high (clear weather pattern) and has no-rain associated with it. The slopes of Tb-R curves are higher for G2, G4, and G6 regions, which imply that those cloud groups are relevant to the convective clouds of low, medium, and high altitudes. For the

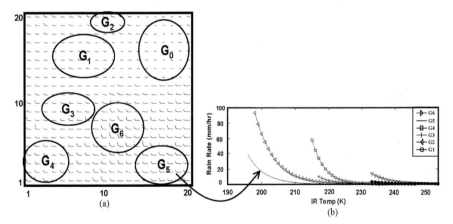

Fig. 4 (a) The Tb-R relationship of 20 × 20 SOFM cloud patch groups and (b) the Tb-R curves with respect to the SOFM groups G0~G6

regions covered by G1, G3, and G5, however, the slopes of *Tb-R* are less steep and they are associated with light or no-rain clouds at low, medium, and high altitudes, respectively.

3 Application Examples

As described in Section 2, PERSIANN-CCS extracts cloud information using three-temperature thresholds (220, 235, and 253 K), which enables it to obtain rainfall rate from different cloud systems. Our experiments show that PERSIANN-CCS can be effective at detecting cirrus clouds and helps to distinguish between different convective rain systems at various heights. A strong convective system is expected to reach the very high troposphere level or enter into the lower stratosphere, while a small local convective may stay in the lower troposphere.

Extreme rainfall events, particularly those associated with strong convective systems, pose a serious flooding threat to many populated areas worldwide. Better representation of the intensity and distribution of these events is extremely important to save human life and property. In the following section, performance of the high resolution PERSIANN-CCS product is evaluated under two extreme precipitation events which resulted in serious damage and fatality.

3.1 Hurricane Ernesto

Formed on August 24 in the eastern Caribbean Sea, Hurricane Ernesto was the costliest tropical cyclone of the 2006 Atlantic hurricane season. Early in the morning of August 30, Ernesto made landfall on the Florida mainland in southwestern Miami-Dade County. It moved across eastern Florida and hit North Carolina coast on August 31, just below hurricane strength.

Compared to ground radar, PERSIANN-CCS is found to be capable of estimating rain rates under regions of cold clouds, but fails to detect rainfall under relatively warm clouds areas, resulting in underestimation of both areal extent and volume of the rainfall (see Fig. 5 for example). Although some discrepancies are observed in hourly 0.04° PERSIANN-CCS rain estimates, overall evaluations have been encouraging. Such a rain estimate with high spatial resolution is a unique feature of PERSIANN-CCS and is beneficial for many applications including hydrological modeling. At coarser time and space resolution PERSIANN-CCS provides fairly reasonable estimation (see Fig. 6a–d). Daily accumulation of 0.04° rain estimate results in capturing rain details fairly well (Figs. 6a, b).

For further statistical assessment of such an extreme rain event, scatter plots of PERSIANN-CCS versus radar rain intensity estimate, accompanied by evaluation statistics, are presented in Fig. 7. In addition, Table 2 demonstrates the evaluation statistics for top 25 and 50 percentile rain rates to assess the capability of PERSIANN-CCS to capture heavy rainfall.

PERSIANN Satellite Product

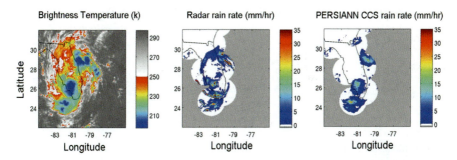

Fig. 5 Demonstration of IR (10.8 μm) image (*Left panel*) captured at 15:15 UTC on August 30 over Florida and corresponding hourly 0.04° rain rate obtained from ground radar (middle panel) and PERSIANN-CCS (Right panel). The event is captured from the hurricane Ernesto, passing over the region

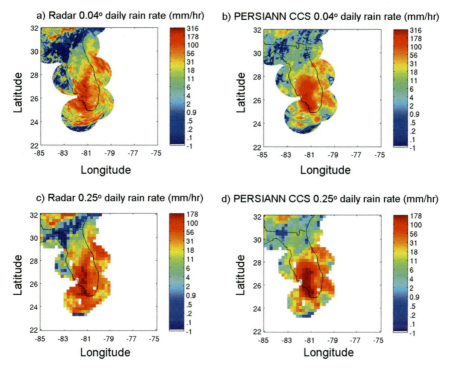

Fig. 6 Analyzing the performance of daily 0.04° and 0.25° PERSIANN-CCS rain rate estimate as compared to ground radar for hurricane Ernesto passing over Florida on August 30, 2006. White grid boxes represent the areas, either not covered by radar or have missing data in all or at least one of the hourly rain rate maps during the period of study

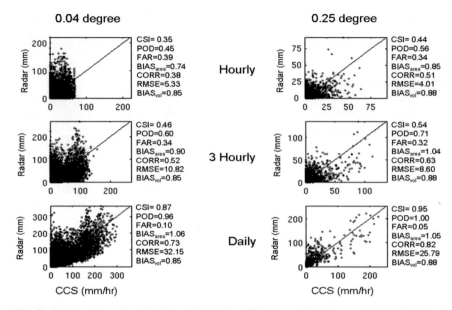

Fig. 7 Scatter plots and evaluation statistics for different combination scenarios of time and space resolution to analyze the performance of PERSIANN-CCS rain rate estimate as compared to ground radar for Hurricane Ernesto on August 30, 2006. Perfect BIAS$_{volume}$ and BIAS $_{area}$ should be 1

Table 2 Evaluation statistics for PERSIANN-CCS top 50 and 25 percentile rain rates for Hurricane Ernesto on August 30, 2006. The ground radar rain estimate is used as reference

Time resolution	Space resolution	Top percentile	CORR	RMSE (mm)	BIAS volume	CSI
Hourly	0.04°	50	0.28	12.46	0.54	0.56
	0.04°	25	0.25	16.46	0.43	0.65
	0.25°	50	0.42	8.14	0.66	0.71
	0.25°	25	0.39	10.54	0.60	0.82
3 Hourly	0.04°	50	0.39	21.70	0.65	0.75
	0.04°	25	0.37	27.86	0.56	0.82
	0.25°	50	0.54	15.74	0.74	0.87
	0.25°	25	0.52	19.99	0.70	0.94
Daily	0.04°	50	0.63	46.50	0.78	0.99
	0.04°	25	0.57	60.23	0.75	1.00
	0.25°	50	0.76	35.94	0.84	1.00
	0.25°	25	0.73	46.43	0.83	1.00

3.2 Hurricane Katrina

Hurricane Katrina is well known as one of the most destructive and deadliest hurricanes in the history of the United States. It formed over the Bahamas on August

23, 2005 and rapidly strengthened in the Gulf of Mexico and caused severe destruction, mainly due to the storm surge, along the Gulf coast. The greatest fatalities and destruction occurred in New Orleans, Louisiana where, Katrina made its second landfall as a Category 3 hurricane, in the morning of August 29, 2005 after the first landfall in Florida on August 24.

Similar to the previous case study, evaluation statistics of different combination scenarios of time and space resolutions are presented in Figs. 8 and 9 and Table 3 for Hurricane Katrina on August 29, 2005. Figure 8 displays daily 0.04° and 0.25° rain rate maps obtained from ground radar and PERSIANN-CCS estimates. Detailed statistics and scatter plots are shown in Fig. 9 and finally, the skill of PERSIANN-CCS to capture higher rain intensities is presented in Table 3.

The statistics demonstrate that both 0.04° and 0.25° PERSIANN-CCS estimates show reasonable results, particularly at coarser time resolution. From Table 3 and compared to the pervious example, it is inferred that PERSIANN-CCS is less skilful in capturing intense rainfall at finer time resolutions. However, at coarser time resolution, while with high spatial resolution, it is found to be relatively reliable in estimating intense rainfall.

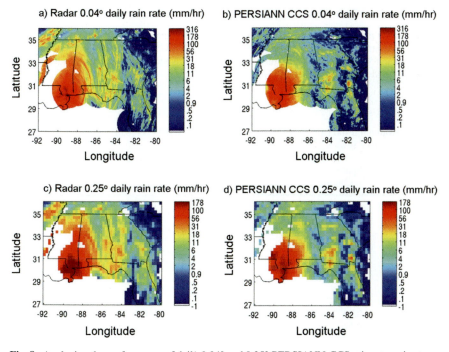

Fig. 8 Analyzing the performance of daily 0.04° and 0.25° PERSIANN-CCS rain rate estimate as compared to ground radar for hurricane Katrina, made its landfall on August 29, 2005. White grid boxes represent no data at all or at least within one of the hourly rain rate maps during the period of study

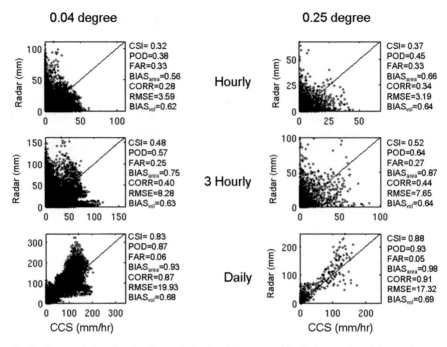

Fig. 9 Scatter plots and evaluation statistics for different combination scenarios of time and space resolution to analyze the performance of the PERSIANN-CCS rain rate estimate compared to ground radar for Hurricane Katrina on August 29, 2005. Note that, at the best condition, both BIAS$_{volume}$ and BIAS$_{area}$ should be 1

Table 3 Evaluation statistics for PERSIANN-CCS top 50 and 25 percentile rain rates: the ground radar rain estimate is used as reference for Hurricane Katrina on August 29, 2005

Time resolution	Space resolution	Top percentile	CORR	RMSE (mm)	BIAS volume	CSI
Hourly	0.04°	50	0.14	8.46	0.39	0.50
	0.04°	25	0.09	11.12	0.32	0.58
	0.25°	50	0.19	7.16	0.45	0.59
	0.25°	25	0.16	9.30	0.37	0.69
3 Hourly	0.04°	50	0.25	16.88	0.48	0.75
	0.04°	25	0.20	22.07	0.40	0.83
	0.25°	50	0.30	15.0	0.52	0.83
	0.25°	25	0.26	19.59	0.44	0.90
Daily	0.04°	50	0.84	29.87	0.66	0.97
	0.04°	25	0.77	39.97	0.71	0.99
	0.25°	50	0.89	25.45	0.68	1.00
	0.25°	25	0.84	33.99	0.73	1.00

4 Real-Time High Resolution Global Precipitation Server

UNESCO's International Hydrological Program (IHP) and the Center for Hydrometeorology and Remote Sensing at the University of California, Irvine (CHRS), have been collaborating to build capacity for the forecasting and mitigation of hydrologic disasters. The focus of these collaborations has been on the development of the means to extend the benefits of space and weather agencies' vast technological resources to the world wide community of hydrologists and resources managers, particularly in areas lacking hydrometeorological observational infrastructure. These collaborations have resulted in the development of several online precipitation mapping resources. Among these resources is the real-time global precipitation server. The server, which can be accessed at (http://hydis.eng.uci.edu/gwadi/), provides visualization and mapping of satellite-based precipitation estimates at $0.04° \times 0.04°$ spatial resolution obtained using real-time implementation of the PERSIANN-CCS. The GeoServer interface to PERSIANN-CCS data accounts for the following requirements: (a) ease of access, (2) rapid image rendering, (c) interface simplicity, and (4) hydrologically relevant functionalities, and (5) portability to other mirror sites. The latter requirement dictates that all development be conducted using public domain software. The server utilizes the open-source University of Minnesota MapServer technology (http://mapserver.org//) to provide real-time visualization of precipitation accumulation for most recent 3, 6, 12, 24, 48, and 72 h periods. The server also utilizes the open source FWTools (http://fwtools.maptools.org/) implementation of the Geospatial Data Abstraction Library (GDAL) and OpenGIS simple features Reference implementation (OGR) library to update precipitation summaries for watersheds including minimum, mean, maximum, and average rainfall within rain-area and to update and populate data-base tables with these information. These libraries are also used to facilitate rapid subsetting of data based on user defined region or a pre-defined polygon areas such as country, administrative unit, and/or watershed boundaries. Figure 10, shows the main page interface which consists of (1) map layer control menu, (2) interactive reference map, (3) map navigation and query menu, (4) map canvas, and (5) location bar. The layer control menu includes 4 sections which allow the user to select for display one or more overlay vector layers, a PERSIANN-CCS data layer, and a baseline image. A query type pull down menu is available with several choices as described below. Figure 11 illustrates the key map navigation functionalities

4.1 Map Navigation

Within the main page, map navigation consists of layer selection using the layer control menu and zooming and panning. Through web forms, users can load into the map canvas (1 in the figure) several vector layers including countries,

Fig. 10 Components of the G-WADI GeoServer web-interface including: (*1*) map layer control menu, (*2*) interactive reference map, (*3*) map navigation and query menu, (*4*) map canvas, and (*5*) location bar

administrative units, and watershed boundaries at different scales (continental-small watershed), streams, and water bodies. The user can also load either precipitation totals for the most recent 3-to-72 h as described above, or maps of extreme rainfall as foreground (2 in the figure). The canvas background layers include satellite image background (from NASA-MODIS), aridity, International Geoshpere-Biosphere Programme (IGBP) land use, and/or elevation maps. The user can zoom in to explore details of precipitation accumulation and view the spatial patterns of rainfall. The speed of image generation is enhanced by using image-tiling techniques to allow for rapid generation of area-views at multiple scales. Panning allows the user to retain the current zoom level while freely moving the area of interest either by sliding the map or by selecting a new map center. In addition, users can jump from one area into another on the globe, while maintaining the current zoom level by clicking the new area on the reference map available in the layer control menu.

PERSIANN Satellite Product

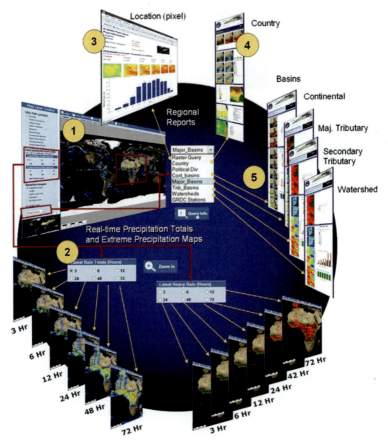

Fig. 11 GWADI-GeoServer map navigation and query functionalities. (*1*) layer selection and navigation, (*2*) available precipitation information, (*3*) precipitation information at a point, (*4*) country and administrative unit's reports, and (*5*) watershed report

4.2 Query Functions

Users can select pixel query to obtain a location report that includes the most recent precipitation accumulation for the selected location along with long-term (monthly average) precipitation. Elevation, land cover and coordinates of the pixel are also reported with rain maps of the $0.5° \times 0.5°$ area surrounding the selected pixel. Additional queries can be performed using geographic boundaries. The results of these queries are precipitation reports for countries and for administrative units. Each report includes the most recent 3, 6, 12, 24, 48, and 72 h rainfall and extreme rainfall maps. In addition, each report includes maps of land cover (based on the IGBP classification), elevation from the United State Geological Survey (USGS) Hydro1k (1 km), and Aridity map. Each country report also includes a link where users can download a suite of vector layers extracted from the Digital Chart

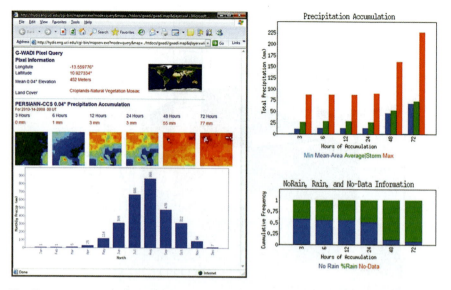

Fig. 12 *Left*: Pixel query, *Right*: Precipitation summary charts available through watershed reports

of the World data set. The hydrologically relevant watershed queries, which are available for four consecutive levels (continental, tributary, major watershed, and watershed) or watersheds add summary charts of minimum, maximum, average, and average/rain-area precipitation values for 6 accumulation periods. Watershed reports also include charts of rain/no-rain/no-data distribution for each of the 6 accumulation periods (Fig. 12).

4.3 Data Access

Users can obtain data either directly from the main interface or through the dynamically generated report pages with the exception of the single pixel report. In the first case, by selecting *get data*, a data access page is generated automatically with the current map extent being used as the default data extraction region. Users can change the default coordinate by selecting a new bounding box with the resulting page and proceeding to download the data. Data access is also available from within all polygon-based query results. For each of the above-described reports, the user can request the data associated with the area represented by the country, administrative unit, or selected watershed using the bounding box of the relevant polygon as the target area. In both cases however, the data is immediately extracted and compressed for all 6 accumulation periods and the user can obtain the data directly through the http server without needing to provide email address. Precipitation data is provided in the Arcview ASCII format which is easily readable by most commercial and public domain GIS.

5 Conclusions and Future Directions

This chapter describes the development of PERSIANN-CCS for rainfall estimation and the data visualization and service through the G-WADI Geo-Server. Two case studies were provided to demonstrate the capability of PERSIANN-CCS in the measurement of precipitation from two hurricane events. The results were evaluated using ground-based radar measurements at various spatial (0.04° and 0.25°) and temporal (hourly to daily) scales over the Southeastern United States. Case studies show that, at 0.04° resolution, accumulation of PERSIANN-CCS estimation from 3-h to daily scale has improved correlation coefficients from 0.52 to 0.73 for Hurricane Ernesto and from 0.40 to 0.87 for Hurricane Katrina. Likewise, at 0.25° resolution, correlation coefficients are improved from 0.63 to 0.82 for Hurricane Ernesto and from 0.44 to 0.91 for Hurricane Katrina (see Figs. 7 and 9). The results demonstrated great potential of using PERSIANN-CCS estimates for hydrologic applications.

Although PERSAINN-CCS performs fairly well under cold clouds (Tb < 253 K), extending the algorithm's capability to capture warmer rainfall requires more investigation. Employing multi-spectral images can be an alternative for this purpose. During daytime a visible channel provides added information relevant to cloud thickness, which is beneficial for rain detection and estimation (Behrangi et al., 2009a, b; Cheng et al., 1993; Griffith et al., 1978; Osullivan et al., 1990). A water vapor channel was also found useful for rain retrieval purposes (Ba and Gruber, 2001; Behrangi et al., 2009a; Desbois et al., 1982; Kurino, 1997). These two channels are currently available globally from a constellation of geosynchronous satellites hence, have potential to be implemented in operational rain estimation algorithms. With the advent of modern imagers on the recent and future geostationary satellites, in particular the Spinning Enhanced Visible and InfraRed Imager (SEVERI) on Meteosat Second Generation (MSG) and the Advanced Baseline Imager (ABI) on GOES-R, more spectral channels with higher temporal and spatial resolution are becoming available.

Recent developments using GEO cloud motion vectors to propagate and smooth MW rainfall estimates have also demonstrated excellent skills in rain retrieval. The Climate Prediction Center Morphing (CMORPH) algorithm, for example, applied the difference between average cloud advection and the motion of surface rainfall and then linearly interpolated MW rainfall estimates along advection streamlines (Joyce et al., 2004). A new multi-platform multi-sensor satellite rainfall algorithm, named Lagrangian Model (LMODEL), has been developed to evaluate the effectiveness of the cloud development modeling/model updating approach (Bellerby et al., 2009; Hsu et al., 2009). The LMODEL synergizes recent developments in cloud development modeling, satellite cloud-feature extraction, cloud image tracking, geostatistics and sequential filtering theory to develop microwave and IR combination algorithm for rainfall retrieval. Investigation of extreme rainfall estimation using multi-spectral imagery and LMODEL cloud tracking approaches are ongoing and will be reported in the near future.

Acknowledgement Partial support for this research is from NASA-EOS (Grant NA56GPO185), NASA-PMM (Grant NNG04GC74G), NASA NEWS (Grant NNX06AF93G) and NSF STC for "Sustainability of Semi-Arid Hydrology and Riparian Areas" (SAHRA) (Grant EAR-9876800). Authors appreciate the satellite data processing from Dan Braithwaite and manuscript editing from Diane Hohnbaum.

References

Atlas, D., D. Rosenfeld, and D. B. Wolff, 1990: Climatologically turned reflectivity-rainrate relationship and links to area-time integrals. *Journal of Applied Meteorology*, **29**, 1120–1135.

Ba, M. B. and A. Gruber, 2001: GOES multispectral rainfall algorithm (GMSRA). *Journal of Applied Meteorology*, **40**, 1500–1514.

Behrangi, A., K. Hsu, B. Imam, S. Sorooshian, and R. J. Kuligowski, 2009a: Evaluating the utility of multi-spectral information in delineating the areal extent of precipitation. *Journal of Hydrometeorology*, In **10**, 684–700.

Behrangi, A., K. L. Hsu, B. Imam, S. Sorooshian, G. J. Huffman, and R. J. Kuligowski, 2009b: PERSIANN-MSA: A precipitation estimation method from satellite-based multi-spectral analysis. *Journal of Hydrometeorology*, In press.

Bellerby, T., M. Todd, D. Kniveton, and C. Kidd, 2000: Rainfall estimation from a combination of TRMM Precipitation Radar and GOES multi-spectral satellite imagery through the use of an artificial neural network. *Journal of Applied Meteorology*, **39**, 2115–2128.

Bellerby, T., 2004: A feature-based approach to satellite precipitation monitoring using geostationary IR imagery. *Journal of Hydrometeorology*, **5**, 910–921.

Bellerby, T., K. Hsu, and S. Sorooshian, 2009: LMODEL: A satellite precipitation algorithm using cloud development modeling and model updating. Part 1: Model development and calibration. *Journal of Hydrometeorology*, In press.

Capacci, D. and B. J. Conway, 2005: Delineation of precipitation areas from MODIS visible and infrared imagery with artificial neural networks. *Meteorological Applications*, **12**, 291–305.

Cheng, M., R. Brown, and C. G. Collier, 1993: Delineation of precipitation areas using Meteosat infrared and visible data in the region of the United Kingdom. *Journal of Applied Meteorology*, **32**, 884–898.

Desbois, M., G. Seze, and G. Szejwach, 1982: Automatic Classification of Clouds on METEOSAT imagery: Application to high-level clouds. *Journal of Applied Meteorology*, **21**, 401–412.

Dobrin, B. P., T. Viero, and M. Gabbouj, 1994: Fast watershed algorithms: analysis and extensions. *Nonlinear Image Processing V*, 2180, 209–220, SPIE.

Griffith, C. G., W. L. Woodley, P. G. Grube, D. W. Martin, J. Stout, and D. N. Sikdar, 1978: Rain estimation from geosynchronous satellite imagery – visible and infrared studies. *Monthly Weather Review*, **106**, 1153–1171.

Hsu, K., X. Gao, S. Sorooshian, and H. V. Gupta, 1997: Precipitation estimation from remotely sensed information using artificial neural networks. *Journal of Applied Meteorology*, **36**, (9), 1176–1190.

Hsu, K., H. Gupta, X. Gao, and S. Sorooshian, 1999: Estimation of physical variables from multichannel remotely sensed imagery using a neural network: Application to rainfall estimation. *Water Resources Research*, **35**, 1605–1618.

Hsu, K., Y. Hong, and S. Sorooshian, 2007: Rainfall estimation using a cloud patch classification map. *Measurement of Precipitation from Space: EURAINSAT and Future*. Edited by V. Levizzani, P. Bauer, and F. J. Turk, Springer Publishing Company, 745 pages, Hardcover, ISBN#978-1-4240-5834-9, pp. 329–343.

Hsu, K., T. Bellerby, and S. : Sorooshian, 2009. LMODEL: A Satellite Precipitation Algorithm Using Cloud Development Modeling and Model Updating. Part 2: Model Updating. In press.

Huffman, G. J., R. F. Adler, D. T. Bolvin, G. Gu, E. J. Nelkin, K. P. Bowman, Y. Hong, E. F. Stocker, and D. B. Wolff, 2007: The TRMM Multisatellite Precipitation Analysis (TMPA): Quasi-Global, Multiyear, Combined-Sensor Precipitation Estimates at Fine Scales. *Journal of Hydrometeorology*, **8**, 38–55.

Hong, Y., K. Hsu, S. Sorooshian, and X. Gao, 2004: Precipitation estimation from remotely sensed imagery using an artificial neural network cloud classification system. *Journal of Applied Meteorology*, **43**, 1834–1852.

Joyce, R. J., J. E. Janowiak, P. A. Arkin, and P. Xie, 2004: CMORPH: A method that produces global precipitation estimates from passive microwave and infrared data at high spatial and temporal resolution. *Journal of Hydrometeorology*, **5**, 487–503.

Kidd, C., D. R. Kniveton, M. C. Todd, and T. J. Bellerby, 2003: Satellite rainfall estimation using combined passive microwave and infrared algorithms. *Journal of Hydrometeorology*, **4**, 1088–1104.

Kohonen, T., 1995: *Self-Organizing Map*. Springer-Verlag, New York

Kurino, T., 1997: A satellite infrared technique for estimating "deep/shallow" precipitation. *Satellite Data Applications: Weather and Climate*, **19**, 511–514.

Marzano, F. S., M. Palmacci, D. Cimini, G. Giuliani, and F. J. Turk, 2004: Multivariate statistical integration of Satellite infrared and microwave radiometric measurements for rainfall retrieval at the geostationary scale. *IEEE Transactions on Geoscience and Remote Sensing*, **42**, 1018–1032.

National Weather Service, NOAA, 2006: Service Assessment: Hurricane Katrina, August 23–31, 2005.

Nicholson, S. E., B. Some, J. McCollum, E. Nelkin, D. Klotter, Y. Berte, B. M. Diallo, I. Gaye, G. Kpabeba, O. Ndiaye, J. N. Noukpozounkou, M. M. Tanu, A. Thiam, A. A. Toure, and A. K. Traore, 2003a: Validation of TRMM and other rainfall estimates with a high-density gauge dataset for West Africa. *Journal of Applied Meteorology*, **42**, Part I, 1337–1354.

Nicholson, S. E., B. Some, J. McCollum, E. Nelkin, D. Klotter, Y. Berte, B. M. Diallo, I. Gaye, G. Kpabeba, O. Ndiaye, J. N. Noukpozounkou, M. M. Tanu, A. Thiam, A. A. Toure, and A. K. Traore, 2003b: Validation of TRMM and other rainfall estimates with a high-density gauge dataset for West Africa. *Journal of Applied Meteorology*, **42**, Part II, 1355–1368.

Osullivan, F., C. H. Wash, M. Stewart, and C. E. Motell, 1990: Rain estimation from infrared and visible GOES satellite data. *Journal of Applied Meteorology*, **29**, 209–223.

Sorooshian, S., K. Hsu, X. Gao, H. V. Gupta, B. Imam, and D. Braithwaite, 2000: Evaluation of PERSIANN system satellite-based estimates of tropical rainfall. *Bulletin of the American Meteorological Society*, **81**, 2035–2046.

Todd, M., C. Kidd, D. R. Kniveton, and T. J. Bellerby, 2001: A combined satellite infrared and passive microwave technique for the estimation of small scale rainfall. *Journal of Atmospheric and Oceanic Technology*, **18**, 742–755.

Turk, F. J. and S. D. Miller, 2005: Toward improving estimates of remotely-sensed precipitation with MODIS/AMSR-E blended data techniques. *IEEE Transactions on Geoscience and Remote Sensing*, **43**, 1059–1069.

Ushio, T., T. Kubota, S. Shige, K. Okamoto, K. Aonashi, T. Inoue, N. Takahashi, T. Iguchi, M. Kachi, R. Oki, T. Morimoto, and Z. Kawasaki, 2008: A Kalman filter approach to the Global Satellite Mapping of Precipitation (GSMaP) from combined passive microwave and infrared radiometric data. *Journal of Meteorology Society Japan*, **87A**, 137–151.

Vincent, L. and P. Soille, 1991: Watersheds in digital spaces: an efficient algorithm based on immersion simulations. *IEEE Transactions on Pattern Analysis and Machine Intelligence*, **13**, (6), 583–598.

Vincent, G., R. A. Scofield, and W. P. Mensel, 1998: The operational GOES infrared rainfall estimation technique. *Bulletin of the American Meteorological Society*, **79**, 1883–1898.

Xu, L., X. Gao, S. Sorooshian, and P. A. Arkin, 1999: A microwave infrared threshold technique to improve the GOES precipitation index. *Journal of Applied Meteorology*, **38**, 569–579.

The Combined Passive Microwave-Infrared (PMIR) Algorithm

Chris Kidd and Catherine Muller

Abstract The retrieval of satellite rainfall estimates from multi-platform Earth observations has received much attention over the last decade. The Passive Microwave – InfraRed algorithm, developed at the University of Birmingham, has been operating in a quasi-operational mode since 2002. The algorithm combines the temporally-rich information from the infrared geostationary observations with the more quantitative, but less frequent, rainfall information from the passive microwave polar-orbiting satellites. Co-located infrared and passive microwave information is entered into a database which is used to generate the relationship between the surface rainfall and infrared cloud top temperatures at a centred-weighted 5×5 scale. The technique produces rainfall estimates at a temporal resolution of 30 min and a spatial resolution of 0.1×0.1: the user can then aggregate these results to suit their requirements.

Keywords Satellite rainfall · Infrared · Passive microwave · Rainfall verification

1 Background

Precipitation is highly variable both spatially and temporally occurring over only a few percent of the Earth's surface at any one time. Conventional means of observation rely primarily on gauges and more recently on radar, although over the world's oceans these observations are often non-existent, and over the land areas coverage is uneven and often sparse. Quantitative precipitation estimates from satellite data can provide spatially and temporally consistent coverage over both land and ocean. With increasing demand for improved precipitation estimates from satellite systems over a range of scales in space and time, new techniques have been developed.

C. Kidd (✉)
School of Geography, Earth and Environmental Sciences, University of Birmingham,
Edgbaston Birmingham, UK
e-mail: c.kidd@bham.ac.uk

Techniques often rely upon visible (Vis) and infrared (IR) observations of the cloud-tops and, although the relationship with surface rainfall is not direct, they benefit from frequent observations. More direct observations can be obtained from passive microwave (PM) observations, although they suffer from poorer less-frequent sampling. This dichotomy is shown in the results from the third Global Precipitation Climatology Project (GPCP) Algorithm Intercomparison Project (AIP-3; Ebert and Manton 1998), which showed that the PM estimates produced the best instantaneous results and the IR-based estimates provided the best long-term estimates.

Although the relationship between the cloud-top characteristics, as observed by VIS and IR sensors, and rainfall can be somewhat tenuous, many techniques have been developed. Examples include the GOES precipitation index (GPI), developed by Arkin and Meisner (1987), and designed to generate monthly rainfall products using the fractional coverage of cloud colder than 235 K in the IR with a fixed rain rate of 3 mmh^{-1}. The use of a simple fixed cloud-top temperature threshold and fixed rain rate has proved to be robust, although can lead to regional biases in the rainfall product. More complex techniques such as the operational GOES IR rainfall estimation technique, or Auto-Estimator, and the GOES Multispectral Rainfall Algorithm (GMSRA), are described by Vicente et al. (1998, 2001) and Ba and Gruber (2001) respectively. These techniques utilize relationships established between the satellite observations and surface data sets together with moisture correction factors to account for evaporation. However, all Vis/IR-based techniques estimate surface rainfall indirectly through the observation of the cloud-top: more direct rainfall measurements are possible through PM observations which respond to the hydrometeor particles, rather than the cloud tops. Precipitation-sized hydrometeors can emit or scatter microwave radiation, depending upon the size of the particle and the frequency of the microwave radiation. At low frequencies (<c.40 GHz) emission from rain droplets provide the main source of information for rainfall retrievals, while above c.40 GHz, the same droplets will scatter microwave radiation. Although the PM observations are more direct than the Vis/IR, the emission signal is only discernable over surface water, and information on rainfall derived from a scattering signal is related to ice particles in the upper regions of the cloud system.

The combination of multi-satellite observations is therefore key to advancing the accuracy of precipitation products, not only from improved sampling, but also from the combination of different sources of information on the precipitation systems. Results from the Precipitation Intercomparison Projects (PIP) (Barrett et al. 1994; Smith et al. 1998; Alder et al. 2001) and the Algorithm Intercomparison Programme (AIP) (summarised by Ebert 1996) demonstrated that PM-only algorithms are more accurate than IR-only algorithms in terms of instantaneous rainfall estimates. However, IR-only techniques provide better long-term estimates than the PM-only techniques due to better temporal sampling. IR techniques generally use data from geostationary satellites, providing a nominal 48 images per day, although more recent satellite systems provide more frequent imagery. It should also be noted that rainfall occurrence is less than cold-cloud occurrence, therefore cloud-based IR techniques provide a smoother, more time-integrated product than rain particle-based PM techniques. Less frequent imagery from PM measurements is obtained from a range of satellite systems including the Defense Meteorological Satellite Program (DMSP) Special Sensor Microwave Imager (SSM/I) and Special Sensor Microwave Imager-Sounder (SSM/IS), the Tropical Rainfall Measuring Mission (TRMM) Microwave Imager (TMI), the Advanced Microwave Scanning Radiometer (AMSR) and the Advanced Microwave Sounding Unit (AMSU).

Adler et al. (1993) first suggested the combination of IR and PM observations to improve precipitation estimates on the basis of retaining the strengths of individual techniques to overcome the other's weaknesses. They compared the results of the GPI and the convective-stratiform technique (CST; Adler and Negri 1988) with an 85-GHz-based PM-technique over monthly timescales to generate modified the rain-rate values for the IR algorithms. Similarly, Kummerow and Giglio (1995) tested both fixed-IR/variable-rainrate and variable-IR/fixed-rainrate techniques over the Pacific atolls, again based upon monthly relationships. The Universally Adjusted GPI (UAGPI; Xu et al. 1999) used the scattering index (SI; Ferraro and Marks 1995) to produce an optimal IR rain/no-rain threshold and optimal conditional rainrates in order to reduce the total error between the IR-based and the PM-based rainfall estimates. Adjustment of algorithm parameters like this is employed by a number of techniques, including the GPCP multi-satellite product (see Huffman et al. 1997, 2001), which combines rainfall estimates from numerous sources by using weightings based upon error estimates assigned to the individual components derived from monthly rainfall products. Other methods use PM rainfall retrievals to calibrate the IR temperatures to derive a generalised IR-rainrate regression. Kidd (1999) calibrated IR temperatures over the AIP-3 region while Todd et al. (2001) describes the calibration of IR temperatures over Africa using a moving $1° \times 1°$ window to generate fine-scale calibrations at $0.25° \times 0.25°$. Techniques combining IR and PM data through neural networks include the Precipitation Estimation from Remotely Sensed Information Using Artificial Neural Networks (PERSIANN) technique (Sorooshian et al., 2000) which generates rainfall estimates from a single IR channel at a nominal resolution of $0.25° \times 0.25°$ every 30 min using five feature parameters based upon a mosaic of overlapping training regions. More recently, techniques, such as CMORPH (Joyce et al. 2004) have been developed whereby the IR information from several time-observations is used to move the areas of PM-derived rainfall.

The choice of the spatial and temporal calibration domain for the IR-PM combined techniques is often subjective. Some techniques rely upon monthly 2.5×2.5 degree domains to provide robust calibrations to overcome the regional biases in the IR-based algorithms (e.g. Adler et al. 1993; Xu et al. 1999). Although the climatological variations in the IR–PM relationship will be reflected in the monthly calibrations they will not respond to the sub-monthly variations in the meteorology. Miller et al. (2001) and Turk et al. (2000) used instantaneous calibrations based upon coincident IR–PM values to reflect the changes in the calibration over short-term periods. However, the use of instantaneous calibration data results in relatively few data and consequently a larger spatial domain is required to ensure an adequate sampled-size: Turk et al. (2000) generated calibrations on a $5° \times 5°$ basis within a $15° \times 15°$ moving window. A trade-off therefore exists between the temporal and the spatial sizes of the domains used for the IR-PM calibration: the choice of the domain can be critical in the accuracy of the final rainfall product (Todd et al. 2001).

Validation studies of these IR–PM techniques are somewhat variable due to the different regions over which they have been tested. Comparison with the GPI technique indicates that most of the combined products showed reduced root-mean squared error (RMSE) and bias (Alder et al. 1993; Kummerow and Giglio 1995; Xu et al. 1999) and similar or slightly worse correlations (Xu et al. 1999; Miller et al. 2001). Todd et al. (2001) highlighted the importance of surface data sets: correlations between the EPSAT gauges and the Microwave-IR Rainfall Algorithm (MIRA) technique exceeded those of GPI and the UAGPI techniques, although when compared with the Global Precipitation Climatology Centre (GPCC) gauge data set, the reverse was true.

2 Algorithm Description

The Passive Microwave-InfraRed (PMIR) technique was initially formulated using data over the AIP-3 study region in the western Pacific (Kidd, 1999). The PMIR has been developed first, to exploit a number near real-time datasets that are available over the Internet and second, to provide a simple yet robust multi-sensor precipitation technique: it is described in detail in Kidd et al. (2003). Precipitation products from this technique have been generated on a quasi-operational basis since January 2002 every 30 min at a resolution of c.12 × 12 km for regions between 60°N and 60°S. The technique, shown in Fig. 1, can be broken down into three separate processing stages: (a) data ingest and preparation, (b) database management and (c) results generation.

2.1 Data Ingest and Preparation

Data for the PMIR technique are derived from two main on-line data sources. The IR dataset is available from the Climate Prediction Center and is described by Janowiak et al. (2001). The data is acquired from five geostationary satellites, currently the two US GOES, the Japanese MT-Sat and the two European MSG satellites. Data from these satellites are remapped to a nominal 4 km spatial resolution and 30 min temporal resolution, with corrections for parallax, viewing angle and inter-satellite calibration (Janowiak et al. 2001).

PM data, from the Defense Meteorological Satellite Program (DMSP) Special Sensor Microwave/Imager (SSM/I) sensors is obtained from the Global Hydrology Climate Center of the National Aeronautics and Space Administration (NASA) Marshall Space Flight Center/University of Huntsville, Alabama. Typically there are three available sensors resulting in between four and six overpasses each day. Rainfall estimates from this data are generated by the combined 85/19 GHz frequency difference technique as submitted to the AIP-3 and PIP-3 inter-comparison

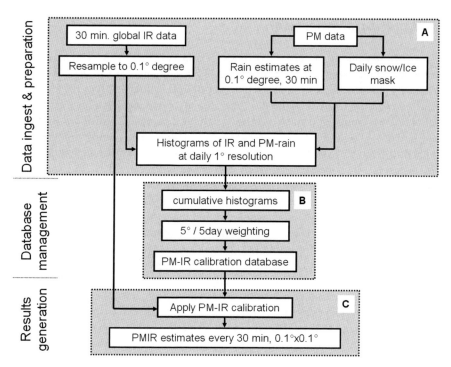

Fig. 1 Processing stages of the PMIR algorithm: data is ingested in *Part A* and histograms prepared of the cloud top temperatures and rainfall rates for each 1° × 1° latitude/longitude. *Part B* generates a temporally and spatially-weighted histograms from which the calibration database is derived. This database, for each 1° × 1° square is then applied (*Part C*) to the IR data to generate 0.1° × 0.1°, 30 min estimates of rainfall

projects by the University of Bristol (see Ebert 1996). The original algorithm was calibrated against UK FRONTIERS radar data but has subsequently been recalibrated against the TRMM PR rainfall data to provide a more globally representative PM-rainfall relationship. Surface contamination is removed through the use of the screening methodology of Ferraro and Marks (1995): surface screening is performed on a daily basis to eliminate surface variations, although some residual contamination due to snow and ice does remain.

The IR data is resampled to a 0.1° grid by using a 3 × 3 filter to average the 4-km data and generate a mean cloud-top temperature over a 12 × 12 km area to approximate the resolution of the PM rainfall estimates generated at the 85 GHz resolution. The resulting rainfall estimates from all PM observations are remapped to a 0.1° grid (approximately 12 km) for each 30-min period centered on the hour and half-hour. The mean IR cloud top temperatures and the PM rainfall estimates are now co-located for each 30 min at a resolution of 0.1° × 0.1°. The PM-IR matched histograms are generated at a scale of 1° × 1° recording IR temperatures (75–329 K) and PM rainfall estimates (0.0–51.1 mm h^{-1}). The histograms for each 30 min are then saved to disk ready for the generation of the database.

2.2 Database Management

The PMIR technique is capable of operating in two modes: a "climatological-historical" mode which uses data from d-2 (day minus two) to d+2 using an arbitrarily derived linear weighting function (0.6, 0.8, 1.0, 0.8, 0.6, respectively) centered on d-0 ("today"), while the "operational" mode uses data for d-0 back to d-4 and is accumulated using a similar weighting function (i.e., d-0 has a weight of 1.0, d-1 × 0.8, d-2 × 0.6, etc.): both sets of temporal weighting functions are applied to frequency of occurrences in the histograms of the IR-temperatures and PM-rainrates. After the data has been aggregated temporally it is then smoothed spatially through the use of a 5° × 5° Gaussian filter. These separate histograms of co-located IR temperatures and PM rainfall rates are then are converted into cumulative histograms and are then matched through the use of a cumulative histogram matching approach such that the coldest IR temperatures are assigned the highest rainfall. The main benefit of using the cumulative histogram matching approach is that it does not assume that the IR and PM data sets are precisely co-located in space and time: co-registered and temporally coincident data sets are rarely achievable. Furthermore, cumulative histogram matching overcomes the problem of the rainrates being heavily skewed towards zero: the technique allows the PM frequency distribution of rainrates to be reflected in the resulting combined product. One drawback is, however, that the IR-calibration will perpetuate any errors in the rainfall distribution generated by the PM-derived rainrates. Thus, the precise screening of non-raining features is critical to the success of the technique. The relationships for each 1° × 1° area are saved as a lookup table for subsequent application to generate the final rainfall product.

2.3 Results Generation

The final step of the technique is the application of the lookup-tables to the IR data sets. Data from the IR is available for each 30 min and is used at a resolution of 0.1° × 0.1°. The corresponding calibration curve for the 1° × 1° area in which the IR value is located is selected and the IR temperature converted to a rainfall rate. Thus the final product is generated at a temporal resolution of 30 min and a spatial resolution of 0.1° × 0.1°, allowing users to average the product to temporal/spatial scales to suit their application.

Examples of the input data sets are shown in Fig. 2. Figure 2(a) shows the global-IR composite imagery from the 5 main geostationary satellites for 11 January 2009 at 12:00 UTC. The PM data is represented by the 85 GHz image (for 11 January 2009) in Fig. 2(b): the highlighted swaths indicate the coverage of the SSM/I between 11:45 and 12:15 UTC. The 85 GHz channels, along with other PM channels, are used to generate the daily snow/ice surface mask shown in Fig. 2(c): the different tones of grey indicate the number of overpasses during the day when snow or ice was retrieved by the algorithm. The PM rainfall product is shown in Fig. 2(d) with these regions of snow and ice masked out in mid-grey.

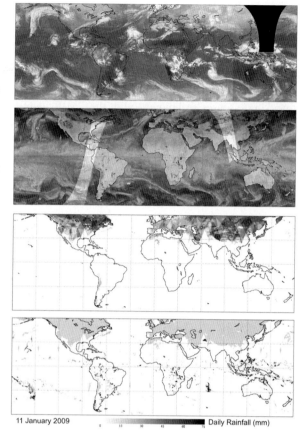

Fig. 2 Component elements used in the PMIR technique: (**a**) the infrared geostationary composite imagery; (**b**) passive microwave 85 GHz imagery (highlighted region is temporally co-incident with the IR imagery); (**c**) snow-ice mask, and; (**d**) the combined PMIR product at 30-min, $0.1° \times 0.1°$ resolution. This example is for 11 January 2009 and illustrates the extent of the snow/ice mask in the northern hemisphere

3 Application and Results

The basis of the technique is described by Kidd et al. (2003, 2007) through the implementation of the PMIR technique over Africa: monthly calibrations of the IR were generated through comparison with the PM-derived rainfall rates. The resulting maps of cloud-top temperatures associated with the rain/no-rain threshold showed considerable regional variations ranging from 210 K, associated with deep convective clouds, through to 290 K in regions of little rainfall. These warm rain/no-rain thresholds and low conditional rainrates were associated with low-level clouds, such as trade-wind cumulus, or sub-resolution rain-cells. Study of the calibration periods revealed significant day-to-day variations in the PM-IR calibration, results which are consistent with the study of Kidd (1999) over the TOGA-COARE region.

Daily rainfall products compared by Kidd et al. (2003) included from the GPI, the Universally-Adjusted GPI (Xu et al. 1999), a variable rain-rate version of the

UAGPI, a PM-only frequency difference algorithm and two PMIR products, one the operational-mode PMIR and the climatological-mode PMIR technique. The results showed that the GPI produced greatest rainfall area, while the PM-only technique generated the least rainfall extent although, statistically, the PM-only technique had the lowest bias of the techniques. Comparisons between the satellite techniques and surface gauge data showed that all the combined algorithms had better correlations than single-sensor techniques, with the PMIR identifying the occurrence of rainfall well. Kidd et al. (2003) however, concluded that the combined techniques did not necessary provide better statistical performance overall due to two main factors. First, the PM-technique that is used to calibrate the IR temperatures is critical in determining the quality of the final product: any inherent weakness in the PM algorithm will be transferred via the calibration to the combined technique and reflected in the final product. Second, the calibrated-IR techniques are still reliant upon the premise that cold cloud tops will faithfully represent the underlying rainfall. Furthermore, it was noted that the calibration itself may not be representative of the meteorological regime, particularly in regions with a strong diurnal cycle and hence different cloud-rain life cycle relationships.

An example of the PMIR precipitation product at 30 min/$0.1° \times 0.1°$ resolution is shown in Fig. 3. Data for 20 July 2008 from 00:00 UTC and 11:30 UTC has been processed and extracted for the region from 120°W to 90°W and from the Equator to 30°N. In the centre left of the sequence of images is Hurricane Fausto, just before it reached its maximum strength: the strengthening of this hurricane through this sequence is evident by the tightening and intensification of the rain bands around the storm centre. To the East are a number of convective clusters moving westwards from the coast of central America into the Eastern Pacific. Also of note is the convective activity over the land areas to the North, over Mexico, associated with the diurnal heating and associated convective storms – the imagery from 00:00 to 11:30 UTC being 17:00–04:30 local time. This case study exemplifies the utility of such a technique for application at fine temporal and spatial scales. While the products from the technique can be averaged to provide coarser resolution estimates, many applications require precipitation estimates at scales finer than many currently-available products. In particular, hydrological modelling often requires rainfall information at fine temporal and spatial scales for input into, for example, flow models. Drainage basins often have irregular shapes with heterogeneous surface types, further emphasising the requirement for precipitation products with spatial resolutions commensurate with the landscape features. Likewise, temporal scales for flood events are typically on scales less than 3-h: the 30-min (potentially 15-min) sampling resolution of this technique addresses this requirement.

The PMIR product has undergone extensive testing and evaluation at the reduced-resolution daily/$0.25°\times0.25°$ scales, comparable with numerous other operational and quasi-operational precipitation products. Figure 4 shows a number of precipitation products generated on 11 January 2009 (00-24Z) at a resolution of $0.25° \times 0.25°$. The 3B42RT product (Fig. 4a), described by Huffman et al. (2007), combines numerous satellite rainfall estimates using weightings based upon error estimates assigned to the individual components. The CMORPH technique (Fig. 4b;

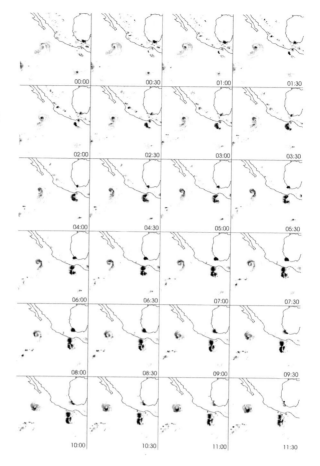

Fig. 3 Example of 0.1° × 0.1° 30 min PMIR precipitation products for rainfall over the Eastern Pacific and Central America on 20 July 2008: grey tones represent rainfall from zero (*white*) to 25 mmh^{-1} (*black*). Region extends from 120°W to 90°W and from 0°N to 30°N

Joyce et al. 2004) is based upon the premise that the estimates from PM-techniques, being superior to those of the IR, are the main source of information with the IR used purely for assessing the movement of the rain systems. The PMIR is shown in Fig. 4c, while the output from the NRL NOGAPS Numerical Weather Prediction (NWP) model is shown in Fig. 4d. It can be noted that the satellite-derived products all show a good degree of similarity, capturing the main precipitation systems, particularly across the Tropics. Some subtle differences can seen, for example, off the coast of South America where the PMIR shows a broader area of heavier rainfall than suggested by the 3B42RT and CMORPH techniques. More pronounced are differences in higher latitude regions: in the southern hemisphere the PMIR shows less rainfall than both the 3B42RT and the CMORPH technique: in the northern hemisphere there are similar differences over the ocean. Over the northern hemisphere land masses the PMIR suggests more rainfall than the other two satellite techniques: both the PMIR and 3B42RT products show some

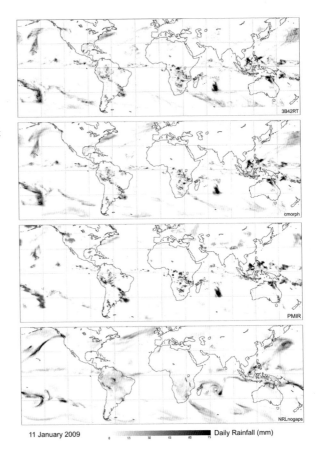

Fig. 4 Comparison of daily (00-24Z) $0.25° \times 0.25°$ precipitation products for 11 January 2009: (**a**) 3B42RT (multi-satellite technique); (**b**) cmorph (IR-advection of PM rainfall); (**c**) PMIR (PM-calibrated IR) and; (**d**) NRL NOGAPS model output

precipitation in the NW of the United States, although the PMIR also shows an area over the Rocky Mountains not shown by the 3B42RT product. The PMIR also shows regions of precipitation over Eurasia not shown in the other two satellite products. The model rainfall product (Fig. 4d) shows broadly similar regions of rainfall, but in less detail and with some of rain features displaced. Of particular note are the regions of convective precipitation over continental Africa and South America: the satellite products capture these precipitation features with a good degree of detail, whereas the model output underestimates the rainfall totals and broadens the extent of the rainfall area.

The PMIR precipitation product is routinely compared with surface data sets and other precipitation products on a daily basis in near real time through the validation efforts of the International Precipitation Working Group (IPWG) (see Ebert et al. 2007): examples of these comparisons are shown in Fig. 5. The Bureau of Meteorology in Melbourne, Australia, oversees the comparison of precipitation products over Australasia: Fig. 5a, shows the PMIR precipitation product compared with surface rain gauge data for 25 December 2008. The form of the precipitation

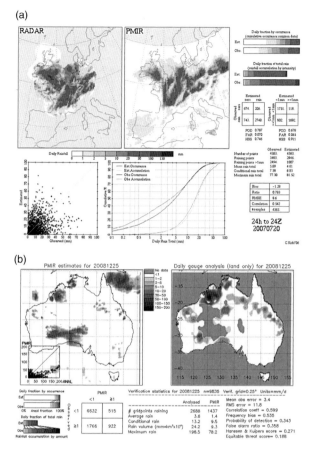

Fig. 5 Examples of the application of the PMIR precipitation product over the international precipitation intercomparison project regions of (**a**) the European validation region on 20 July 2007 and (**b**) Australian validation region on 25 December 2008

regions are generally similar between the satellite and surface data sets, although the satellite product produces a smaller extent than the surface gauge data, while generating only ~60% of the rainfall totals. Figure 5b shows an example of the PMIR over the IPWG European validation region on 20 July 2007. The PMIR performs reasonably well, with a correlation of 0.542, although generating less rainfall extent (85%) and less total rainfall (ratio=0.783) than suggested by the surface radar data. However, as in the example over Australia, the general form of the rainfall pattern in similar, albeit with the PMIR showing some influence from the IR cold cloud patterns.

Longer term performance of the PMIR is shown in Fig. 6: here comparison is made between the PMIR, the NRL NOGAPS model output and five other satellite precipitation products. A 31-day moving-average correlation coefficient has been generated for daily products over the IPWG European validation region. In common with other findings (e.g. Ebert et al. 2007), the satellite precipitation products

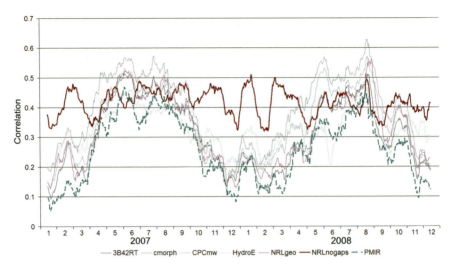

Fig. 6 Statistical performance of satellite and model precipitation products during 2007 and 2008 over the IPWG European validation region. The model output is represented by the thick continuous line while the PMIR is represented by the thick *broken line*

show a marked seasonal cycle in the retrieval performance, while the model shows reasonably consistent performance, although poorer during the summer months and somewhat better during the winter months. This seasonal cycle has been attributed to a number of factors including the difficulty in extracting precipitation signals over cold and snow-covered surface and the variations in the structure of the precipitation systems – winter-time weather systems having less vertical extent and less liquid water than summer-time systems. The overall best-performing satellite algorithm during this period was the CMORPH technique. Although performance of the PMIR was generally poorer than the other techniques, it was on a par with the Hydro-Estimator algorithm (Scofield and Kuligowski 2003) and the NRL geo. It should also be noted that the techniques is designed to operate at finer resolutions than daily/$0.25° \times 0.25°$ and at present the technique is working with only two PM satellite data sets for calibration (the F13 and F15).

4 Conclusions

The exploitation of observations from multiple satellite sensors is key to the improvement of rainfall estimates over single-sensor techniques. Over the last decade there has been much work devoted to the development of such techniques, ranging from combining individual sensor products, to the calibration of the IR with PM rainfall information to the advection of PM rainfall measurements by IR motion-vectors. At present the advection/morphing techniques produce precipitation products with superior statistical performance: however, such techniques are somewhat complicated requiring cloud-tracking routines. The PMIR technique described here is relatively straightforward, currently relying upon two easily available data sets, the global-IR composite and the PM DMSP SSM/I data set.

New observations from future satellite missions and the full exploitation of current satellite observations hold great promise for improvements in quantitative precipitation estimates from satellites. In particular, the full exploitation of imagery from the satellites such as the Meteosat Second Generation (MSG) have yet to be realised: the use of multichannel information should enable improved delineation of precipitation regions and provide information on rainfall "potential" information over the IR channel alone. In addition, newer geostationary satellite systems have high scanning/image capture capabilities enabling precipitation products to be generated at finer temporal resolutions, critical for flood-monitoring capabilities. In particular, the development of the Global Precipitation Mission (GPM) (Flaming 2002) to provide 3-hourly data will prove very useful in IR-calibration techniques. Although the GPM mission would enhance the temporal sampling capability of PM-derived precipitation products, sub-three hourly estimates would still be needed. In addition, the GPM dual-frequency precipitation radar will provide unrivalled independent information on the precipitation structures in the mid- to high-latitudes, helping to resolve some of the shortcomings of the satellite products over the seasonal cycle. However, critical to the success of any combined technique is the performance of the individual algorithm components: to date there has been no comprehensive study to evaluate whether certain techniques perform best as a result of individual components, or through the technique of combining the information.

Acknowledgments The author would like to thank NASA under their Precipitation Measurement Missions programme for their continuing support of this research. Global infrared data is provided courtesy of the Climate Prediction Center and John Janowiak; passive microwave data from the SSM/I is courtesy of the Global Hydrology and Climate Center, NASA/MSFC.

References

Adler RF, and Negri AJ (1988) A satellite technique to estimate tropical convective and stratiform rainfall. Journal of Applied Meteorology, 27, 30–51.

Adler RF, Negri AJ, Keehn PR, and Hakkarinen IM (1993) Estimation of monthly rainfall over Japan and surrounding waters from a combination of low-orbit microwave and geosynchronous IR data. Journal of Applied Meteorology 32, 335–356.

Adler RF, Kidd C, Petty G, Morrisey M, and Goodman MH (2001) Intercomparison of global precipitation products: The Third Precipitation Intercomparison Project (PIP-3). Bulletin of the American Meteorological Society 82, 1377–1396.

Arkin PA and Meisner BN, (1987) The relationship between largescale convective rainfall and cold cloud over the Western Hemisphere during 1982–84. Monthly Weather Review 115, 51–74.

Ba MB and Gruber A (2001) GOES multispectral rainfall algorithm (GMSRA). Journal of Applied Meteorology 40, 1500–1514.

Barrett EC, Adler RF, Arpe K, Bauer P, Berg W, Chang A, Ferraro R, Ferriday J, Goodman S, Hong Y, Janowiak J, Kidd C, Kniveton D, Morrissey M, Olson W, Petty G, Rudolf B, Shibata A, Smith E, and Spencer R (1994) The first WetNet Precipitation Inter-comparison Project: Interpretation of Results. Remote Sensing Reviews 11, 303–373.

Ebert EE (1996) Results of the 3rd Algorithm Intercomparison Project (AIP-3) of the Global Precipitation Climatology Project (GPCP). Research Rep. 55, Bureau of Meteorology Research Centre, Melbourne, Australia, 204 pp.

Ebert EE and Manton MJ (1998) Performance of satellite rainfall estimation algorithms during TOGA COARE. Journal of Atmospheric Science 55, 1537–1557.

Ebert EE, Janowiak JE and Kidd C (2007) Comparison of near real time precipitation estimates from satellite observations and numerical models. Bulletin of the American Meteorological Society 88, 47–64.

Ferraro RR and Marks GF (1995) The development of SSM/I rain-rate retrieval algorithms using ground based radar measurements. Journal Atmospheric and Oceanic Technology 12, 755–770.

Flaming GM (2002) Requirements for global precipitation measurement. Proc. IGARSS'02, Vol. 1, Toronto, ON, Canada, IEEE, 269–271.

Huffman GJ, Adler RF, Arkin R, Chang A, Ferraro R, Gruber A, Janowiak J, McNab A, Rudolf B and Schneider U (1997) The global precipitation climatology project (GPCP) combined precipitation dataset. Bulletin of the American Meteorological Society 78, 5–20.

Huffman GJ, Adler RF, Morrisey MM, Bolvin DT, Curtis S, Joyce R, McGavock B, and Susskind J (2001) Global precipitation at one-degree daily resolution from multi-satellite observations. Journal of Hydrometeorology 2, 36–50.

Huffman GJ, Adler RF, Bolvin DT, Gu G, Nelkin EJ, Bowman KP, Hong Y, Stocker EF and Wolff DB (2007) The TRMM multi-satellite precipitation analysis (TMPA): Quasi-global, multiyear, combined-sensor precipitation estimates at fine scales. Journal of Hydrometeorology 8, 38–55. doi:10.1175/JHM560.1

Janowiak JE, Joyce RJ and Yarosh Y (2001) A real-time global half-hourly pixel-resolution infrared dataset and its application. Bulletin of the American Meteorological Society 82, 205–217.

Joyce RJ, Janowiak JE, Arkin PA and Xie P (2004) CMORPH: A method that produces global precipitation estimates from passive microwave and infrared data at high spatial ad temporal resolutions. Journal of Hydrometeorology 5, 487–503.

Kidd C (1999) Results of an infrared/passive microwave rainfall estimation technique. Proc. Remote Sensing Society, Cardiff, Wales, United Kingdom, Remote Sensing Society, 685–689.

Kidd C, Kniveton DR, Todd MC and Bellerby TJ (2003) Satellite rainfall estimation using a combined passive microwave and infrared algorithm. Journal of Hydrometeorology 4 1088–1104.

Kidd C, Tapiador FJ, Sanderson V and Kniveton D (2007) The University of Birmingham Global Rainfall Algorithms. p.255–268. In Measuring Precipitation from Space: EURAINSAT and the future. Eds. V. Levizzani, P. Bauer and J. Turk. Springer, New York 722 pp.

Kummerow C and Giglio L (1995) A method for combining passive microwave and infrared rainfall observations. Journal of Atmospheric and Oceanic Technology 12, 33–45.

Miller SW, Arkin PA and Joyce R (2001) A combined microwave/ infrared rain rate algorithm International Journal of Remote Sensing 22, 3285–3307.

Scofield RA and Kuligowski RJ (2003) Status and outlook of operational satellite precipitation algorithms for extreme-precipitation events. Weather and Forecasting 18, 1037–1051.

Smith EA, Lamm JE, Adler RF, Alishouse J and Aonashi K (1998) Results of the WetNet PIP-2 project. Journal of Atmospheric Science 55, 1483–1536.

Sorooshian S, Hsu K-L, Gao X, Gupta HV, Imam B and Braithwaite D (2000) Evaluation of PERSIANN system satellite-based estimates of tropical rainfall. Bulletin of the American Meteorological Society 81, 2035–2046.

Todd MC, Kidd C, Kniveton D and Bellerby TJ (2001) A combined satellite infrared and passive microwave technique for estimation of small-scale rainfall. Journal of Atmospheric Oceanic Technology 18, 742–755.

Turk FJ, Hawkins J, Smith EA, Marzano FS, Mugnai A and Levizzani V (2000) Combining SSM/I, TRMM and infrared geostationary satellite data in a near-real time fashion for rapid precipitation updates: Advantages and limitations. Proc. 2000 EUMETSAT Meteorological Satellite Data Users' Conf., Vol.2, Bologna, Italy, EUMETSAT, 705–707.

Vicente GA, Scofield RA and Menzel WP (1998) The operational GOES infrared rainfall estimation technique. Bulletin of the American Meteorological Society 79, 1883–1898.

Vicente GA, Davenport JC and Scofield RA (2001) The role of orographic and parallax correction on real time high resolution satellite rain rate distribution. International Journal of Remote Sensing 23, 221–230.

Xu L, Gao X, Sorooshian S, Arkin PA and Imam B (1999): A microwave infrared threshold technique to improve the GOES precipitation index. Journal of Applied Meteorology 38, 569–579.

The NRL-Blend High Resolution Precipitation Product and its Application to Land Surface Hydrology

Joseph T. Turk, Georgy V. Mostovoy, and Valentine Anantharaj

Abstract In this chapter, we discuss the basic workings of the NRL-Blend high-resolution precipitation product, followed by a validation experiment. We employ satellite omissions to the existing (late 2008) constellation of low Earth orbiting satellite platforms to examine the impact of several proxy Global Precipitation Mission (GPM) satellite constellation configurations when used to initialize land surface models (LSM). The emphasis is on how high resolution precipitation products such as the NRL-Blend are affected by such factors as sensor type (conical or across-track scanning) and nodal crossing time, using a collection of GPM proxy datasets gathered over the continental United States. We present results which examine how soil moisture states simulated by the two state-of-the-art land surface models are impacted when forced with the various precipitation datasets, each corresponding to a different proxy GPM constellation configuration.

Keywords Satellite · Microwave · Surface · Model · Hydrology · GPM

1 High Resolution Precipitation Products (HRPP)

High Resolution Precipitation Products (HRPP) combine a multitude of spaceborne remotely-estimated and ground-based datasets in order to generate a precipitation product that is of a finer spatial and/or temporal resolution than any of the individual input datasets. These HRPPs are relevant to a variety of applications relating to Earth's hydrological cycle. Passive microwave sensors onboard low Earth orbiting (LEO) and geostationary Earth orbiting (GEO) environmental satellite systems provide the basic building blocks of an HRPP, augmented in some cases by surface radar and raingauge information and analyses from numerical

---

J.T. Turk (✉)
Jet Propulsion Laboratory, Radar Science and Engineering/334, Pasadena, CA 91109-8099, USA
e-mail: jturk@jpl.nasa.gov

weather prediction (NWP) models. Examples of commonly used HRPPs are the Tropical Rainfall Measuring Mission (TRMM) Multisatellite Precipitation Analysis (TMPA) (Huffman et al., 2007), the Precipitation Estimation from Remotely Sensed Information using Artificial Neural Networks (PERSIANN) datasets (Hsu et al., 1997), the Climate Prediction Center morphing technique (CMORPH) (Joyce et al., 2004), and the NRL-Blend (Turk and Miller, 2005) among others. Typically, these HRPPs combine multiple satellite datasets (and some add in additional raingauge and other non-satellite data) and produce estimates of three-hourly accumulated precipitation between ±60° latitude, updated every three hours, at a gridded spatial resolution of 0.25°. From these three-hourly accumulations, longer time scale accumulations can be generated. Some of these HRPPs are designed to be operated strictly in near realtime (e.g, NRL-Blend), while others create near realtime as well as a higher quality, post-processed non-realtime datasets.

2 NRL-Blend HRPP Technique

In this section we outline the design and implementation of the NRL blended satellite precipitation technique (referred to as NRL-Blend), which is based upon a real time, underlying collection of time and space-matching pixels from all operational geostationary (GEO) visible/infrared (VIS/IR) imagers and passive microwave (PMW) imagers onboard low Earth orbiting (LEO) satellites. It operates in an autonomous, operational mode with a steadily arriving stream of near real-time data from the operational GEO and LEO satellites (Turk and Miller, 2005). As of January 2009, the current operational GEO satellites are GOES-11, GOES-12, Meteosat-7, Meteosat-9 (MSG-2), and GMS-6 (MTSAT-1R). The current LEO constellation utilizes all 12 available satellites for a total of 12 PMW sensors and one active radar system, including (launch date in parentheses),

- The crosstrack Advanced Microwave Sounding Units (AMSU) onboard the National Oceanic and Atmospheric Administration (NOAA)-15 (May 1998), NOAA-16 (September 2000), and NOAA-17 (June 2002),
- The Microwave Humidity Sounder (MHS) onboard NOAA-18 (May 2005) and the EUMETSAT Meteorological Operational Platform (METOP)-A (October 2006),
- The Special Sensor Microwave Imagers (SSMI) onboard the Defense Meteorological Satellite Program (DMSP) F-13 (March 1995), F-14 (April 1997), and F-15 (December 1999),
- The Special Sensor Microwave Imager Sounder (SSMIS) onboard DMSP F-16 (October 2003) and F-17 (November 2006),
- The Advanced Microwave Scanning Radiometer (ASMR-E) onboard the Earth Observing System (EOS) Aqua (May 2002),

The WindSat polarimetric radiometer onboard Coriolis (January 2003),
The Tropical Rainfall Measuring Mission (TRMM) Microwave Imager (TMI) and its companion Precipitation Radar (PR) (November 1997).

All of these satellites orbit sun-synchronously with the exception of TRMM (TRMM's local observing time repeats approximately every 28 days at the equator). NOAA-15/17, Metop-A, all DMSP, and Coriolis are in orbit patterns which crossover in mostly morning (AM) and evening (within a few hours of the solar terminator), whereas NOAA-16/18 and Aqua are in orbits which crossover in early afternoon (PM) and early morning (NOAA-16 is an operational backup which has drifted from its initial afternoon 1400 local equator crossing time to near 1700 as of late 2008). AMSU and MHS are across-track scanning sounders which can be used also for precipitation estimation, PR is an across-track scanning radar, whereas SSMI, SSMIS, WindSat and AMSR-E all scan conically. Although the NRL-Blend is operated with all of these LEO datasets to minimize overall revisit time, as an option any of these LEO datasets can be omitted such as to study, for example, how the loss of a particular satellite in a particular orbit will affect the overall performance of the NRL-Blend. This option will prove useful when the land surface hydrology modeling is examined in Section 5.

2.1 Time-Space Colocation of LEO and GEO Datasets

Underlying the NRL blended satellite technique is an ongoing, real-time, dynamic collection of colocated (in time and space) intersecting pixels from all GEO VIS/IR and LEO PMW imagers. The operation of the NRL-Blend is essentially described by three procedures. The first procedure involves dataset collocation and is done in the background. As new PMW datasets arrive, the PMW-derived rainrate pixels are paired with the time and space-coincident geostationary 11-μm IR brightness temperature (T_B) data from areas of GEO satellite scan coverage, using a 15-min maximum allowed time offset between the pixel observation times. Figure 1 depicts an example where a TRMM orbit passes over areas covered by each of the five GEO satellites, and matched pairs of IR T_B and LEO-derived rainrate are paired with the associated date and geolocation. This collocation procedure is constantly ongoing with newly arrived datasets. This background collection of collocated data is used to update global histograms of the IR T_B and the PMW rain rate (R) in the nearest 2° latitude-longitude box, as well as the eight surrounding boxes (this overlap assures a fairly smooth transition in the histogram shape between neighboring boxes), depicted on the left side of Fig. 2. The reasoning behind these threshold values for time-collocation and box size are discussed in [25]. As soon as a box is refreshed with new data, a probabilistic histogram matching relationship is updated using the PMW rainrate and IR T_B histograms, and an updated T_B-R lookup table (LUT) is created. The global LUT update process is constantly ongoing for all satellite intersections of the 2° boxes, with operationally arriving global LEO and GEO datasets.

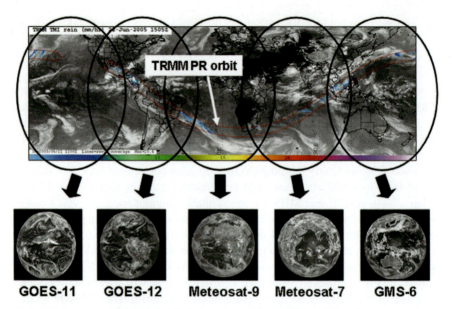

Fig. 1 Illustration depicting an global orbit overpass of the TRMM satellite and the rainrate inside of the PR coverage area (*red box*). The orbit intersects coverage areas of each of the five geostationary satellites (circa late 2008) at different times during its orbit. Colocated pairs of IR temperatures and rainrates are collected along with the associated date and geolocation

2.2 Instantaneous Rainrate Adjustment

The second procedure is initiated with newly arrived GEO datasets and is depicted on the right side of Fig. 2 (the foreground process). These GEO data are mapped to a common $0.1°$ pixel^{-1} rectangular map projection (1200 lines × 3600 samples, within ±60° latitude) and assigned a rainrate through bicubic interpolation of the rainrate derived from the four surrounding LUT values. The consultation of the LUT's generated by the background is illustrated by the arrow pointing between the background and foreground sides of Fig. 2. Bicubic interpolation assures smooth transitions in rain rates across box boundaries. If any LUT is more than 24-h old relative to the GEO dataset time, that LUT is not used (if all four LUTs are bad, then a missing value is assigned for the rainrate). A final step involves the use of a numerical weather prediction (NWP) model data to account for underlying environmental conditions that are not detected (or not accounted for) in a satellite-only analysis. Using the Navy Operational Global Atmospheric Prediction System (NOGAPS) forecast model fields (interpolated to the satellite time), the 850-hPa wind vectors, temperature, humidity, and total column precipitable water (TPW) are combined with a high-resolution topographic database. A threshold based upon the product of the humidity and TPW (Vicente et al., 1998) is used together to screen false rain identification. The TPW and terrain slope are used to apply a scaling factor in regions of likely orographic effects on both the upslope and downslope sides (Vicente et al., 2002).

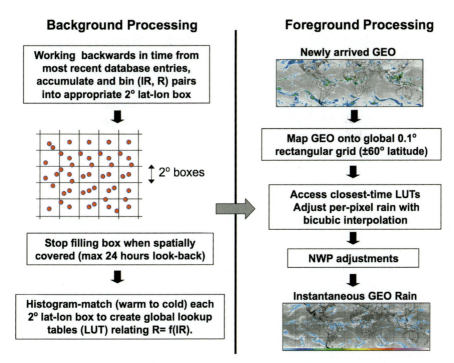

Fig. 2 Block diagram of the generation of the global lookup tables (LUT) for all 2° latitude-longitude boxes (background process, *left side* of *vertical line*), and the adjustment of GEO datasets into instantaneous rain (foreground process, *right side* of *vertical line*)

2.3 Accumulations Procedure

At each 3-hourly synoptic time (00, 03, ...21 UTC), the precipitation accumulations are updated by backwards time-integrating the instantaneous LEO and GEO datasets from the previous 24 h, and outputting an accumulations dataset at 3, 6, 12 and 24-h interval. In the accumulations procedure, each GEO instantaneous rain rate pixel is weighted according to its time proximity to the nearest PMW overpass. The PMW estimates are always fully weighted and the GEO estimate receives a smaller weight the closer it occurs to a PMW overpass, as depicted in Fig. 3 which shows an example for the 12-h accumulations ending at 12 UTC. For accumulations intervals

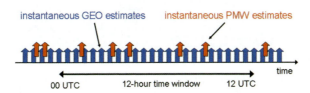

Fig. 3 Depiction and example of how the NRL-Blend accumulations are generated by weighted backwards time integration of the instantaneous GEO and LEO rainrates

beyond 24-h (such as monthly or seasonally), the 3-hourly accumulations are further time integrated (for computational efficiency). Although the computations are done on the 0.1° grid, the final products are averaged to a global 0.25° grid (480 lines × 1440 samples) for size and stored in a basic binary format. Each pixel is stored as a 2-byte short integer, where the integer represents the average rainrate (mm hour^{-1}) over the time interval scaled by 100. To get back the accumulations totals (mm), the integer value is therefore divided by 100 and multiplied the number of hours in the accumulations interval. Although the NRL-Blend was operated intermittently beginning in 2002, official data collection of the global precipitation accumulation products began in January 2004.

2.4 Comparisons with Numerical Weather Prediction Models

An example of the NRL-Blend product is shown in Fig.4a. In this figure, the 12-h precipitation accumulations ending at 12 UTC on 6 September 2006 are shown (color scale in units of mm), with the grayshade background depicting a composited longwave infrared (IR) satellite image at the accumulations end time. Visually, the depiction and location of precipitation appears consistent with known precipitation

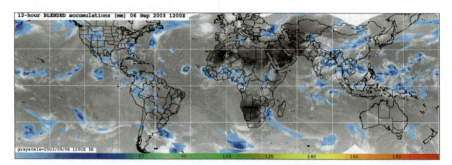

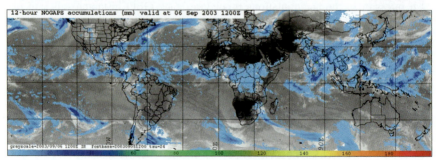

Fig. 4 (**a**) 12-h accumulated precipitation (color scale in mm) ending at 12 UTC on 6 September 2006 from the NRL-Blend HRPP (**b**) As in Fig. 4a, but the 12-h accumulated precipitation (color scale in mm) ending at 12 UTC on 6 September 2006 from the 24-h forecast of the NOGAPS model

patterns at this time of year, e.g., several tropical cyclones in the Atlantic Ocean and tropical disturbances in the western Pacific Ocean. However, each pixel is associated with an error whose structure has various components. For example, the majority of the error is likely associated with the accumulated errors from the instantaneous rainrate estimates provided by each component satellite. Another error may be from precipitation evolution that was not captured by the intermittent revisit schedule of the component satellites.

An additional source of global precipitation data is available from many global numerical weather prediction (NWP) forecast models. While many NWP models do not yet specifically carry rain as a prognostic variable, the conversion of cloud liquid water to rain is typically based on various temperature and humidity thresholds, including parameterizations of convective processes. Since models evolve the moisture with the dynamical state of the atmosphere, the models generally do a better job of following the movement of precipitation associated with frontal systems. This may explain why the motion-based HRPPs (such as CMORPH) tend to perform better in the middle latitudes than other types of HRPPs which carry little or no knowledge of the atmospheric dynamical state. Although an NWP model precipitation dataset is a forecast and not a true observation, it nevertheless is an additional source of data that in some cases may outperform the HRPPs. An example from the Navy Operational Global Atmospheric Prediction System (NOGAPS) NWP forecast model is shown in Fig. 4b, valid at the same time as Fig. 4a. While NOGAPS does capture the same basic patterns of precipitation as noted in Fig. 4a, there are regions of discrepancy. For example, the model tends to spread out too much light precipitation and doesn't capture the intense, smaller scale precipitation (e.g., tropical cyclones). On the other hand, the model pick up light rain associated with mid-latitude frontal systems and doesn't show evidence of discontinuities between land and water backgrounds (which is an issue with passive-microwave based precipitation estimates). Nevertheless, while there are situations where HRPP-derived precipitation is likely to be superior, there are other situations where the opposite is true. Furthermore, NWP model data could complement HRPP data in a combined model-plus-satellite approach. It is therefore constructive to include NWP model forecasted precipitation datasets as part of any HRPP validation effort, as discussed in the next section.

3 Ground Validation

Different applications can accommodate different types of errors and uncertainties in the HRPPs. For example a drought analysis is interested in locations where little or no precipitation has fallen over long time intervals. A flood warning system needs to know how much rainfall has fallen over a period time ranging from hours to weeks. In this section, we summarize previous and ongoing ground validation efforts involving the NRL-Blend. A unique series of satellite-denial experiments are presented which are aimed towards understanding how the performance of HRPPs such as the NRL-Blend is affected by the types of satellite sensors employed in the underlying LEO satellite constellation.

3.1 Verification Efforts of the International Precipitation Working Group (IPWG)

Nearly all of the verification of the NRL-Blend has taken place with ground-based raingauge networks as part of the HRPP verification activities of the International Precipitation Working Group (IPWG) (Turk and Bauer, 2006). Ebert et al. (2007) presented a summary of the over-land validation of 12 HRPPs (including the NRL-Blend) and four NWP models (including NOGAPS) done on a daily time scale and at a 25-km spatial resolution, using the Australian Bureau of Meteorology daily raingauge analysis (Weymouth et al. 1999). The results demonstrated that the NRL-Blend (and other HRPPs) precipitation occurrence and amount were more accurate than NOGAPS (and other NWP) models during summer months and lower latitudes (where mainly convective type precipitation is present). Conversely, NOGAPS (and other NWP) models exhibited superior performance compared to the NRL-Blend (and other HRPPs) during winter months and higher latitudes (mainly lighter, stratiform precipitation). Sapiano and Arkin (2008) analyzed the performance of several three-hourly HRPP accumulations over the central United States, and with an ocean buoy network in the tropical Pacific Ocean. The results showed that the NRL-Blend did resolve the diurnal cycle of precipitation, but with a high bias over land. The recent workshop of the Program for the Evaluation of High Resolution Precipitation Products (PEHRPP), sponsored by the IPWG, provides further summary and recommendations on future validation activities (Turk et al., 2008).

3.2 Satellite Omission Experiments

The Global Precipitation Mission (GPM) is an upcoming joint mission between the National Aeronautics and Space Agency (NASA) and the Japanese Aerospace Exploration Agency (JAXA). It builds upon the heritage of the Tropical Rainfall Measuring Mission (TRMM) with an advanced core spacecraft augmented by a constellation satellite and other international satellite systems with precipitation-sensing capabilities. With changes to satellite missions and sensor capabilities, it is unlikely that the satellites forming the GPM constellation configuration will be known until close to the launch of the core spacecraft, and will change during the lifetime of the GPM. Therefore, it is instructive to study how the retention or loss of a particular satellite platform and/or sensor type will affect the performance of the GPM precipitation products and other applications that utilize GPM products. In this study, we use the existing (late 2008) PMW satellite constellation to examine the impact of several proxy GPM satellite constellation configurations. The focus is on how HRPPs such as the NRL-Blend are affected by such factors as sensor type (conical or across-track scanning) and orbit equator crossing time (many of the satellites in the GPM constellation will be in sun-synchronous orbits), using a collection of GPM proxy datasets gathered over the continental United States. In this section, we examine how the performance of the NRL-Blend is impacted when one or more satellite systems are omitted, using an existing surface gauge network analysis (Chen et al., 2008) for ground validation.

As mentioned in Section 2, the NRL-Blend can be configured to run with any or all of a number of LEO satellites with precipitation-sensing capabilities. In order to examine the impact of particular satellite types, equator crossing times and sensor types (conical or crosstrack), the NRL-Blend was configured for ten parallel runs, each run employing different combinations of satellites and sensor types, beginning in June 2007. Each of these runs employed a different set of satellites relative to the "all satellites" configuration. The all-satellites run, as the name suggests, utilized all 12 available low Earth-orbiting satellites for a total of 13 sensors, previously described in Section 2.

The other parallel runs of the NRL-Blend were configured to specifically study the impact of omitting either the morning (AM) or afternoon (PM) satellites, and the impact of omitting the crosstrack microwave sounding sensors. The reasoning behind the latter is that owing to changes to satellite programs and sensors could lead to the omission of the preferred conically-scanning microwave imagers, leading to a increased role for the crosstrack microwave sounders (designed mainly for temperature and humidity profiling and not precipitation sensing) than originally envisioned in GPM. Microwave sounding sensors such as AMSU, MHS and the future Advanced Technology Microwave Sounder (ATMS) are typically placed in AM or PM sun-synchronous orbits to satisfy observational requirements for NWP data assimilation applications. Several recent studies and algorithms have demonstrated that the AMSU channel suite, with its suite of sounding and window channels, is capable of improved detection of precipitation at high latitudes (Surussavadee and Staelin, 2008). With these encouraging results, it is therefore instructive to determine how the performance of an HRPP (in this case, the NRL-Blend, but the concept could be extended to any HRPP) is impacted by the loss of satellites systems with microwave sounding sensors, and their associated equator crossing times.

The ground truth data used is the optimal interpolation (OI) daily 0.5° gridded gauge analysis provided by the National Oceanic and Atmospheric Administration (NOAA) Climate Prediction Center (CPC) (Chen et al., 2008) over the continental United States during two 3-month periods, Jun–Aug 2007 (JJA) and Dec 2007–Feb 2008 (DJF).

For ease of interpretation with previous studies, the panels in Fig. 5 use the identical box-and-whiskers type presentation as used in Ebert et al. (2007), using the same 1 mm day^{-1} threshold for rain detection. The panels from top to bottom are bias, equitable threat score (ETS), probability of detection (POD) and false alarm rate (FAR). The left figure only considers data west of 100 W longitude (generally higher elevation terrain, colder surface backgrounds), and the right figure is for data

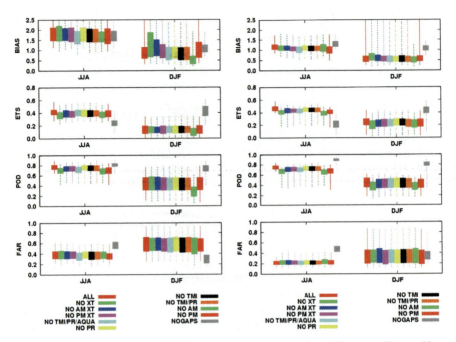

Fig. 5 Seasonal performance of HRPP precipitation estimates using different satellite combinations over the continental United States west of 100 W longitude (*left*), and east of 100 W longitude (*right*), using a threshold of 1 mm day^{-1}. From *top*, bias, equitable threat score (ETS), probability of detection (POD) and false alarm rate (FAR) from ten NRL-Blend versions and one NWP model. Each color refers to a different set of satellites that was omitted from the "all satellites" baseline product from the NRL-Blend. "No XT" refers to no crosstrack sounders, "No AM XT" refers to no morning nodal crossing crosstrack sounders, "No TMI+PR+Aqua" refers to no TRMM TMI and PR and no AMSR-E from Aqua

east of 100 W longitude. The colors refer to different runs of the NRL-Blend each with a different set of satellites. For example, "No AM XT" refers to the NRL-Blend precipitation estimates when all morning (time of ascending node near 1800 local) satellites with crosstrack sounders were omitted from the "all satellites" configuration of the NRL-Blend. "No PM XT" refers to the NRL-Blend precipitation estimates when all afternoon (time of ascending node near 1330 local) satellites with crosstrack sounders were omitted. Only one NWP model (NOGAPS) is shown (gray color). By intercomparing the bias, ETS, POD and FAR side-by-side, one can easily notice where one satellite combination performed better or worse than another, and how this ensemble of HRPPs performed relative to the NOGAPS model during summer and winter seasons, and at lower and higher elevations. In general, there is overall performance degradation for all HRPPS over the western United States (US) compared to the eastern US. This is consistent with studies that have shown the poor performance of PMW scattering-based techniques when used over high elevation and complex terrain (Bennartz and Bauer, 2003). At first glance there is not much difference amongst the various satellite omission runs for the NRL-Blend "adjustment-based" HRPP technique, but closer inspection (green box) illustrates largest performance impact is the omission of the morning overpass crosstrack sounders ("No AM XT" and "No AM" configurations).

4 Sensitivity of Land Surface Parameters

The improved spatio-temporal coverage envisioned for the GPM-era satellite precipitation estimates will facilitate and enable a suite of operational applications that will routinely utilize the estimated satellite rainfall data, especially in ungauged areas, as well as in regions that lack adequate radar coverage. In fact, GPM is promoted as a science mission with broad societal applications, (GPM, 2008a) that will address societal benefits related to human health (soil moisture, climate and disease outbreak), homeland security (removal of chemical/biological/nuclear agents), flooding potential and warning, water availability, water quality, and agriculture and food security. Hence the GPM mission will potentially extend its scope beyond meeting the scientific objectives of advancing precipitation measurement capabilities, improving understanding of the global energy and water cycle variability, and advancing climate and weather prediction (GPM, 2008b).

In addition, the science objectives of GPM also include "hydrometeorological prediction" capabilities for flood-hazard and fresh-water resource modeling and prediction, as well as quantifying the improvements in the estimation and partitioning of land surface parameters of runoff/infiltration/storage and latent/sensible heat fluxes. Hence it is helpful to quantify and understand the sensitivities of some of the commonly used land surface models (LSM) to uncertainties in the satellite precipitation estimates. We use the suite of GPM proxy data from the satellite omission experiments to force the land surface models and then estimate the sensitivities of the different models to differences in the precipitation forcing data.

5 Land Surface Model Response

Impact and validation efforts also include the use of land surface models (LSM) and other types of hydrological observations (other than raingauge as was done above) to examine the impact of these GPM proxy data upon streamflow, discharge, soil moisture and other runoff measurements. By employing the Noah (Ek et al., 2003) and the Mosaic (Koster and Suarez, 1992) LSMs, incorporated with the NASA Land Information System (LIS) (Kumar et al., 2008) to simulate land surface and hydrological states, the performance impact of different GPM constellations can be examined. A similar methodology was used by Gottschalck et al. (2005) to study the impact of different precipitation products on soil states simulated by the Mosaic LSM over the continental United States. Precipitation is considered as an important factor in controlling spatial and temporal patterns of the soil moisture, especially in arid and semi-arid regions (Grayson et al., 2006). The analysis domain covers the south-central United States where there is a large number of well-instrumented watersheds in the Arkansas and Red River basins (Duan and Schaake, 2003).

5.1 Configuration of Land Surface Models

The Noah LSM (version 2.7.1) was used for retrospective soil moisture simulations with four standard soil layers having thickness (from top to bottom) of 10, 30, 60, and 100 cm. Also, the Mosaic model had a standard configuration with 3 soil layers having the following thickness: 2, 158, and 200 cm. Both models were configured at 0.1° × 0.1° latitude-longitude grid over the experimental domain.

There are certain differences in the physics and structure between Noah and Mosaic models that are relevant to this study. In both LSMs, a tile approach is adopted to represent the surface heterogeneity within the model grid cells. The total heat and water vapor fluxes from the individual grid cell are estimated as an area-weighted average of the fluxes produced by the each tile within the grid cell. Every tile in the Mosaic model is completely homogeneous (e.g. it is represented either by complete vegetation cover of a prescribed type or by the bare soil surface). On the other hand, the tiles in the Noah LSM are heterogeneous and treated as a mixture of vegetation canopy and bare soil surfaces. The total tile flux is a weighted average of the fluxes coming from vegetation and bare soil surfaces with weights equal to their corresponding area fractions within the tile. Heat and moisture fluxes are calculated separately over these surfaces. Four different tiles were permitted within each model grid cell in the current simulations of the surface states with the Noah and Mosaic LSMs. It is assumed that both in Noah and Mosaic models the tiles are described by only one surface temperature, which is a reasonable assumption for the Mosaic homogeneous tile, but it can be considered as an oversimplification within the Noah model having heterogeneous tiles composing of vegetation canopy and bare soil surfaces.

Soil temperature prediction is based on the heat diffusion equation, which is numerically solved for four soil layers in the Noah model, but a simpler approach based on the force-restore method (Deardorff, 1978) is adopted in the Mosaic model to predict the soil temperature only in two layers (surface and deep). Note that the surface temperature (T_s), which is equal to the canopy temperature, is estimated from the prognostic energy balance equation in the Mosaic model. Conversely in the Noah model, T_s is evaluated diagnostically from an equation of the surface energy budget, linearized in T_s. Therefore, the Mosaic LSM implies some heat storage (the term $\partial T_s/\partial t$ is nonzero) within the thin surface layer having a finite thickness, which includes both the soil substrate and the canopy air. Due this reason in part, the Mosaic model has a relatively thinner top soil layer of only 2 cm. But in the Noah model the term $\partial T_s/\partial t$ is assumed to be zero, and this fact suggests a zero thickness of the surface layer and therefore no heat storage within it. More details about differences in the surface layer description have been presented by Smirnova et al. (1997).

In both LSMs, soil moisture is predicted from a numerical solution of the diffusion-type equation (e.g., Chen et al. 1996). Water movement in the soil depends on hydraulic properties (saturated conductivity and matrix potential, porosity, and others) of the soil substrate. It assumed that the entire soil column having 2-m depth in the Noah model is homogeneous and each of four Noah model's layers within this column has the same hydraulic properties. Conversely, the Mosaic LSM allows vertical heterogeneity within the 3.5 m soil column and each soil layer is characterized by its own set of hydraulic properties. The explicit use of soil vertical heterogeneity makes the Mosaic model more realistic/adequate in describing water flows in natural conditions in comparison with the Noah LSM, which accepts simplified assumption of vertical homogeneity. A free drainage condition is adopted at the bottom boundary of the soil column in both LSMs.

Surface static fields (vegetation fraction, leaf and stem area indices, soil porosity and texture, sand/clay/silt fraction, elevation, slope, and others), which are necessary for the land state simulations with the LSMs, are bilinearly interpolated or aggregated from their native grids (most of these fields are available at $0.01°$ grid spacing) to the $0.1° \times 0.1°$ latitude-longitude grid using routines available from the NASA LIS. In order to produce realistic soil moisture fields, both LSMs were integrated for a 2.5 year period (from 1 Jan. 2005 to 1 June 2007) using the North American Land Data Assimilation System (NLDAS) forcing fields including precipitation (Cosgrove et al., 2003). Initially, a constant 30% volumetric moisture content was assigned at all model grid points. After the 2.5 year spin-up time, the Noah and Mosaic LSMs were additionally integrated for 14 months (until 31 August 2008) using six different GPM proxy precipitation products and the NLDAS atmospheric fields (except for precipitation). Hourly soil moisture values outputted from both LSMs were averaged to produce daily mean values and these daily values of soil moisture were used as a basis for the analysis described in the next section.

5.2 Soil Water Content Sensitivity

A top 1-m soil water content (SWC) was used as an integral soil moisture measure to compare simulation capabilities between Noah and Mosaic LSMs in reproducing a response of the soil moisture to variations in precipitation. Usually, the total SWC within a column is measured in length units (mm or cm) and can be considered as the amount of water stored in the control volume represented, in our case, by the 1-m soil column having the unit cross-section area. Due to this definition, the SWC is also known as the water storage (Mitchell et al., 2004; Schaake et al., 2004). Either the water storage (or the column SWC) and its temporal change/range at monthly and seasonal scales are broadly adopted for the LSM intercomparison. Previous studies (Mitchell et al., 2004; Schaake et al., 2004) have shown that the local maximum water holding capacity of the soil substrate (defined as the SWC difference between saturation and wilting points, which depend on the soil texture) has a little impact upon the observed and simulated water storage range. Rather,

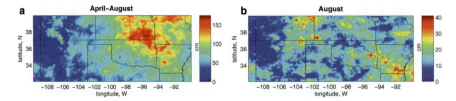

Fig. 6 (a) Geographical distribution of accumulated precipitation (cm) from the NRL-Blend "All-Satellites" configuration during 1 April to 31 August 2008. **(b)** Same as above, but during August 2008 only

they have suggested that the monthly/seasonal water storage changes are controlled by the description of the model's evaporation and runoff.

In order to better understand the seasonal variations of the impact of the various NRL-Blend precipitation products on the simulated SWC, we focused on a monthly analysis of SWC difference *relative to the "all-satellites" NRL-Blend configuration* during Mar–Aug 2008. Figure 6a depicts the estimated precipitation from the NRL-Blend "All Satellites" configuration between 1 April and 31 August 2008, and Fig. 6b represents August 2008 only. It would be instructive to consider monthly mean SWC fields (climatology) simulated using the all-satellites precipitation product before performing an intercomparison between SWC fields produced by the six GPM proxy precipitation products. Figure 7a illustrates geographical distribution of top 1-m SWC simulated with Noah and Mosaic LSMs and averaged for August 2008. Although there is a general agreement in geographical patterns of 1-m SWC simulated by the two LSMs (the corresponding correlation between the Noah and Mosaic SWC fields is as high as 0.67), the SWC values produced by the Mosaic model are substantially underestimated in comparison to those simulated by the

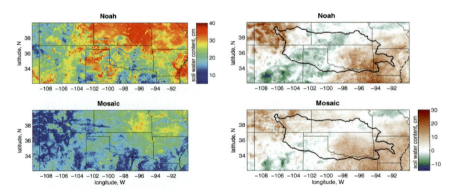

Fig. 7 Geographical distribution of (**a**, *left*) top 1-m soil water content (averaged for August 2008) and (**b**, *right*) its change from 1 April to 31 August 2008, simulated with Noah (*upper frame*) and Mosaic (*lower*) land surface models, using the "all-satellites" NRL-Blend precipitation product. Positive values shown by brown color in right frames stand for soil drying and negative (*green/blue*) for soil moistening. The thick line in Fig. 7b outlines the boundary of the Arkansas-Red River basin

Noah LSM. The mean top 1-m SWC difference (bias) between Noah and Mosaic (Noah minus Mosaic value) simulations is 8.3 cm for August 2008. This rather large SWC bias demonstrated by the Mosaic model might be attributed to more efficient surface evaporation and drainage through the lower boundary of the soil column, accounting for the slope of the model grid cell, as compared to that in the Noah LSM. Comparisons between observed and simulated top 2-m SWC performed over the state of Illinois (Schaake et al., 2004) also indicated a low SWC bias predicted by the Mosaic model, especially for SWC $\leq$ 50 cm, and almost zero bias for the Noah LSM. Figure 7b shows the Apr–Aug SWC (top 1-m water storage) change simulated by the Noah and Mosaic LSMs. Despite differences in the model physics, both LSMs produce highly-correlated spatial patterns of the water storage change. These patterns include areas of drying (positive values of SWC change) in the NW and SE parts of the domain, and a distinct zone of moistening (negative values) stretching from the SW to the NE corner of the domain.

The impact of precipitation in a LSM is dependent upon many physical factors, such as soil type, vegetation, etc. and a soil moisture analysis at a given time is likely to be the cumulative result of precipitation amount and variability from weeks or months prior. For example, note how the soil moisture pattern in Fig. 7a is better matched with the pattern of the precipitation that fell for several months prior (Fig. 6a). To accommodate this, Fig. 8 shows the geographical distribution (August 2008 average values) of top 1-m SWC difference relative to the all-satellites configuration for the six different proxy constellations. Results for both Noah (upper frame) and Mosaic (lower frame) LSMs are depicted in Fig. 8. This monthly mean SWC difference can be considered as a typical bias in the soil moisture produced by the omission of the specified satellites and sensors. We note that the biggest impact (largest absolute values of the top 1-m SWC biases) is due to the omission of either the crosstrack microwave sounders, or the morning crossing (AM) satellites from the NRL-Blend. On the other hand, omission of afternoon (PM) satellites in the NRL-Blend resulted in the smallest impact upon the soil moisture simulated both with the Noah and Mosaic models. Although the relative magnitude of these SWC changes is small (generally, they are in the ± 5 cm range), and the spatial scales are different, these results are consistent with the raingauge-only validation presented earlier in Fig. 5.

Note that all the scenarios except the TMI+PR+Aqua omission case provide a spatially coherent response in the SWC difference relative to the all-satellites configuration. Indeed, a positive SWC difference prevails in the western half of the domain (west of 100 W) and negative over the eastern half (east of 100 W longitude) as shown in Fig. 8. These marked features of the spatial SWC response simulated with the two LSMs might be associated with east-west gradients of hydrometeorological variables (Duan and Schaake, 2003). Generally over this basin, annual precipitation and runoff decrease and potential evaporation increases from east to west, resulting in wetter areas east of 100 W and drier to the west. Also, the accuracy of satellite precipitation estimates are generally lower over high elevation terrain concentrated west of 100 W longitude. It is important to note a rather high similarity in spatial

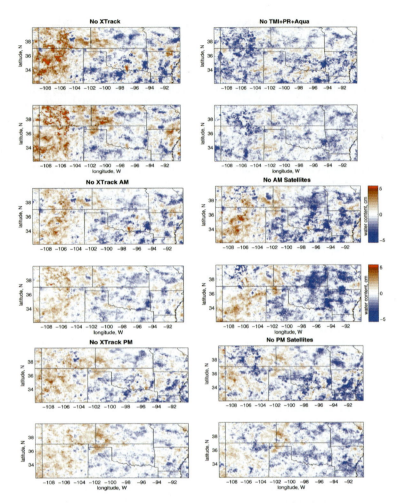

Fig. 8 Geographical distribution of top 1-m soil water content (SWC) difference relative to the all-satellites configuration for the six different GPM proxy constellations noted at the top of each frame (upper part of each frame corresponds to Noah and lower to Mosaic model simulations). SWC represents August 2008 averaged values

patterns of SWC differences simulated with two LSMs with quite different structure and physics. Indeed, correlations of SWC difference between the Noah and Mosaic LSMs are high for all different satellite omission scenarios. Corresponding correlation coefficients range between 0.7–0.8 during summer months and 0.6–0.7 in spring with a little variation among the six satellite omission cases. Observed high coherence in the top 1-m spatial response (estimated as a difference relative to some baseline case) simulated with different LSMs may suggest that precipitation is a more important factor in controlling the spatial distribution of the SWC relative response in comparison to the model physics.

While this single month (August 2008) analysis is important, these results are further substantiated by analyzing the symbolic boxplots (Wilks, 2006) of the top

1-m SWC difference (relative to the all-satellites scenario) distribution when plotted for an entire spring-summer 2008 and additionally stratified into two regions east/west of 100 W longitude. Figure 9 shows these symbolic boxplots (each of the six panels uses a different constellation configuration, identical to Fig. 8) for both Noah and Mosaic LSMs and for the entire spring-summer 2008. As before, the largest SWC deviations (they are proportional to the range between upper and lower quartiles depicted in Fig. 9) from the all-satellites configuration are observed when the crosstrack microwave sounders or AM crossing satellites are excluded. As in Fig. 8, less response in SWC is produced by both the Noah and Mosaic models west of 100 W longitude over high elevation terrain and drier regions. Also, both LSMs indicate that SWC deviations have a tendency to be positive (soil moistening in comparison to the all-satellites case) west and negative (soil drying) east of

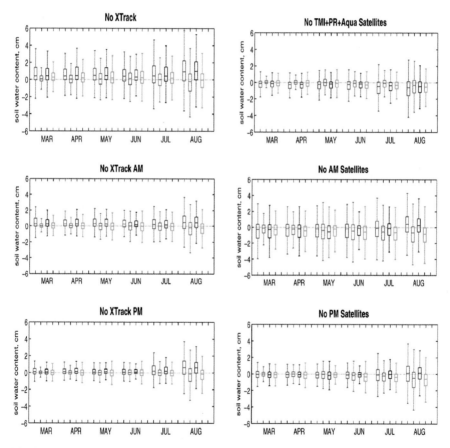

Fig. 9 Symbolic distribution plots (boxplots) of top 1-m soil water content (SWC) difference relative to the "all-satellites" NRL-Blend configuration. Median, upper and lower quartiles, and data range within lower/upper inner fences are shown. SWC differences for the regions west of 100 W are depicted in blue (Noah) and black (Mosaic), and for the region east of 100 W are shown in magenta (Noah) and green (Mosaic)

100 W longitude. In addition to the above mentioned geographic differences, Fig. 9 illustrates a clear seasonal tendency in SWC change from the relatively low SWC response during spring and early summer months, to the high SWC deviation by the end of summer.

Despite spatial scale differences, these results are consistent with the validation impact studies presented in Section 3, demonstrating the importance of the crosstrack microwave sounders and the morning local time crossing satellites. However, we note that these results are unique to the type of HRPP technique (NRL-Blend) used in this study, and the results may differ when used with other types of HRPP techniques. Also, precipitation often has a predominant time-of-day cycle and therefore the local time of the observation is important. Relatively speaking, soil moisture changes over longer time scales than precipitation does. Therefore in the case of the soil moisture simulations, we note that the removal of the morning (AM) satellites likely has less to do with the specific local time-of-day observation than it does with the fact that the bulk of the current (2008) satellites (DMSP, Coriolis and several NOAA) have early morning crossing times.

6 Conclusions

We have examined the impact of omitting certain types of satellites and sensors upon the performance of the NRL-Blend high resolution precipitation product (HRPP). Each separate run of the NRL-Blend omitted one or more sensors relative to the "all satellites" satellite configuration. These omission experiments were designed to examine possible types of satellite constellation configurations that may exist during the GPM era. A specific purpose was to examine the utility of the crosstrack passive microwave (PMW) sounders which, while not designed for quantitative precipitation retrieval, have shown promise in precipitation estimation at higher elevations and latitudes. A second purpose was to examine the local time of observation by the sun-synchronous satellite platforms (many GPM constellation satellites will orbit in sun-synchronous orbits) affects the performance of the satellite precipitation products. The impact was examined two ways. The first was by examining traditional performance metrics (bias, correlation, etc.) amongst the various NRL-Blend products, where the validation data consisted of a dense raingauge analysis over the central United States, and where the validation was separated along the 100 W longitude line (which roughly separates drier, high elevation regions west of this boundary from lower, moist regions to the east). The overall performance of the NRL-Blend degrades over higher elevation areas, where PMW techniques are known have problems. A slight but noticeable degradation in the performance of the NRL-Blend was noted for the case when all crosstrack sounders were removed or the morning local time crossing satellites were removed (change to equitable threat score of about 0.1 and a similar degradation in probability of detection).

The second impact consisted of examining the outputs of land surface models (LSM) where the LSM was forced with these same precipitation datasets. Both the Noah and the Mosaic LSMs, incorporated with the NASA Land Information

System, were used to simulate land surface and hydrological states when these same NRL-Blend precipitation datasets were used (in lieu of the LIS-provided precipitation). The LSM analysis examined seasonal changes in modeled soil moisture. The biggest impact (largest absolute values of the top 1-m SWC biases) was due to the omission of either the crosstrack microwave sounders, or the morning crossing (AM) satellites. On the other hand, omission of afternoon (PM) satellites in the NRL-Blend resulted in the smallest impact upon the soil moisture simulated both with the Noah and Mosaic models. Although the relative magnitude of these SWC changes is small (in the ± 5 cm range), these results are consistent with the raingauge-only validation conclusions, suggesting the importance of the crosstrack microwave sounders in future GPM constellations.

GPM is currently planned to be active during the NASA Soil Moisture Active Passive (SMAP) mission. Although each mission is designed to function independently of each other, there exists significant GPM-SMAP overlap in terms of science goals and applications, specifically towards the utilization of frequent precipitation estimates. For example, one of the biggest obstacles to improved over-land precipitation estimation is the large variability in land surface emissivity, which is affected by near-surface soil moisture. Conversely, GPM can potentially benefit SMAP soil moisture retrievals with its capability for improved tracking of precipitation evolution between SMAP revisits. It is constructive to examine the connection between these two missions at this stage as to maximize the utility of the overall data towards achieving mission goals.

References

Bennartz, R. and P. Bauer, 2003: Sensitivity of microwave radiances at 85–183 GHz to precipitating ice particles. *Radio Sci.*, **38**, 4, 8075, doi:10.1029/2002RS002626.

Chen, F., et al., 1996: Modeling of land surface evaporation by four schemes and comparison with FIFE observations. *J. Geophys. Res.*, **101**, 7251–7268.

Chen, M., W. Shi, P. Xie, V. B. S. Silva, V. Kousky, R. Higgins, and J. Janowiak, 2008: Assessing objective techniques for gauge-based analyses of global daily precipitation. *J. Geophys. Res.*, **113**, D04110, 1–13.

Cosgrove, B. A., et al., 2003: Real-time and retrospective forcing in the North American Land data Assimilation System (NLDAS) project. *J. Geophys. Res.*, **108**, 8842, doi: 10.1029/2002JD003118.

Deardorff, J. W., 1978: Efficient prediction of ground surface temperature and moisture, with inclusion of a layer of vegetation. *J. Geophys. Res.*, **83**, 1889–1903.

Duan, Q. and J. C. Schaake, Jr., 2003: Total water storage in the Arkansas-Red River basin. *J. Geophys. Res.*, **108**, 8853, doi:10.1029/2002JD003152.

Ebert, E., C. Kidd, and J. Janowiak, 2007: Comparison of near real-time precipitation estimates from satellite observations and numerical models. *Bull. Amer. Meteor. Soc.*, **88**, 47–64.

Ek, M. B., K. E. Mitchell, Y. Lin, E. Rogers, P. Grunmann, V. Koren, G. Gayno, and J. D. Tarplay, 2003: Implementation of Noah land surface model advances in the National Centers for Environmental Prediction operational mesoscale Eta model. *J. Geophys. Res.*, **108**, 8851, doi:10.1029/2002JD003296.

GPM, 2008a: GPM Science Serving Society. *Global Precipitation Measurement, NASA Goddard Spaceflight Center* http://gpm.gsfc.nasa.gov/features/servingsociety.html

GPM, 2008b: GPM Science Objectives. *Global Precipitation Measurement, NASA Goddard Spaceflight Center* http://gpm.gsfc.nasa.gov/science.html

Gottschalck, J., J. Meng, M. Rodell, and P. Houser, 2005: Analysis of multiple precipitation products and preliminary assessment of their impact on Global Land Data Assimilation System land surface states. *J. Hydrometeor.*, **6**, 573–598.

Grayson, R. B., A. W. Western, J. P. Walker, D. D. Kandel, J. F. Costelloe, and D. J. Wilson, 2006: Controls on patterns of soil oisture in arid and semi-arid systems, Chapter 7, in *Ecohydrology of Arid and Semi-Arid Ecosystems*. Eds. P. D'Ordorico and A. Porporato, Springer, The Netherlands, 341p.

Huffman, G. J., R. F. Adler, D. T. Bolvin, G. Gu, E. J. Nelkin, Y. Hong, D. B. Wolff, K. Bowman, and E. F. Stocker, 2007: The TRMM multisatellite precipitation analysis (TMPA); Quasi-global, multiyear, combined-sensor precipitation estimates at fine scales. *J. Hydrometeor.*, **8**, 38–55.

Hsu, K., X. Gao, S. Sorooshian, and H. V. Gupta, 1997: Precipitation estimation from remotely sensed information using artificial neural networks. *J. Appl. Meteor.*, **36**, 1176–1190.

Joyce, R. J., J. E. Janowiak, P. A. Arkin, and P. Xie, 2004: CMORPH: A method that produces global precipitation estimates from passive microwave and infrared data at high spatial and temporal resolution. *J. Hydromet.*, **5**, 487–503.

Koster, R. D. and M. J. Suarez, 1992: Modeling the land surface boundary in climate models as a composite of independent vegetation stands. *J. Geophys. Res.*, **108**, 2697–2715.

Kumar, S. V., C. D. Peters-Lidard, J. L. Eastman, and W. -K. Tao, 2008: An integrated high-resolution hydrometeorological modeling testbed using LIS and WRF. *Environ. Model. Soft.*, **23**, 169–181.

Mitchell, K. E., et al., 2004: The multi-institution North American land Data Assimilation System (NLDAS): Utilizing multiple GCIP products and partners in a continental distributed hydrological modeling system. *J. Geophys. Res.*, **109**, D07S90, doi: 10.1029/2003JD003823.

Sapiano, M. R. P. and P. A. Arkin, 2008: An inter-comparison and validation of high resolution satellite precipitation estimates with three-hourly gauge data. *J. Hydromet.*, in press. doi: 10.1175/2008JHM1052.1.

Schaake, J. C., et al., 2004: An intercomparison of soil moisture fields in the North American Land data Assimilation System (NLDAS). *J. Geophys. Res.*, **109**, D01S90, doi:10.1029/2002JD003309.

Smirnova, T. G., J. M. Brown, and S. G. Benjamin, 1997: Performance of different soil model configurations in simulating ground surface temperature and surface fluxes. *Mon. Wea. Rev.*, **125**, 1870–1884.

Surussavadee, C. and D. H. Staelin, 2008: Rain and snowfall retrievals at high latitudes using millimeter wavelengths, *Proc IGARSS 2008*, 6–11 July, Boston.

Turk, J. and P. Bauer, 2006: The international precipitation working group and its role in the improvement of quantitative precipitation measurements. *Bull. Amer. Meteor. Soc.*, **87**, 643–647.

Turk, F. J. and S. Miller, 2005: Toward improving estimates of remotely-sensed precipitation with MODIS/AMSR-E blended data techniques. *IEEE Trans. Geosci. Rem. Sens.*, **43**, 1059–1069.

Turk, F. J., P. Arkin, E. Ebert, and M. Sapiano, 2008: Evaluating High Resolution Precipitation Products. *Bull. Amer. Meteor. Soc.*, December issue.

Vicente, G., R. A. Scofield, and W. P. Menzel, 1998: The operational GOES infrared rainfall estimation technique. *Bull. Amer. Meteor. Soc.*, **79**, 1883–1898.

Vicente, G., J. C. Davenport, and R. A. Scofield, 2002: The role of orography and parallax correction on real time high resolution satellite rainfall estimation. *Int. J. Remote Sens.*, 23, 221–230.

Weymouth, G., G. A. Mills, D. Jones, E. E. Ebert, and M. J. Manton, 1999: A continental-scale daily rainfall analysis system. *Aust. Meteorol. Mag.*, **48**, 169–179.

Wilks, D. S., 2006: Statistical Methods in the Atmospheric Sciences. 2nd Ed. Elsevier, New-York, NY, 627 p.

Kalman Filtering Applications for Global Satellite Mapping of Precipitation (GSMaP)

Tomoo Ushio and Misako Kachi

Abstract GSMaP (Global Satellite Mapping of Precipitation) is a project aiming (1) to produce high-precision and high-resolution global precipitation maps using satellite-borne microwave radiometer data, (2) to develop reliable microwave radiometer algorithms, and (3) to establish precipitation map techniques using multi-satellite data for the coming GPM era. The GSMaP_MVK system uses a Kalman filter model to estimate precipitation rate at each $0.1°$ with 1-h resolution on a global basis. The input data sets are precipitation rates retrieved from the microwave radiometers and infrared images to compute the moving vector fields. Based on the moving vector fields calculated from successive IR images, precipitation fields are propagated and refined on the Kalman filter model, which uses the relationship between infrared brightness temperature and surface precipitation rate. This Kalman filter – based method shows better performance than the moving vector – only method, and the GSMaP_MVK system shows a comparable score compared with other high-resolution precipitation systems.

Keywords Kalman filter · Infrared radiometer · Precipitation map · Microwave radiometer

1 Introduction

Estimation of the global distribution of precipitation with high accuracy has long been a major scientific goal. The making of precipitation maps on a global basis is important for modeling of the water cycle, maintaining the ecosystem environment, agricultural production, improvements of weather forecast precision, flood warning, and so on. Because most rain gauges are distributed in the Northern Hemisphere, and

T. Ushio (✉)
Osaka University, Suita, Osaka 565-0871, Japan
e-mail: ushio@comm.eng.osaka-u.ac.jp

there are extremely few rain gauges on the sea, it is difficult to measure the temporal and spatial changes of the rain rate on a global scale. In addition to this, precipitation is basically a rapid process and has a great variability in space and time. Due to these features, it is quite difficult to capture the global distribution of precipitation with high resolution and enough accuracy for the scientific and practical applications.

As a result of recent progress in technology, many satellites for observation of meteorological phenomena have been launched, enabling us to observe precipitation on a global basis with high resolution and accuracy. The passive observation of precipitation at microwave frequencies has been shown to be very effective for estimating the rainfall rate (mm/h) with enough accuracy, and we now have several active satellites with on-board microwave radiometers. Following that success, future satellite-based programs like GPM (Global Precipitation Measurement) are planned. The GPM project is a follow-on mission of the TRMM (Tropical Rainfall Measuring Mission) under international cooperation including the United States, Japan, and other countries, which will extend TRMM observation to higher latitudes with 3-h sampling at any given point on the earth. Since the microwave radiometers are all on low earth orbit satellites, the problem of sampling error is unavoidable, even if all the microwave radiometers aboard the satellite are used. Therefore it is necessary to utilize a gap-filling technique to generate precipitation maps from only the microwave radiometer data if temporal resolution of 3 h or less is required, which is important for the operational applications like flash flood warning systems.

Worldwide, damages by floods account for more than two-thirds of the total damage caused by natural disasters. Floods are usually caused by heavy rainfalls and/or tropical cyclones, such as typhoons, hurricanes, and cyclones, and occur almost every season and year all over the world. Accurate estimations of precipitation, water vapor, and clouds from satellite observations are needed in higher resolution in time and space for this purpose. Long-term operation and near-real-time availability are also required. Great expectations for development of early warning and alert systems for flood events using satellite-derived precipitation information have been recently raised internationally.

On the basis of the requirements for global precipitation estimates, there are numerous global precipitation systems, for example TMPA-RT (Huffman et al. 2007), NRLgeo (Turk and Miller 2005), PERSIANN (Sorooshian et al. 2000), CMORPH (Joyce et al. 2004), and PMR (Kidd et al. 2003), that provide data in near real time through the Internet. Among them, the WCRP (World Climate Research Program) GPCP (Global Precipitation Climatology Project) has been a pioneer in this field and the most successful system that provides precipitation estimates on a monthly 2.5° grid with two decades of data (1979 to present) (Adler et al. 2003). The most recent system named TMPA (The Tropical Rainfall Measuring Mission Multi-satellite Precipitation Analysis) provides 3-h real-time rainfall analysis on a 0.25° grid (Huffman et al. 2007). The TMPA method utilizes precipitation estimates from the various microwave radiometer data and the surface precipitation estimates by the ground rain gauges, based on histogram matching for the long term. TMPA's estimates are generated from infrared radiometer data of the geosynchronous meteorological satellites with precipitation estimates from the microwave radiometer. While the GPCP type system uses the direct conversion method from brightness temperature at infrared wavelengths to the rainfall rate on the microwave radiometer data as a calibrator, CMORPH (CPC Morphing technique) (Joyce et al. 2004) takes a different approach to produce precipitation maps that have higher resolution of 30 min and 0.073°. CMORPH calculates atmospheric motion vectors from two successive infrared images at 30-min intervals and then the precipitation pixels are propagated according to the moving vector fields. In this approach, no direct conversion between the infrared brightness temperature and rainfall rate is used, which is a distinct feature compared with precipitation systems such as CMORPH. In addition to the TMPA and CMORPH products, the PERSIANN (Sorooshian et al. 2000) system works on a neural network based procedure to compute a precipitation estimates with 0.25° resolution, and the NRL blended technique (Turk and Miller 2005) is built on the statistical relationships from co-located passive microwave and infrared pixels, and is operated in near real time. In spite of the physical simplicity of CMORPH, this moving method shows excellent scores from the daily 0.25° comparisons with the radar-rain gauge networks in several regions on the globe in PEHRPP (Pilot Evaluation of the High Resolution Precipitation Products).

The Global Satellite Mapping of Precipitation (GSMaP) project started in 2002 with the support of the Japan Science and Technology Agency (JST). Its aims are (1) to produce high-precision and high-resolution global precipitation maps from satellite-borne microwave radiometer data, (2) to develop reliable microwave radiometer algorithms, and (3) to establish a precipitation map technique by using multi-satellite data for the coming GPM era. In this project, in order to make global precipitation maps with 0.1°/1 h resolution, we take an approach that takes advantage of both the moving vector method and direct conversion of brightness temperatures to rain rates. In order to combine these advantages, Kalman filter theory is applied to the precipitation rate propagated along with the atmospheric motion vector. We refer to this Kalman filter–based system as GSMaP_MVK, short for GSMaP moving vector with Kalman filter method, while the moving vector–only approach is labeled GSMaP_MV. Using this technique, a near-real-time system named GSMaP_MVK_RT, which simply contains only the propagation process forward in time, was developed and opened to the public via the Internet. The full system (GSMaP_MVK), which contains both the forward and backward propagation processes, has been developed and implemented (Ushio et al. 2009). In this chapter, the GSMaP_MVK system is summarized and reviewed.

2 Data

This section describes the input data sets that are used in this study. As is stated above, the precipitation estimation from the microwave radiometer is better than that of IR-only approaches. However, all these microwave radiometers aboard the satellite are in low earth orbit (LEO) and there is a problem of sampling error. On the other hand, infrared radiometers on geostationary orbit can give information on cloud top layers and equivalent blackbody temperatures. Since the Geo-IR sensors sees the cloud top pattern with hourly to sub-hourly resolution, the sampling interval is less than that of LEO microwave radiometers. However, the IR channel is not as sensitive to precipitation rates. Therefore, the microwave radiometer (MWR) on low earth orbit (LEO) and the infrared radiometer (IR) on geostationary (Geo) orbit are quite complementary with each other for monitoring the precipitation. Hence in this system, both the LEO-MWR and the Geo-IR data sets are used.

At the moment of this writing, the available microwave sensors are SSM/I (Special Sensor Microwave/Imager), TMI (TRMM Microwave Imager), and AMSR-E (Advanced Microwave Scanning Radiometer for EOS), whose characteristics are listed in Table 1. The algorithm to convert the brightness temperature observed at the several microwave channels to surface rain rate is described in Aonashi et al. (1996) and Kubota et al. (2007), and the surface rain rate is retrieved by finding the optimal rain intensity that matches the brightness temperature calculated from the radiative transfer model with the observed brightness temperature. At the time of this writing, no AMSU data sets are used in this product, because development of the AMSU rainfall product from the Aonashi's algorithm have not yet been completed. In the near future, the AMSU data set will be included in this GSMaP_MVK product.

Table 1 Characteristics of microwave radiometers used in this study

Name	Altitude (km)	Sensor	Frequency (GHz)
TRMM	402	TMI	10, 19, 21, 37, 85
AQUA	705	AMSR-E	7, 10, 19, 24, 37, 89
DMSP-F13	803	SSM/I	19, 37, 85
DMSP-F14	803	SSM/I	19, 37, 85
DMSP-F15	803	SSM/I	19, 37, 85

The IR data sets used in the current version of the system are from the CPC (Climate Prediction Center) (Janowiak et al. 2001) through the Man – computer Interactive Data Access System (McIDAS; Lazzara et al. 1999). The latitude and longitude resolution of the data are 0.03635° (4 km at the equator). The latitude range is 60° N–60° S. The temporal resolution is about 30 min. While the standard system uses these IR data sets as input, the near-real-time system named GSMaP_NRT uses slightly different IR data for the purpose of near-real-time supply of the system as described below.

3 Methodology

Figure 1 shows the algorithm flow of the GSMaP_MVK system. The basic procedures are (1) to propagate the rain pixels along with the atmospheric motion vector derived from successive IR images in 1-h intervals, (2) to refine the precipitation rate on a Kalman gain, and (3) to combine the backward precipitation fields. Since the propagation process is similar to the morphing technique in CMORPH, an overview of the moving vector technique is first given here and then the method for applying the Kalman filter is presented.

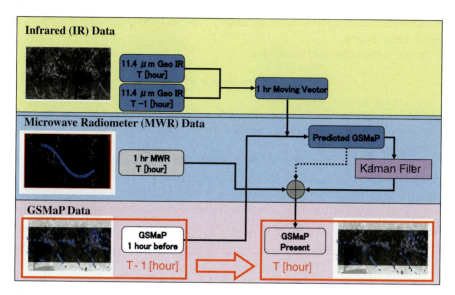

Fig. 1 Flow chart of the algorithm developed in this study. Adapted from Ushio et al. (2009)

The procedure to calculate the atmospheric motion vector is simple. That is, a sub-region is selected, the region is correlated, and the peak correlation is determined. Consider the two successive IR images at times t and $t + 1$ h. If the cloud area at time $t + 1$ moves horizontally a few pixels from time t, the two images match at the offset of a few pixels. This offset value can be obtained by (1) calculating the two-dimensional correlations between the two images for the zero offset, (2) repeating the calculation for various offsets, and then (3) looking for the maximum correlation. The offset value in longitudinal and latitudinal directions that shows the maximum correlation is the atmospheric motion vector that we want. In this system, a FFT (Fast Fourier Transformation) based algorithm is applied to compute the correlation coefficient in order to speed up the computation.

According to the method stated above, the moving vector fields are obtained globally with several degrees resolution. Based on the vector fields, the precipitation pixels that were retrieved from the microwave radiometer data are propagated forward in time and the precipitation rate at the propagated pixel is reformed using the Kalman filter. This procedure is repeated until the next microwave radiometer overpass.

The Kalman filter is a theory that provides an efficient recursive means to estimate the state of a process from a series of noisy measurements. In this application, we pay much attention to the relationship between the IR brightness temperature and the surface precipitation rate to refine the precipitation rate propagated by the moving vector. Although the brightness temperature at IR wavelengths is not so sensitive to surface precipitation rate, the data have no sampling error with 1 h resolution and are statistically correlated with the surface precipitation rate with large variances. This noisy measurement in terms of precipitation rate provides better feedback information to more accurately represent the temporal variation of a precipitation system.

In this system, the state equation is

$$X_k = X_{k-1} + w \qquad (1)$$

where X_k is the precipitation rate at time k, which is the propagated precipitation rate forward in time; and w is the process noise, which shows the variation of the precipitation system.

Figure 2 shows a histogram of the w values, which indicate the uncertainty in precipitation rate after 1 h from its propagation compared with the precipitation rate retrieved from the microwave radiometer. It is shown that the variation of precipitation rate in 1 h is normally distributed with zero mean, enabling application of the Kalman filter theory.

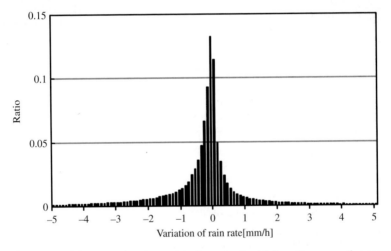

Fig. 2 Histogram of the uncertainty in precipitation rate after 1 h from its propagation compared with the rain rate retrieved from the microwave radiometer. Adapted from Ushio et al. (2009)

Figure 3 shows the geostationary satellite IR brightness temperature relative to precipitation rate estimated from microwave radiometers. We see that the IR brightness temperature is non-linearly correlated with surface precipitation rate with large variance and a noisy measurement of the true precipitation is made from the IR observation. Based on this relationship, at time k a measurement y_k of the true state X_k is made according to

$$y_k = HX_k + v \qquad (2)$$

where H is a constant coefficient and v is the measurement error expressed as bars in Fig. 3. To apply the Kalman filter on a linear basis, the linearization approximation is needed and has been performed for every 1 mm h^{-1} increment in the present study.

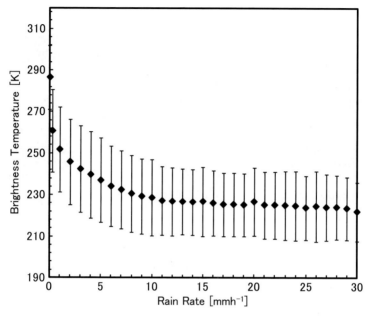

Fig. 3 Geostationary satellite IR brightness temperature (ordinate) relative to precipitation estimated from microwave radiometers (abscissa) during July 2005. Adapted from Ushio et al. (2009)

Based on these equations, the Kalman gain is computed to refine the precipitation rate after its propagation. After thus applying the Kalman filter, the same procedure is applied to the precipitation pixels from the next microwave radiometer traversal. Following the CMORPH strategy, this procedure is illustrated in Fig. 4. Consider the case that the microwave radiometer passes over a certain area at time t. According to the procedure described above, the rainy pixels are propagated from the moving vector derived from successive IR images and the precipitation rate is refined via the Kalman filter. This process is repeated until the next microwave radiometer traversal. When the next microwave radiometer arrives, the revisited precipitation pixels are also propagated backward in time and the Kalman filter refines the precipitation rate as illustrated in the middle column in Fig. 4. This process is repeated until the most recent microwave radiometer traversal. The optimal precipitation rate is calculated from the weighted average between the forward and backward estimates.

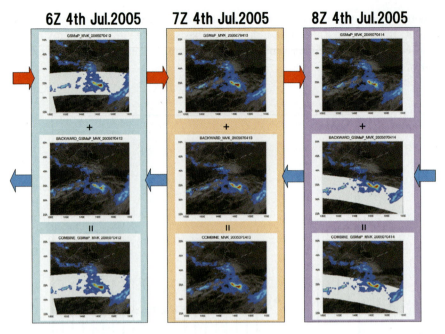

Fig. 4 Schematic illustration combining the precipitation fields forward and backward in time. The white belts denote the coverage of the microwave sensors. Adapted from Ushio et al. (2009)

4 Current Status of the System

Since the GSMaP_MVK system is from a newly developed algorithm, the processing and data supply systems are being upgraded. To meet user requirements of the high-resolution precipitation data in near real time, the JAXA Earth Observation Research Center (EORC) has developed and operated a near-real-time data processing system, which is called "Global Rainfall Map in Near Real Time (GSMaP_NRT)" based on the GSMaP algorithms described in previous sections. Core algorithms of the system are based on those provided by the GSMaP project, microwave radiometer retrieval by GSMaP_MWR, and IR data merged by GSMaP_MVK.

GSMaP_NRT uses TRMM/TMI, AMSR-E, three SSM/I, and geostationary satellite information as inputs. Input data are almost the same as GSMaP standard systems, but differ slightly in terms of their real-time availability. Table 2 summarizes the differences of input data including ancillary data between GSMaP_NRT and GSMaP standard systems. Data available within three hours from observation are utilized in the GSMaP_NRT system. Infrared (IR) data observed by each geostationary satellite, such as the MTSAT satellite by the Japan Meteorological Agency, GOES satellites by NOAA, and Meteosat satellites by EUMETSAT, are provided through the Japan Weather Association within 1 h after observation. Note that the SSM/I data on DMSP F-15 satellite have been used for rainfall retrieval only over

Table 2 Difference between GSMaP_NRT and standard systems

Input data	Sensor	GSMaP_NRT	GSMaP standard
Passive microwave radiometer	TRMM TMI	NASA/GSFC Realtime version	NASA/GSFC Standard version
	Aqua AMSR-E	JAXA/EORC	JAXA/EORC
	DMSP SSM/I (F13, 14, 15)	NOAA/national weather service	Remote sensing systems
GEO Infrared radiometer	MTSAT, METEOSAT-7/8, GOES-11/12	Globally-merged pixel-resolution data by JWA	Full-resolution IR data by NASA/GSFC and NOAA/climate prediction center (CPC)
Atmospheric information	–	JMA global analysis (GANAL) realtime version	JMA global analysis (GANAL)
Sea surface temperature	–	JMA Merged satellite and in situ data global daily sea surface temperatures in the global ocean (MGDSST)	JMA MGDSST

the ocean because of interference problems in the 22 GHz Vertical polarization channel since August 2006. Also note that SSM/I data on DMSP F-14 satellite have not been available since 23 August 2008.

The current GSMaP_NRT system operationally produces frequent and accurate hourly global rainfall maps within four hours after observation. Figure 5 is a schematic flow of the system. Note that the system uses only the forward processing of the GSMaP_MVK, because it gives greater importance to data availability than quality, to meet user requirements such as application to flood prediction. A system combining forward and backward processing in the operational system, which will be available three days after observation, is in preparation. Full-resolution IR data, which are provided by NASA/GFSC and NOAA/CPC, and microwave radiometer data, which have not been included in near-real-time processing because of data delay, will be used for reprocessing three days after observation.

Currently, GSMaP_NRT rainfall systems are distributed with browse images and KMZ files for Google Earth software via the Internet (http://sharaku.eorc.jaxa.jp/GSMaP/). The system has been processing by means of the Version 1 system since November 2007, and a major update of the algorithm to Version 2 has been applied since October 2008. Table 3 is a summary of GSMaP_NRT Version 2 systems. A satellite information flag denotes all satellite sensors that are used in estimation of rainfall at each pixel during one-hour time period. An observation time flag indicates the relative time of the latest microwave

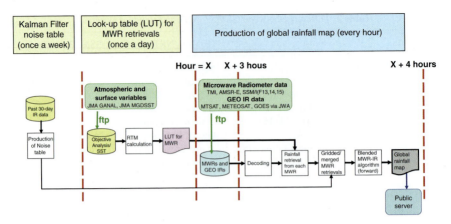

Fig. 5 Schematic flow of the GSMaP_NRT processing system

Table 3 Summary of GSMaP_NRT Systems (Version 2)

Parameter [unit]	Coverage	Horizontal resolution	Temporal resolution
Hourly rain rate [mm/h] Satellite information flag Observation time flag	Global (60°N–60°S)	0.1° grid box	Hourly
Hourly rain rate in text format [mm/h]	Global (60°N–60°S) but separated to regional files		
Daily accumulated rainfall [mm/day]	Global (60 N–60S)	0.25° grid box	Daily (accumulation from 00Z to 23Z of the specified day) Daily (accumulation from 12Z of previous day to 11Z of the specified day)

radiometer observation at each pixel. Those two flags will help users evaluate the reliability of the rainfall estimates. Daily and 0.25° latitude/longitude grid averaged rainfall systems are also produced by the system for the International Precipitation Working Group (IPWG) satellite precipitation validation/intercomparison studies. Currently GSMaP_NRT results, as well as those of other satellite-based rainfall systems, are compared with regional ground rain gauge and radar network data on a near-real-time basis over United States (University of Maryland), Australia (Bureau of Meteorology, Australia), South America (University of Maryland), and Japan (Osaka Prefecture University).

GSMaP_NRT is also characterized as prototype of GPM. The GPM mission consists of two categories of satellites. One of is the TRMM-type core (GPM core satellite) satellite jointly developed by the United States and Japan that will carry an active precipitation radar (Dual-frequency Precipitation Radar: DPR) and a passive microwave radiometer as a calibrator to other satellites. The other is a constellation of several satellites developed by each international partner (space agency) that will carry passive microwave radiometers and/or microwave sounders. The core satellite will make detailed and accurate estimates of precipitation structure and microphysical properties from the GSMaP_MWR algorithm using TRMM's Precipitation Radar (PR) and TRMM Microwave Imager (TMI), while the constellation of satellites will provide suitable temporal sampling of highly variable precipitation systems enabled by GSMaP_MVK.

Figure 6 is a comparison of precipitation observations by a single satellite with high horizontal resolution and accuracy (TRMM) and by combined multi-satellites with high temporal resolution and global coverage (GSMaP_NRT). The

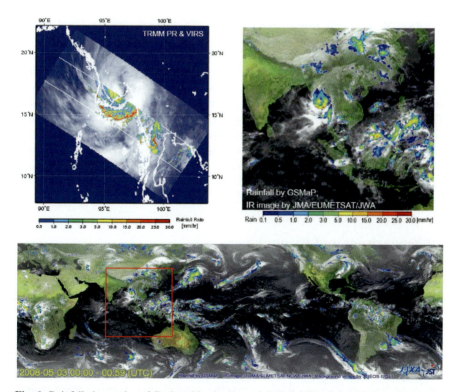

Fig. 6 Rainfall observation of Cyclone Nargis. *Upper left*: Rainfall and cloud image observed by TRMM PR and visible infrared scanner (VIRS) at 00:43 UTC 3 May 2008. *Upper right*: Zoomed image of red rectangle area in *lower panel*. Rainfall (*color*) estimated by GSMaP algorithm in near-real-time system, and cloud image (*grayscale*) observed by geostationary satellites. *Lower*: Global rainfall at 00:00–00:59 UTC 3 May 2008

upper left panel is an observation of Cyclone Nargis just after its landfall on the coast of Myanmar at 00:43 UTC 3 May 2008. Nargis caused serious damage in the Irrawaddy Delta of Myanmar and became a wide-scale disaster. Color denotes rainfall observation by the Precipitation Radar (PR), and the grayscale denotes simultaneous observation of cloud images by the Visible Infrared Scanner (VIRS), both onboard the TRMM satellite. The footprint of PR is about 5 km, and detailed structures of strong rain bands are indicated in yellow and red. Although PR observes a three-dimensional structure, the observation swath is narrower (about 250 km) than that of VIRS (about 850 km). The upper right and lower panels are hourly global rainfall by the GSMaP_NRT systems and cloud image observed by geostationary satellites at 00:00–00:59 UTC 3 May 2008. The upper right panel is a zoomed image of the red rectangle area in lower panel. The well-organized rainfall area in the middle of the image is Cyclone Nargis at the coast of Myanmar. The horizontal resolution of the precipitation map (GSMaP_NRT) is a 0.1° latitude/longitude grid, coarser than that of PR (about 5 km), but it has major advantage in global coverage and temporal resolution.

Research efforts on utilization of satellite precipitation data for flood prediction have been underway recently. Although the TRMM satellite has achieved highly accurate precipitation observations over tropical and subtropical regions, its observations are infrequent because it is only a single satellite in low earth orbit. Production of high temporal resolution global rainfall maps by a single satellite is difficult, but has been achieved by combining multiple satellites and sensors with TRMM as GSMaP system demonstrated in Fig. 6. Using large-scale rainfall information provided by high-resolution rainfall systems, application in flood prediction areas will be able to evolve into operational uses. Coordination research with flood communities, in order to use GSMaP_NRT systems in their flood alert system and tools, has been in progress toward GPM era.

5 Comparisons and Examples

5.1 Example

In order to demonstrate the performance of the GSMaP_MVK system, Fig. 7 shows examples of the system in hourly to seasonal time scales. The top panel shows a representative example of the GSMaP_MVK on an hourly scale on July 15, 0 UTC 2005. As is seen from this panel, the global precipitation distribution in a certain hour is produced by interpolating the precipitation area from the IR and MWR data sets using Kalman filtering. While convective precipitation systems appear mainly in the tropics, band type precipitation systems can be seen in middle latitudes. A typhoon system with heavy rainfall rate is also seen in this panel far south of Japan. This typhoon (Typhoon Haitang) actually brought heavy rainfall to the Taiwan area a few days after this observation.

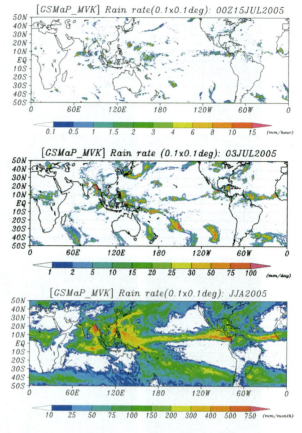

Fig. 7 Examples of the GSMaP_MVK systems for various time scales. From top to bottom, global precipitation distributions in 1 h, daily, seasonal, and climatological scales are shown. Adapted from Ushio et al. (2009)

The second panel is an example of the GSMaP_MVK on a daily scale on 3 July 2005. On this time scale, some features of precipitation systems of longer duration are evident. For example, a Baiu (Japanese summer monsoon) front is seen over Japan; this system brought significant amounts of precipitation not only to Japan but also to the Pacific Ocean off Japan. With this front, a historically heavy rain rate exceeding more than 100 mm/day was recorded in the western part of Japan on July 3 by the Japanese Meteorological Agency, which caused hazardous flooding at that time. Looking at an example in the seasonal scale in the bottom panel, some strong precipitation areas are evident. For example, across the Pacific Ocean along the ITCZ, a relatively heavy precipitation belt appears with a peak at the eastern part. In the Bay of Bengal, a strong precipitation pattern associated with an Asian monsoon is evident.

A series of images for a particular event in every hour is effective to show the usefulness of this high-resolution system. In Fig. 8, consecutive images of the Typhoon Nabi with 1 h/0.1° resolution are shown. This typhoon arose in 29 August 2005 at the Mariana Island, hit Japan, and then turned into an extratropical cyclone in 8 September 2005 at the Kuril Islands. Consecutive 12 h of images just before hitting Japan Island are shown in this figure. With the combining method described here,

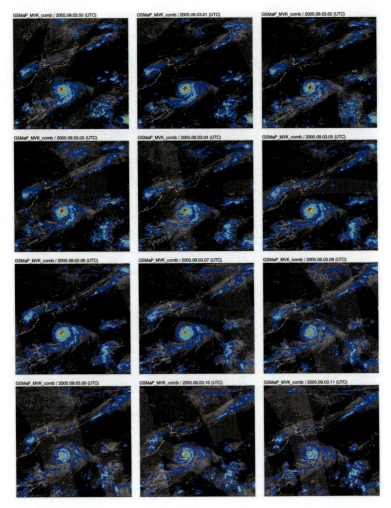

Fig. 8 Consecutive images of 12 h with 1 h/0.1° resolution from GSMaP_MVK for July 3, showing the progression of Typhoon Nabi. For each image, the color superimposed on the IR cloud image denotes the rain rate, while the grey belts denote the coverage by the various microwave sensors. Adapted from Ushio et al. (2009)

the precipitation pattern associated with the typhoon is successfully demonstrated. The precipitation system is spiraling cyclonically and the distribution consists of the eyewall as well as the principal and secondary rain bands that occur outside the eyewall. Changes of the structure are observed every hour as it propagates, suggesting that the GSMaP_MVK system can be useful for typhoon tracking and monitoring, although the reproduction of the precipitation distribution on an hourly/10 km scale is generally difficult and possesses low reliability. It is also suggested that the GSMaP_MVK system is useful for monitoring large-scale phenomena other than typhoons.

5.2 Comparison and Validation

In this section the Kalman filter method itself is first evaluated and then the GSMaP_MVK system is compared with other ground-based data sets. In the method described here, precipitation estimates retrieved from the microwave radiometer data are propagated with moving vectors and refined from the Kalman filter. In the current algorithm, on the next microwave radiometer overpass, the refined precipitation rate at the pixel is overwritten by the precipitation rate from an updated microwave scan. At the time of this overwrite, the two precipitation estimates from the Kalman filter output and from the microwave radiometer are obtained simultaneously, and the Kalman filtered result in this system can be assessed through the comparison with the updated microwave scan data. Thus Fig. 9 shows how the correlation drops as the time from the last microwave radiometer overpass increases. As expected the correlation gradually decreases after the microwave radiometer overpass in cases both with and without Kalman filtering. One or two hours after the updates, no significant difference with and without Kalman filter can be seen. But after three hours or more, the moving vector approach with Kalman filtering tends to be superior to the moving vector only method without Kalman filtering.

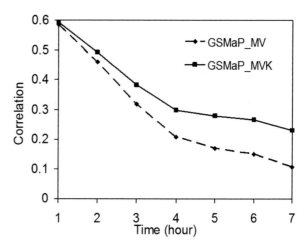

Fig. 9 Correlation coefficient between GSMaP and precipitation estimated from microwave radiometers for the period of July 2005 in the tropics and the extra-tropics over land and ocean. The solid line denotes the Kalman filter method, while the dotted line denotes the moving vector only method

One of the best ways to validate the satellite precipitation system is to use the ground-based radar data calibrated by the dense rain gauge network. In this section, some comparisons with the data from the radar rain gauge network in Japan called RADAR-AMeDAS are presented for validation of the GSMaP_MVK system. Detailed analyses are described in Ushio et al. (2009) and Kubota et al. (2009). Scatter plots comparing observed precipitation distributions for one month with 1 h/0.1° and for 3 h/0.25° resolutions are shown in Fig. 10. In this figure, the unit is 10 log(rain rate in mm h^{-1}) [dBR]. Since the original resolution of the GSMaP_MVK is 1 h/0.1°, in the left panel a comparison with this resolution is shown. As a whole, the precipitation estimates from GSMaP_MVK and ground radar data match well, with a correlation coefficient of 0.44, though generally speaking, precipitation estimates on a global basis with such a high resolution are difficult. The high-density region spreading vertically around −4 dBR on the abscissa arises from an observation problem and should be ignored. The monthly mean precipitation rate is 0.17 mmh^{-1} for the GSMaP_MVK and 0.23 mmh^{-1} for the RADAR-AMeDAS, and the difference is about 0.06 mmh^{-1}, suggesting that the satellite system slightly underestimates the precipitation rate. The lower resolution comparison (3 h/0.25°) is presented in the right panel. It is anticipated that the correlation coefficient rises to 0.65 and a more linear relationship appears in the comparison with this resolution.

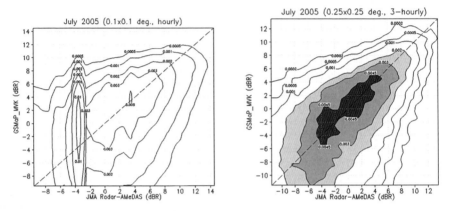

Fig. 10 Scatter plots comparing the hourly precipitation estimates from GSMaP_MVK with RADAR-AMeDAS with 1 h/0.1° (*left panel*) and 3 h/0.25° (*right panel*) resolution. Adapted from Ushio et al. (2009)

As for the daily comparison, it is very helpful to use the IPWG/PEHRPP validation data (Ebert et al. 2007). The IPWG/PEHRPP provides most of the statistical parameters for the evaluation of the several high-resolution satellite precipitation systems at several regions including the United States, Australia, Europe, Japan and so on. In this study, the statistical validation results are shown from the Japanese validation site. Figure 11 shows a time series of monthly averaged daily correlation and root mean square error (RMSE) for some satellite precipitation systems such as 3B42RT, CMORPH, and GSMaP_MVK. The top panel shows that the correlation rises in the summer season and gradually decreases toward the winter season in all the satellite systems. In July the correlation shows its highest value of 0.7 and in January the minimum value of 0.2 appears. At this writing, this trend arises partly because the algorithm of the microwave radiometer fails to retrieve solid precipitation rate data correctly. Generally, the GSMaP_MVK and CMORPH systems perform better in terms of correlation and RMSE than the 3B42RT, indicating that the moving vector type approach works well. The GSMaP_MVK and CMORPH tend to have similar scores, and it seems that the CMORPH has a slightly better score than the GSMaP_MVK in spite of the Kalman filter. Probably this is because the

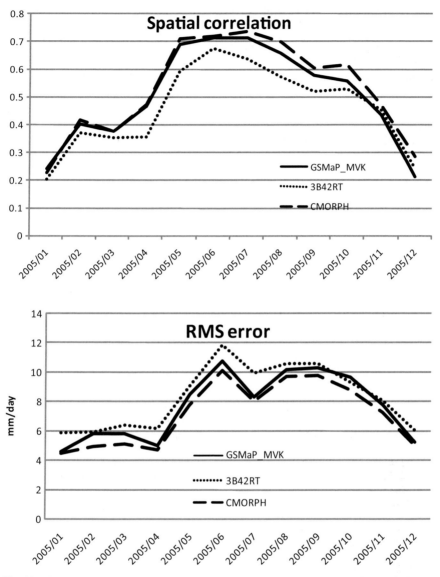

Fig. 11 Time series of correlation coefficient between the GSMaP_MVK, CMORPH, 3B42RT, and RADAR-AMeDAS data during 2005. Adapted from Ushio et al. (2009)

CMORPH uses not only the microwave radiometer but also the microwave sounder data sets, while the current version of the GSMaP_MVK uses only the microwave radiometer data. But it is anticipated that the future version of the GSMaP will include the AMSU data, and the development of the precipitation retrieval algorithm of the microwave sounder data is ongoing. The initial comparison of the GSMaP_MVK including the microwave sounder data shows the best performance.

6 Future Plans and Conclusions

In this chapter, concept, methodology, and comparisons of the GSMaP_MVK were described, and a data distributing and processing system of this system was also presented. Using the Kalman filter theory, the GSMaP_MVK takes advantage of both the moving vector type and IR T_b to rain rate conversion type approaches. The evaluation of the Kalman filter approach clearly shows a better score than the moving vector only approach. Some comparisons with ground-based data sets show that the GSMaP_MVK has one of the best scores for estimating the precipitation rate.

Acknowledgements This work is partly supported by the JAXA/TRMM and GPM program and the JST/CREST. We thank Dr. Takuji Kubota to produce some figures.

References

Adler, R. F., G. J. Huffman, A. Chang, R. Ferraro, P. P. Xie, J. Janowiak, B. Rudolf, U. Schneider, S. Curtis, D. Bolvin, A. Gruber, J. Susskind, P. Arkin, and E. Nelkin, 2003: The version-2 Global Precipitation Climatology Project (GPCP) monthly precipitation analysis (1979–present). *J. Hydrometeor.*, **4**, 1147–1167.

Aonashi, K., A. Shibata, and G. Liu, 1996: An over-ocean precipitation retrieval using SSM/I multi-channel brightness temperature. *J. Meteor. Soc. Japan*, **74**, 617–637.

Ebert, E., J. Janowiak, and C. Kidd, 2007: Comparison of near-real-time precipitation estimates from satellite observations and numerical models. *Bull. Amer. Meteor. Soc.*, **88**, 47–64.

Huffman, G., R. F. Adler, D. T. Bolvin, G. Gu, E. J. Nelkin, K. P. Bowman, Y. Hong, E. F. Stocker, and D. B. Wolff, 2007: The TRMM Multisatellite Precipitation Analysis (TMPA): Quasi-global, multiyear, combined-sensor precipitation estimates at fine scales. *J. Hydrometeor.*, **8**, 38–55.

Janowiak, J., R. J. Joyce, and Y. Yahosh, 2001: A real-time global half-hourly pixel-resolution IR dataset and its applications. *Bull. Amer. Meteor. Soc.*, **82**, 205–217.

Joyce, R., J. E. Janowiak, and G. J. Huffman, 2001: Latitudinally and seasonally dependent zenith-angle corrections for geostationary satellite IR brightness temperatures. *J. Appl. Meteor.*, **40**, 689–703.

Joyce, R., J. E. Janowiak, P. A. Arkin, and P. Xie, 2004: CMORPH: A method that produces global precipitation estimates from passive microwave and infrared data at high spatial and temporal resolution. *J. Hydrometer.*, **5**, 487–503.

Kidd, C., D. Knoveton, M. Todd, and T. Bellerby, 2003: Satellite rainfall estimation using combined passive microwave and infrared algorithm. *J. Hydrometeor.*, **4**, 1088–1104.

Kubota, T., S. Shige, H. Hashizume, K. Aonashi, N. Takahashi, S. Seto, M. Hirose, Y. Takayabu, K. Nakagawa, K. Iwanami, T. Ushio, M. Kachi, and K. Okamoto, 2007: Global precipitation map using satellite-borne microwave radiometers by the GSMaP project: Production and validation. *IEEE Trans. Geosci. Remote Sens.*, **45**, 2259–2275.

Kubota, T., T. Ushio, S. Shige, S. Kida, M. Kachi, and K. Okamoto, 2009: Verification of high resolution satellite-based rainfall estimates around Japan using a gauge-calibrated ground radar dataset. *J. Meteor. Soc. Japan*, Vol. 87A, 203–222.

Lazzara, M. A., J. M. Benson, R. J. Fox, D. J. Laitsch, J. P. Rueden, D. A. Santek, D. M. Wade, T. M. Whittaker, and J. T. Young, 1999: The man computer interactive data access system: 25 Years of interactive processing. *Bull. Amer. Meteor. Soc.*, **80**, 271–284.

Sorooshian, S., K. Hsu, G. Gao, H. Gupta, B. Imama, and D. Braitwaite, 2000: Evaluation of PERSIANN system satellite-based estimates of tropical rainfall. *Bull. Amer. Meteor. Soc.*, **81**, 2035–2046.

Turk, J. and S. Miller, 2005: Toward improving estimates of remotely sensed precipitation with MODIS/AMSR-E blended data techniques. *IEEE Trans. Geosci. Remote Sens.*, **43**, 1059–1069.

Ushio, T., K. Sasashige, T. Kubota, S. Shige, K. Okamoto, K. Aonashi, T. Inoue, N. Takahashi, T. Iguchi, M. Kachi, R. Oki, T. Morimoto, and Z. -I. Kawasaki, 2009: A Kalman filter approach to the Global Satellite Mapping of Precipitation (GSMaP) from combined passive microwave and infrared radiometric data. J. Meteor. Soc. Japan, Vol. 87A, 137–151.

Part II
Evaluation of High Resolution Precipitation Products

Neighborhood Verification of High Resolution Precipitation Products

Elizabeth E. Ebert

Abstract High resolution satellite-derived precipitation fields may be quite useful for many applications even if they do not exactly match with the observations. To try to assess their quality, verification techniques known collectively as *neighborhood* techniques have been developed. These techniques compare the estimates and observations within space/time neighborhoods and measure their "closeness" according to various criteria such as the similarity of estimated and observed precipitation intensity distributions, occurrence of precipitation exceeding critical thresholds, fractional precipitation area, and so on. By changing the size of the space/time neighborhoods it is possible to assess at which scales the satellite estimates have sufficient accuracy for a particular application. This chapter demonstrates the neighborhood verification approach using two satellite-based high resolution precipitation products, and interprets their accuracy according to four different "closeness" criteria.

Keywords Verification · Neighborhood verification · Evaluation · Closeness

1 Introduction

High space and time resolution precipitation products from satellites and numerical weather prediction models are now widely available and are used for a variety of applications from weather analysis and prediction, climate analysis, emergency management, stream flow prediction, water resource management, agriculture, etc. These applications have different requirements regarding accuracy and timeliness. Therefore, the source of precipitation data that best suits one application may not be ideal for another. Among hydrological applications, flash flood forecasting calls

E.E. Ebert (✉)
Centre for Australian Weather and Climate Research, Melbourne, Australia
e-mail: E.ebert@bom.gov.au

for rapid data transmission but can accommodate moderate accuracy in the estimates, while water resource management needs unbiased precipitation estimates even though they may take some time to acquire.

Satellite-based estimates in particular can be very useful when it is necessary to know the spatial distribution of precipitation in near real time, and radar-based precipitation estimates are either not available or not of suitable quality to use quantitatively. The finest scales at which operational satellite precipitation estimates are normally made are ~ 8 km and 30 min for estimates using geostationary satellite data (e.g., HydroEstimator (Scofield and Kuligowski, 2003), CMORPH (Joyce et al., 2004), and PERSIANN (Sorooshian et al., 2000)), and ~ 25 km and 3-hourly for estimates using mainly passive microwave data (e.g., TMPA-RT (Huffman et al., 2007) and NRL-PMW (Turk and Miller, 2005)). The high spatial and temporal resolution offers users the flexibility to either make direct use of the estimated fine-scale precipitation distributions, or average or accumulate the estimates to larger space and time scales as required.

In order to make appropriate use of high resolution satellite precipitation estimates it is necessary to understand the nature of their estimation errors. This is done by verifying the satellite estimates against reference data such as rain gauge or radar analyses, where these are assumed to give a reasonable representation of the true precipitation distribution. Due to the sparseness of gauge networks measuring sub-daily precipitation, the finest scales at which gauge analyses are normally used for satellite precipitation validation are ~ 25 km and 24-hourly (e.g., Ebert et al., 2007). Verification of sub-daily satellite precipitation estimates usually relies on radar precipitation composites or radar-gauge analyses. This is starting to be done more frequently (e.g., Hossain and Huffman, 2008; Sapiano and Arkin, 2008), and will be done in this chapter as well.

When satellite estimates and surface reference data are available on a common grid, the traditional verification approach is to compare the value of the satellite in a given pixel with the corresponding observation in that pixel. This matching strategy makes it very difficult for high resolution satellite estimates to demonstrate good skill, for several reasons. In addition to retrieval error, which is related to the conversion of satellite-measured radiances to surface precipitation, other non-trivial sources of error contribute to the total error. Sampling error can occur when there are spatial and temporal mismatches between the satellite products and the reference data. The satellites give estimates of instantaneous precipitation rates over the pixel-sized areas, while gauges provide temporal accumulations at a point. Gauge analyses that map point data to the pixel scale contain error associated with the interpolation. Gauge measurements may be affected by under-catch errors. Radar provides instantaneous areal precipitation estimates that are analogous to the satellite estimates, albeit at much higher resolution than the satellite. However, the conversion of radar reflectivity to surface precipitation is complex and itself contains a variety of errors (Collier, 1996). For a review of precipitation measurement methodology and errors see Michaelides (2008). It is no wonder, then, that it is so difficult to achieve perfect verification results when the reference data give an imperfect estimate of the truth.

Upscaling to coarser space and time resolution is a well known approach for reducing the sampling error, but this process loses important information on the precipitation intensity distribution. For many applications it may not be essential to get the precipitation position and timing exactly right; instead, "close enough" may be good enough.

An emerging verification approach called "neighborhood verification" has been described by Ebert (2008)[1]. Instead of requiring an exact space/time match, all pixel-scale values within a spatial and/or temporal neighborhood of the observation are considered to be equally likely estimates of the "true" value, thus representing a probabilistic view of verification. Some neighborhood verification methods compare the estimated values within a neighborhood to the observation in the center, while others compare the estimates to observed values within the same neighborhood. The advantage is that useful skill can be demonstrated even if perfect correspondence is not achieved at the pixel scale. By varying the size of the neighborhoods and performing the verification at multiple scales and for multiple intensity thresholds, it is possible to determine at which scales the satellite estimates have useful skill. This strategy evolved from the need for more appropriate verification approaches for high resolution model precipitation forecasts, but is equally applicable to high resolution satellite precipitation estimates.

This chapter describes the general neighborhood verification approach (Ebert, 2008), and focuses on four methods that are particularly suited for evaluating precipitation mean values, precipitation frequency, occurrence of extreme values, and similarity of the intensity distribution. The verification methodology is demonstrated on precipitation estimates from the CMORPH algorithm of Joyce et al. (2004) and the near real time TRMM Multisatellite Precipitation Analysis (TMPA-RT) of Huffman et al. (2007), using gauge analyses and radar estimates over Australia as reference data. The information that can be gained from each of the methods is highlighted.

2 Neighborhood Verification Methods

Neighborhood verification computes error metrics for the set of all neighborhoods, or space/time windows, in a domain, rather than the set of all individual pixels. The use of continuous, rather than discrete, sampling leads to more robust statistics. The size of the local spatial neighborhood around a pixel is increased linearly or exponentially from 1×1, 3×3, 5×5, etc., to some upper bound on the window size that reflects the maximum distance that may still be considered relevant for precipitation guidance. For hydrological applications this might be one or two times the typical catchment size, or the size of a synoptic scale rain system. If a temporal domain is used then t time windows are increased in the same way.

---

[1] Although she called it "fuzzy verification" the term "neighborhood verification" is preferred as it more clearly describes the nature of the approach.

Since many neighborhood verification methods use the concept of an "event", i.e., the occurrence of a value greater than or equal to some threshold value, the precipitation intensity threshold for an event is also varied from small to large values, $R_1,\ldots,R_m$. Thus, instead of the single score that is normally reported for pixel-scale validation using a rain/no rain threshold, neighborhood verification provides an $m \times n \times t$ array of scores for varying scales and thresholds. It is then possible to examine the array of scores to determine which space and time scales have useful skill for precipitation exceeding various intensities.

A well known neighborhood verification approach is *upscaling*, in which the estimates and observations at pixel scale are averaged to larger scales before being compared using the usual continuous and categorical verification metrics. Zepeda-Arce et al. (2000) and Yates et al. (2006) describe the use of upscaling to evaluate model output against radar data and gauge analyses. More recently, Hossain and Huffman (2008) upscaled precipitation estimates from four different high resolution satellite products to show how several commonly used verification metrics improved with increasing spatial scale from $0.04°$ to $1.0°$.

In recent years many new techniques that verify neighborhoods of pixels have been proposed in the meteorological literature (Brooks et al., 1998; Zepeda-Arce et al., 2000; Atger, 2001; Casati et al., 2004; Germann and Zawadzki, 2004; Theis et al., 2005; Rezacova et al., 2007; Roberts and Lean, 2008). Ebert (2008) describes twelve neighborhood verification methods, four of which are demonstrated here. Each method is characterized by a decision model regarding what constitutes a useful forecast or estimate. For example, the upscaling method considers a useful estimate to be one that has the same average value as the observations. This is one criterion for judging whether precipitation estimates at the catchment scale are useful for hydrological purposes.

The *fractions skill score* (FSS) method of Roberts and Lean (2008) considers a perfect estimate to be one with the same frequency of events as was observed within a neighborhood. This neighborhood method implicitly acknowledges that the observations are likely to contain random error at the pixel scale, and asserts that a better approach to comparing estimates with observations is to assess their similarity in terms of their fractional coverage of raining pixels. The fractions skill score is probabilistic in nature, and is based on a variation of the Brier score used to verify probability forecasts:

$$\text{Fractions Brier Score} = FBS = \frac{1}{N} \sum_N (P_{est} - P_{obs})^2 \qquad (1)$$

P_{est} and P_{obs} are the fractional coverages of estimated and observed precipitation pixels, respectively, in each of the N neighborhoods in the domain. The FBS is the mean squared error in probability space, with lower values of FBS indicating more accurate satellite estimates. To transform the score into a positively oriented metric the FBS is referenced to the corresponding value for the unmatched case, giving a fractions skill score of

$$FSS = 1 - \frac{\text{FBS}}{\frac{1}{N}\left[\sum_N P_{est}^2 + \sum_N P_{obs}^2\right]} \qquad (2)$$

The FSS varies between 0 for a complete mismatch (i.e., negatively correlated) and 1 for a perfect match.

Roberts and Lean (2008) show that the target value of FSS above which the estimates are considered to have useful (better than a uniform probability forecast of f_{obs}, the observed rain fraction in the domain) skill is given by

$$FSS_{useful} = 0.5 + \frac{f_{obs}}{2} \qquad (3)$$

where f_{obs} is the fraction of observed raining pixels in the full domain. This leads to the concept of a "skillful scale", namely the smallest scale at which the FSS exceeds FSS_{useful}. This is a more meaningful concept for many users. In fact, the skillful scale is now used in the United Kingdom to help weather forecasters at the Met Office understand the quality of high resolution model forecasts (M. Mittermaier, personal communication, 2008).

The *multi-event contingency table* (MECT) method of Atger (2001) considers an estimate to be useful if at least one occurrence of an event is estimated close to an observed event. "Close" can refer to space, time, intensity, or any other important aspect. This is an important criterion for emergency managers and disaster relief agencies using satellite estimates to detect heavy precipitation in a remote region, or for weather forecasters using model output to prepare warnings of heavy precipitation. The MECT method compares a neighborhood of estimates to an observation in the center using traditional categorical metrics such as frequency bias, probability of detection, false alarm ratio, and so on. These are derived from the four elements of the contingency table, namely the number of hits (observed precipitation correctly detected), misses (observed precipitation not detected), false alarms (detections of precipitation where none occurred), and correct negatives (no precipitation detected or observed)[2]. According to the MECT method, whenever precipitation is observed in the central pixel of the neighborhood and also detected by the satellite in at least one pixel in the neighborhood, this counts as a hit. If there is no precipitation observed in the central pixel but one or more neighborhood pixels with detected precipitation, a false alarm is counted. As the neighborhood increases in size, it is easier to get a hit, but also easier to get a false alarm. Although any categorical score can be computed, the one most relevant to accuracy assessment in this case is the Hanssen and Kuipers discriminant HK, which measures the difference between the probability of detection (rewarding hits) and the probability of false detection (penalizing false alarms). In terms of the contingency table elements HK is given by

---

[2] See Jolliffe and Stephenson (2003) or JWGV (2008) for more detailed information on contingency tables and categorical verification scores.

$$HK = \frac{hits}{hits + misses} - \frac{false\ alarms}{correct\ rejections + false\ alarms} \quad (4)$$

A score of 0 indicates no skill, while a score of 1 signifies perfect performance.

Germann and Zawadzki (2004) proposed a neighborhood verification method that uses as its criterion for goodness, "A forecast is useful if it has a high probability of matching the observed value." Called the *conditional square root of RPS* (CSRR), it explicitly uses a probabilistic approach to compare precipitation estimates to observed precipitation in the center of each neighborhood. They compute the ranked probability score (RPS), which measures the domain-averaged mean squared difference in cumulative probability space, for precipitation in M logarithmically increasing intervals[3]:

$$RPS = \frac{1}{N}\left[\frac{1}{M-1}\sum_{m=1}^{M}(CDF_{est,m} - CDF_{obs,m})^2\right] \quad (5)$$

where $CDF_{est,m}$ is the cumulative probability of the estimates exceeding the intensity threshold for category m, and $CDF_{obs,m}$ is the observed cumulative probability, equal to 1 if the observed value exceeds the threshold for category m, and 0 if not. The RPS rewards estimates with an intensity distribution that peaks sharply near the observed value. The square root of the RPS can be interpreted as the standard error of the probability estimates across the full range of precipitation intensities. This quantity is normalized by the observed precipitation fraction in the domain to enable performance to be compared for different cases:

$$CSRR = \frac{\sqrt{RPS}}{P_{obs}} \quad (6)$$

This is a negatively oriented score, i.e., a perfect estimate would have a value of CSRR=0. Unlike the three methods described previously, CSRR values are not computed for varying precipitation thresholds since the score evaluates the full intensity distribution of the estimates. Likely users of CSRR information would include emergency managers and other decision makers concerned about the effects of local heavy precipitation.

The first two neighborhood methods described in this section, upscaling and FSS, compare neighborhoods of satellite estimates against neighborhoods of observations. Ebert (2008) calls this strategy "model oriented", meaning that the observations are manipulated to represent the scales that can be resolved by the numerical

---

[3]The M intervals are chosen here to have bounds identical to the thresholds used for the other methods. Since the RPS is sensitive to the choice of intervals, this score should be used in a relative sense (i.e., to compare performance) rather than an absolute sense.

weather prediction model (in that paper); "product oriented" might be a better term to use with satellite estimates. This gives a fair assessment of the satellite products in the sense that they are being evaluated only on scales that they claim to resolve. The last two neighborhood methods, MECT and CSRR, compare each neighborhood of satellite estimates against the single observation in the center of the neighborhood. Although this may seem "unfair", many users wish to know the accuracy of the satellite estimate at a particular location. Note that this "user oriented" philosophy is more demanding than the "product oriented" one, but not as tough as the traditional pixel-to-pixel verification, since skill can still be demonstrated when the satellite estimate detects precipitation close to the observation.

3 Neighborhood Verification of CMORPH and TMPA Precipitation Estimates

The upscaling, FSS, MECT, and CSRR methods were used to evaluate four aspects of the precipitation distribution, namely the mean value, the frequency of precipitation pixels, the occurrence of an estimated event close to an observed event, and the estimated intensity distribution, within spatial and temporal neighborhoods of increasing size. Several intensity thresholds were used with particular emphasis on evaluating the higher precipitation rates.

To better understand the neighborhood verification we can compare verification results from two different hours on the same day. November 3, 2007 was characterized by a deepening low pressure system over southeastern Australia which produced heavy precipitation in the vicinity of Melbourne. Figure 1 shows the precipitation from the Rainfields merged radar-gauge analyses (Seed and Duthie, 2007) and CMORPH satellite estimates (Joyce et al., 2004), valid for the hours ending at 10 and 16 UTC. The first case shows CMORPH estimates to be relatively unbiased, but with the precipitation area in the northeast slightly displaced from the radar observations (Fig. 1a). In the second case the CMORPH estimate had excellent placement but its intensities were much lighter than observed (Fig. 1b).

To prepare the data for verification, the original 2 km resolution radar pixels were averaged to the 8 km scale of the satellite product. The neighborhood verification used six spatial scales increasing logarithmically from 1×1 to 29×29 pixels, the largest that the radar analysis could accommodate, and eight intensity thresholds ranging from 0.1 to 20 mm h^{-1}. No time window was used.

The results are shown in Figs. 2 and 3 as a function of the precipitation intensity threshold (x-axis) and spatial scale (y-axis). In these plots the shading and the number show the value of the score. The value in the lower left corner is the score that would be achieved using traditional pixel matching and a very low threshold, essentially rain/no rain.

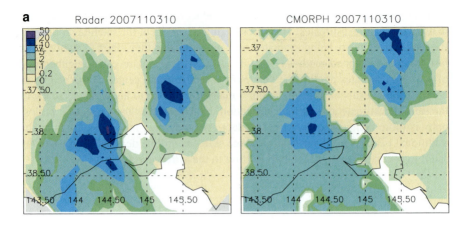

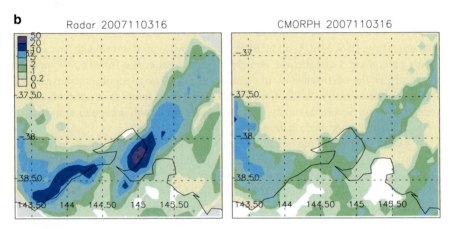

Fig. 1 Hourly precipitation estimated from a radar-gauge analysis (*left*) and the CMORPH satellite precipitation product (*right*), at (**a**) 10 UTC and (**b**) 16 UTC on 3 November 2007

For the upscaling approach the equitable threat score[4] was chosen as the error metric since it penalizes both misses and false alarms. Other metrics such as mean

---

[4]The equitable threat score is a categorical verification metric used widely by the meteorological community to verify precipitation forecasts and estimates. It is defined as the fraction of all events forecast and/or observed that were correctly diagnosed, accounting for the hits that would occur purely due to random chance:

$$ETS = \frac{hits - hits_{random}}{hits + misses + false\ alarms - hits_{random}}$$

where $hits_{random} = \frac{1}{N}\left(N_{obs.rain} \times N_{fcst.rain}\right)$. The ETS can be interpreted as the detection skill relative to random chance, and varies between -1/3 and 1.

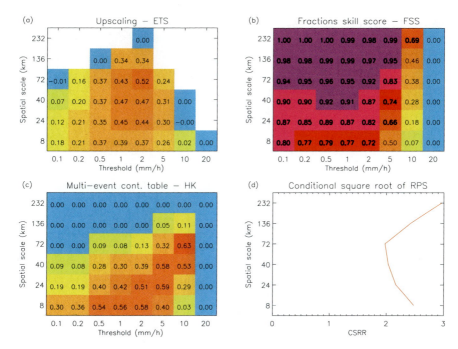

Fig. 2 Neighborhood verification scores for hourly CMORPH estimates verified against radar-gauge analyses in southeastern Australia at 10 UTC on 3 November 2007. In (**a**)–(**c**) the spatial scale is varied along the y-axis with larger scales at the top of the plot, while the precipitation intensity threshold is varied along the x-axis with larger values to the right. The value in the *lower left* corner represents the usual pixel-scale score for a rain/no rain threshold. Score values of 0 occur where there were no satellite estimates for that intensity-scale combination but there were observed values. Blank entries represent intensity-scale combinations for which there were no observations or the score was undefined. The bold numbers in the fractions skill score plot indicate where useful skill was achieved according to Eq. (3). The CSRR in (**d**) has no threshold dependency and is plotted simply as a function of spatial scale

error, root mean square error, and Nash-Sutcliffe efficiency coefficient can also be computed as a function of scale to evaluate the mean value; a categorical score was chosen here to illustrate the varying performance with precipitation intensity.

In both cases, CMORPH showed optimal skill at a spatial scale of 72 km. For the 10 UTC case this occurred for an intensity threshold of 2 mm h^{-1}, indicating that the moderate precipitation was located more accurately than precipitation of other intensities (Fig. 2a). Although there were pixel scale observations exceeding 20 m h^{-1}, there were no corresponding satellite detections, resulting in a score of 0 as seen in the right side of the diagram. At 16 UTC the optimal ETS occurred for much lower precipitation thresholds, reflecting the bias errors (Fig. 3a). In general the 16 UTC case scored more highly than the 10 UTC case due to its excellent precipitation detection.

The opposite was true when comparing the estimated and observed fractional coverage of precipitation pixels using the fractions skill score. Useful detections were made for a greater range of scales and intensities for the 10 UTC case

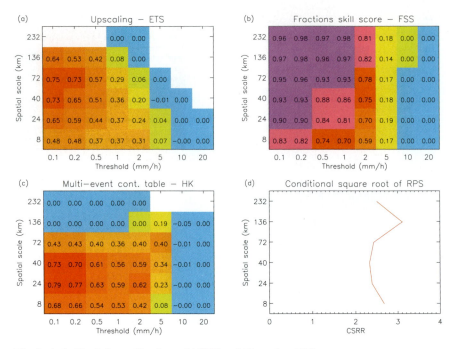

Fig. 3 As in Fig. 2, for verification at 16 UTC on 3 November 2007

compared to the 16 UTC case (Figs. 2b, 3b). By considering the fractional coverage of pixels exceeding a particular intensity, the FSS puts greater emphasis on the precipitation intensity distribution than does the upscaling approach. This explains why the unbiased case outperformed the biased case with better detection. FSS generally improves with increasing spatial scale and decreasing precipitation intensity, and tends to be characterized by higher numerical values than the ETS and HK.

The ability of CMORPH to detect precipitation of a given intensity close to where it was observed was quite different at 10 and 16 UTC, according to the MECT verification (Figs. 2c, 3c). A good way to use the results from the MECT method is to choose a threshold of interest, say 10 mm h^{-1}, and scan vertically to see which scales had a large HK score. In a warning context this would tell the user how large the neighborhood surrounding the point of interest should be to provide a useful indication of precipitation, with many hits and not too many false alarms. For example, at 10 UTC a neighborhood of 40–72 km (effective radius of 23–40 km) gave useful detections of precipitation exceeding 10 mm h^{-1}, while no detections of 10 mm h^{-1} were found in the neighborhood of any observed values at 16 UTC due to the low bias of the CMORPH estimate. However, for light precipitation only small neighborhoods were required for very good detections. The diagonal pattern of higher HK scores seen in Figs. 2c and 3c is typical for this method, as the difficulty in detecting higher precipitation rates (due to their greater temporal variability and associated higher sampling errors) means that larger neighborhoods are generally needed to find a match.

The final evaluation of the CMORPH estimates was done using the probabilistic CSRR method, which has a low (good) value when the estimated intensity distribution peaks near the observed value. The focus of the CSRR on the precipitation rate distribution is similar to that of the FSS, but neighborhood estimates are compared to point values rather than the neighborhood distribution of observed precipitation rates. The CSRR showed better performance for CMORPH at 10 UTC, when there was little bias, than at 16 UTC when the estimates contained significant bias error (Figs. 2d, 3d). The scales that best represented the observed precipitation distribution were 24–72 km, in agreement with the upscaling and MECT results. The return to lower CSRR values at the largest scale for the 16 UTC case is related to the inclusion of moderate CMORPH precipitation rates from the southeastern corner of the domain in the precipitation rate distribution being compared to observed pixels near the center of the domain (Fig. 1b).

The CMORPH example focused on the interpretation of neighborhood verification results, where the goal was to interpret the behavior of the scores in light of the errors that can be seen in the radar and satellite precipitation maps. The more typical use of verification is to detect systematic errors over some period of time using a much larger dataset. The aggregated verification results can then be used to guide the improvement of detection algorithms and the appropriate interpretation of existing satellite precipitation products.

To demonstrate the use of neighborhood verification for providing information on systematic errors, we verify satellite estimates from the TMPA-RT algorithm (also known as 3B42RT; Huffman et al., 2007). Estimated 3-hourly precipitation accumulations at 0.25° spatial resolution were summed to 24 h accumulations and verified against operational daily rain gauge analyses at the same scale over Australia (Weymouth et al., 1999) during two seasons, December 2005–February 2006 (summer) and June–August 2006 (winter). The neighborhood verification scores for each method were computed for each day separately, then aggregated for the full season. This involved summing the daily hits, misses, and false alarms (upscaling and MECT methods) or squared errors (FSS and CSRR methods) and computing the scores from the summed components.

Not all possible verification scales are useful in practice – for daily precipitation we will be primarily interested in scales below about a few degrees latitude/longitude. Therefore, the neighborhood sizes used in the neighborhood verification were 1×1, 3×3, 5×5, 9×9, and 17×17. As before, no temporal neighborhoods were used.

Figure 4 shows the neighborhood verification of TMPA-RT daily precipitation estimates during the summer season, while Fig. 5 shows performance for the winter season. Looking first at the general performance, the TMPA-RT performed better in summer than in winter as has been found earlier (Ebert et al., 2007). The maximum precipitation rates were much higher in summer than in winter, and there were fewer instances of non-detection.

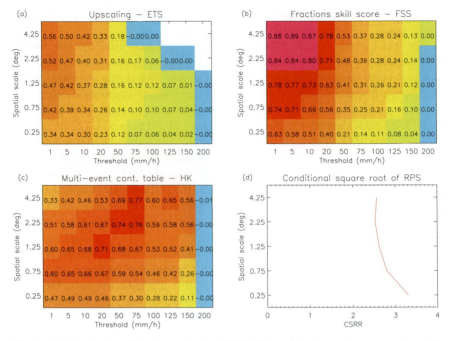

Fig. 4 Neighborhood verification results for TMPA-RT precipitation estimates over Australia during the summer season, December 2005–February 2006

The upscaling method shows that the skill was greatest for low precipitation thresholds (light precipitation) and decreased monotonically as the precipitation rate increased. In summer the ETS values for precipitation thresholds exceeding 1 mm d^{-1} were better than 0.5 for scales of 2 to 4°, while for heavy precipitation ($\geq$50 mm d^{-1}) the ETS was less than 0.2 for all scales (Fig. 4a). Wintertime performance was much worse, with negligible skill for precipitation heavier than 20 mm d^{-1} (Fig. 5a).

To understand these results it is helpful to look at the frequency biases for summer and winter, shown in Fig. 6. The frequency bias is simply the ratio of detected to observed precipitation events, with a value of 1 indicating unbiased estimates. In summer the frequency of light precipitation was relatively unbiased, but the occurrence of heavier precipitation at pixel scale was overestimated by more than a factor of 4 for high thresholds. The high rate of false alarms contributed to the poorer ETS values at these scales. The situation was reversed in winter when precipitation frequency was underestimated at all scales and intensities, and the high rate of misses led to low overall skill. This behavior is in agreement with the pixel-scale results shown by Ebert et al. (2007) for earlier dates.

As with the upscaling approach, the performance according to FSS was greater at smaller precipitation thresholds and larger scales (Figs. 4b, 5b). This means that the estimated precipitation frequencies better matched the observed frequencies occurring at those scales and intensities. To show whether these matches were useful

Neighborhood Verification

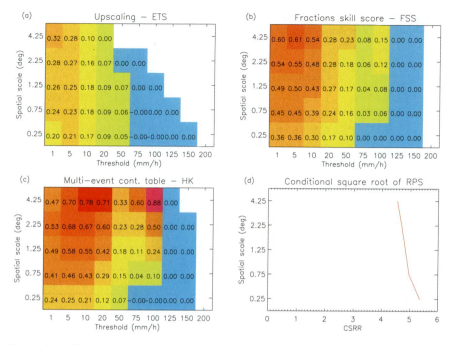

Fig. 5 As in Fig. 4, for the winter season, June–August 2006

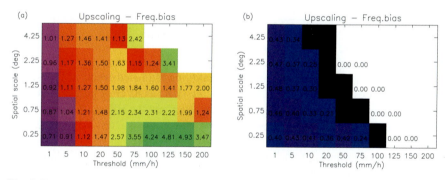

Fig. 6 Frequency bias of TMPA-RT for the upscaling verification approach, for (**a**) summer 2006 and (**b**) winter 2006

according to the Eq. (3), the observed frequency, target skill, and skillful scale (i.e., the minimum scale at which the target skill is met) are listed in Table 1. In some cases these were interpolated from the results in Figs. 4 and 5. In summer the skillful scales were conveniently small, a few pixels or less, for precipitation threshold up to 20 mm d^{-1}, beyond which the estimated frequencies were not sufficiently close the observed frequencies. This is most likely due to the high biases noted earlier. In winter useful skill was shown only for low thresholds and at scales greater than 100 km.

Table 1 Threshold-dependent observation frequency, target FSS, and estimated skillful scale for TMPA-RT estimates during (**a**) summer 2005–2006 and (**b**) winter 2006

(**a**)

Dec 2005–Feb 2006	Precipitation threshold (mm d^{-1})								
	1	5	10	20	50	75	100	125	150	200
Observed frequency, f_{obs}	0.27	0.14	0.08	0.04	0.006	0.002	0.001	0	0	0
Target FSS_{useful}	0.64	0.57	0.54	0.52	050	0.50	0.50	0.50	0.50	0.50
Estimated skillful scale (° lat/lon)	0.25	0.25	0.35	0.63	3.1	–	–	–	–	–

(**b**)

June–Aug 2006	Precipitation threshold (mm d^{-1})								
	1	5	10	20	50	75	100	125	150	200
Observed frequency, f_{obs}	0.09	0.03	0.01	0.004	0	0	0	0	0	0
Target FSS_{useful}	0.54	0.52	0.51	0.50	050	0.50	0.50	0.50	0.50	0.50
Estimated skillful scale (° lat/lon)	2.3	1.6	3.2	–	–	–	–	–	–	–

The MECT method is concerned with identifying estimated events near observed events, and uses the Hanssen and Kuipers score (HK) as its error metric. As events become increasingly rare the false alarm rate drops and the HK score tends toward the probability of detection. According to the MECT, the greatest skill at detecting precipitation in the vicinity of an observed value occurred at moderate to heavy precipitation thresholds of 20–75 mm d^{-1}, and spatial scales of 100–400 km during summer (Fig. 4c). Good performance was seen at these scales even for precipitation exceeding 150 mm d^{-1}, which indicates the potential usefulness of TMPA-RT for flooding precipitation applications (e.g., Hong et al., 2007). During winter good performance was seen for slightly lower precipitation thresholds (10–20 mm d^{-1}) and greater spatial scales ($\geq$400 km) (Fig. 5c).

The seasonal mean CSRR scores were significantly lower for summertime TMPA-RT estimates than for winter (Figs. 4d, 5d). The poor wintertime performance was mainly related to low bias seen in Fig. 6b, with the estimated intensity distribution not matching closely with the observed values. In both seasons quite large neighborhoods were required to get optimal precipitation rate distributions, a result of both sampling and detection errors.

4 Discussion

As the production and distribution of high resolution satellite-based precipitation products becomes increasingly common, the need to evaluate them appropriately becomes more important. Standard pixel-by-pixel verification can suggest that high resolution precipitation products are not as accurate as lower resolution products, yet most users intuitively feel that the high resolution products should be more useful. Neighborhood verification gives credit to estimates that are "close" to the observations, thus offering an alternative to traditional verification approaches. This is achieved by looking in space/time neighborhoods surrounding the observations and evaluating the degree of "closeness" according to various criteria.

Many neighborhood verification methods are now available to answer different questions about the accuracy of high resolution products. The neighborhood verification framework of Ebert (2008) includes twelve methods, four of which have been described and demonstrated here. Upscaling verifies the mean value of the satellite estimates against the mean value of the observations in successively greater neighborhoods. The fractions skill score (FSS) compares estimated and observed distributions in neighborhoods by computing squared errors in probability space. Two methods were shown for comparing estimates to pixel-scale observations. The multi-event contingency table (MECT) looks for estimates that are located nearby observed precipitation events and measures the closeness using the Hanssen and Kuipers score. The conditional square root of RPS (CSRR) is a probabilistic approach, evaluating whether the distribution of estimated precipitation rates in the neighborhood of an observation peaks near the observed intensity. Each of these methods addresses a different aspect of accuracy.

By evaluating the accuracy of the satellite estimates as a function of both intensity and spatial scale, neighborhood verification gives information about which scales have useful skill. This helps users to decide whether to use the estimates at face value at full resolution, or spatially transform the values to give more accurate and useful information.

Most users of precipitation estimates are not very familiar with objective verification techniques and scores, and thus neighborhood verification may seem somewhat daunting. Two new metrics, namely the FSS and the CSRR, have only recently been introduced into the meteorological literature and are not yet found in standard textbooks on verification. Even those who are comfortable with verification methods and scores may find it overwhelming to interpret the results from several neighborhood methods, each of which produces a large array of scores. The key is to first identify which is the most important aspect(s) of the estimated precipitation to get right – is it the spatial average, the precipitation area, the presence of one or more high intensity estimates nearby the location of interest, the precipitation rate distribution, or something else, then choose the neighborhood verification method that addresses this aspect – upscaling, FSS, MECT, or CSRR, respectively. Focusing on an intensity threshold of interest and condensing its scale-dependent performance into a single easily-interpreted value like the "skillful scale" can help make the verification results much more accessible. As neighborhood verification becomes more widely used, new approaches will certainly emerge for interpreting the results in ways that intuitively meet the needs of specific users.

Acknowledgements The author would like to thank Faisal Hossain and George Huffman for their encouragement in applying neighborhood verification methods to evaluate satellite precipitation estimates. Three anonymous reviewers gave helpful suggestions which helped to clarify this paper.

References

Atger, F., 2001: Verification of intense precipitation forecasts from single models and ensemble prediction systems. *Nonlin. Proc. Geophys.*, **8**, 401–417.

Brooks, H. E., M. Kay, and J. A. Hart, 1998: Objective limits on forecasting skill of rare events. *19th Conf. Severe Local Storms, Amer. Met. Soc., Minneapolis, MN, 14–18 September 1998, AMS.* 552–555.

Casati, B., G. Ross, and D. B. Stephenson, 2004: A new intensity-scale approach for the verification of spatial precipitation forecasts. *Meteorol. Appl.*, **11**, 141–154.

Collier, C. G., 1996: *Applications of Weather Radar Systems: A Guide to Uses of Radar Data in Meteorology and Hydrology.*. Wiley, Chichester, 390 pp.

Ebert, E. E., 2008: Fuzzy verification of high resolution gridded forecasts: A review and proposed framework. *Meteorol. Appl.*, **15**, 51–64.

Ebert, E. E., J. E. Janowiak, and C. Kidd, 2007: Comparison of near real time precipitation estimates from satellite observations and numerical models. *Bull. Amer. Met. Soc.*, **88**, 47–64.

Germann, U. and I. Zawadzki, 2004: Scale dependence of the predictability of precipitation from continental radar images. Part II: probability forecasts. *J. Appl. Meteorol.*, **43**, 74–89.

Hong, Y., R. F. Adler, A. Negri, and G. J. Huffman, 2007: Flood and landslide applications of near real-time satellite rainfall products. *Natural Hazards*, **43**, 285–294.

Hossain, F. and G. J. Huffman, 2008: Investigating error metrics for satellite rainfall data at hydrologically relevant scales. *J. Hydrometeor.*, **9**, 563–575.

Huffman, G. J., R. F. Adler, D. T. Bolvin, G. Gu, E. J. Nelkin, K. P. Bowman, Y. Hong, E. F. Stocker, and D. B. Wolff, 2007: The TRMM Multisatellite Precipitation Analysis (TMPA): Quasi-global, multiyear, combined-sensor precipitation estimates at fine scales. *J. Hydrometeor.*, **8**, 38–55.

Jolliffe, I. T. and D. B. Stephenson, 2003: *Forecast Verification. A Practitioner's Guide in Atmospheric Science.* . Wiley and Sons Ltd, Chichester, 240 pp..

Joyce, R. J., J. E. Janowiak, P. A. Arkin, and P. Xie, 2004: CMORPH: A method that produces global precipitation estimates from passive microwave and infrared data at high spatial and temporal resolution. *J. Hydrometeor.*, **5**, 487–503.

JWGV (Joint Working Group on Verification), 2008: Forecast verification: Issues, methods, and FAQ. [Avaliable at http://www.bom.gov.au/bmrc/wefor/staff/eee/verif/verif_web_page.html.]

Michaelides, S. C., 2008: *Precipitation: Advances in Measurement, Estimation and Prediction.*. Springer, New York.

Rezacova, D., Z. Sokol, and P. Pesice, 2007: A radar-based verification of precipitation forecast for local convective storms. *Atmos. Res.*, **83**, 221–224.

Roberts, N. M. and H. W. Lean, 2008: Scale-selective verification of rainfall accumulations from high-resolution forecasts of convective events. *Mon. Wea. Rev.*, **136**, 78–97.

Sapiano, M. R. P. and P. A. Arkin, 2008: An inter-comparison and validation of high resolution satellite precipitation estimates with three-hourly gauge data. *J. Hydromet.*, in press [Available at: http://ams.allenpress.com/perlserv/?request=get-abstract&doi=10.1175%2F2008JHM1052.1].

Scofield, R. A. and R. J. Kuligowski, 2003: Status and outlook of operational satellite precipitation algorithms for extreme-precipitation events. *Wea. Forecasting*, **18**, 1037–1051.

Seed, A. and E. Duthie, 2007: Rainfields: A quantitative radar rainfall estimation scheme. *33rd Conf. Radar Meteorology, Amer. Met. Soc., Cairns, Australia, 6–10 August 2007*.

Sorooshian, S., K. -L. Hsu, X. Gao, H. V. Gupta, B. Imam, and D. Braithwaite, 2000: Evaluation of PERSIANN system satellite-based estimates of tropical rainfall. *Bull. Amer. Met. Soc.*, **81**, 2035–2046.

Theis, S. E., A. Hense, and U. Damrath, 2005: Probabilistic precipitation forecasts from a deterministic model: a pragmatic approach. *Meteorol. Appl.*, **12**, 257–268.

Turk, F. J. and S. D. Miller, 2005: Toward improving estimates of remotely-sensed precipitation with MODIS/AMSR-E blended data techniques. *IEEE Trans. Geosci. Rem. Sensing*, **43**, 1059–1069.

Weymouth, G., G. A. Mills, D. Jones, E. E. Ebert, and M. J. Manton, 1999: A continental-scale daily rainfall analysis system. *Aust. Met. Mag.*, **48**, 169–179.

Yates, E., S. Anquetin, V. Ducrocq, J. -D. Creutin, D. Ricard, and K. Chancibault, 2006: Point and areal validation of forecast precipitation fields. *Meteorol. Appl.*, **13**, 1–20.

Zepeda-Arce, J., E. Foufoula-Georgiou, and K. K. Droegemeier, 2000: Space-time rainfall organization and its role in validating quantitative precipitation forecasts. *J. Geophys. Res.*, **105** , (D8), 10129–10146.

A Practical Guide to a Space-Time Stochastic Error Model for Simulation of High Resolution Satellite Rainfall Data

Faisal Hossain, Ling Tang, Emmanouil N. Anagnostou, and Efthymios I. Nikolopoulos

Abstract Abstract For continual refinement of error models and their promotion in prototyping satellite-based hydrologic monitoring systems, a practical user guide that readers can refer to, is useful. In this chapter, we provide our readers with one such practical guide on a space-time stochastic error model called SREM2D (A Two Dimensional Satellite Rainfall Error Model) developed by Hossain and Anagnostou (*IEEE Transactions on Remote Sensing and Geosciences*, 44(6), pp. 1511–1522, 2006). Our guide first provides an overview of the philosophy behind SREM2D and emphasizes the need to flexibly interpret the error model as a collection of modifiable concepts always under refinement rather than a final tool. Users are encouraged to verify that the complexity and assumptions of error modeling are compatible with the intended application. The current limitations on the use of the error model as well as the various data quality control issues that need to be addressed prior to error modeling are also highlighted. Our motivation behind the compilation of this practical guide is that readers will learn to apply SREM2D by recognizing the strengths and limitations simultaneously and thereby minimize any black-box or unrealistic applications for surface hydrology.

Keywords Satellite rainfall · Infrared · Passive microwave · Uncertainty

1 Introduction

To the surface hydrologist, rainfall remains one of the most complex hydrologic variables exhibiting intermittency across scales of interest. Being a binary phenomenon (e.g. it is either raining or is completely dry), rainfall is one of the few natural variables whose lack of continuity in space and time dominates

F. Hossain (✉)
Department of Civil and Environmental Engineering, Tennessee Technological University, Cookeville, TN 38505-0001, USA
e-mail: fhossain@tntech.edu

as scales become smaller (unlike stream flow or soil moisture). Although, the space-time structure of rainfall directly affects the response of dynamic terrestrial hydrologic processes such as runoff generation and soil moisture evolution, this scale-dependent complexity has remained a challenge to its mathematical modeling and a topic of much research the last few decades.

Models that simulate the rainfall generation process are aplenty. Using various discrete pulse-type probability distributions and/or the physics of the atmospheric process, these models can simulate the evolution of rainfall in the space-time continuum. The modeling of the rainfall process has been a much studied topic since the 1970s (see for example, Anagnostou and Krajewski, 1997; Stewart et al., 1984; Bras and Rodriguez-Iturbe, 1976; Eagleson, 1972). For a review of currently available rainfall models, the reader is referred to Waymire and Gupta (1981) and Fowler et al. (2005).

However, error models on rainfall, which are conceptually different from rainfall models because they simulate the measurement error of rainfall, are relatively less common, particularly if the focus is on space-borne platforms Hossain and Huffman (2008). Satellite rainfall error modeling has a relatively shorter heritage than radar rainfall error modeling (see for example, Ciach et al., 2007 and Jordan et al., 2003). The issue of "error" (hereafter used synonymously with "uncertainty") arises when there is more than one source of data observing the same rainfall process, with one source having typically lower confidence than the other. Satellite rainfall, on account of being indirect "measurements" of the rainfall process are often linked with such lower levels of confidence than the more conventional measurement arising from ground networks such as weather radars and in-situ gages (Huffman, 2005). As satellite rainfall data become more easily available at higher spatial and temporal resolutions from multiple sources, a natural outcome will be an explosion of its use in surface hydrologic applications over regions where it is needed most. For applications that are very critical for society (such as flood/landslide monitoring or drought management), it is important therefore that users understand the uncertainty associated with satellite rainfall data prior to building decision support systems.

The purpose of this chapter is to provide readers with a detailed practical guide on the use of a space-time satellite rainfall error model called SREM2D developed earlier by authors of this chapter – F. Hossain and E.N. Anagnostou ("A Two Dimensional Satellite Rainfall Error Model" *IEEE Transactions on Remote Sensing and Geosciences*, 44(6), pp. 1511–1522, 2006). In another work by Hossain and Huffman (2008), a detailed overview on the history of error quantification of satellite rainfall data and its modeling is provided. Thus, other competing error models are not the subject of interest in this chapter.

Also, due to increased interest on SREM2D from users of various backgrounds, this practical guide is considered timely for advancing the application of high resolution (satellite) precipitation products (HRPPs) in surface hydrology (*hereafter, rainfall is used as a shorthand for precipitation*). At the time of writing this manuscript, users from the following organizations and institutions were identified as having expressed a direct interest or already begun using SREM2D in their analyses: (1) NASA Laboratory of Atmospheres, (2) NASA Data Assimilation Branch,

(3) University of Oklahoma, (4) Mississippi State University's GeoResources Institute, (5) University of Mississippi Geoinformatics Center. Most error models described in literature are written for researchers engaged in development and assessment of satellite rainfall data. There is none, to the best of our knowledge, that aims to guide a user towards its practical use, calibration, limitations and interpretation of error model output. Hence, a motivation behind the compilation of this practical guide is that readers and users alike will learn to apply SREM2D recognizing simultaneously the pros and cons and thereby minimize any black-box or invalid applications for surface hydrology.

The paper is organized as follows. Section Two addresses the question *Why SREM2D?* and provides an overview of the philosophy behind SREM2D. Section Three dwells on the general modeling structure of the SREM2D error model. Section Four describes the formulation of SREM2D error metrics, followed by "Data Quality Control/Quality Assessment (QA/QC) and Error Metric Calibration" in Section Five. This section (Five) explains readers the computation of various error metrics of SREM2D from the data and the potential limitations that may be associated with the calibration approach. Section Six describes issues of SREM2D simulation and reproducibility of error statistics via ensemble generation of synthetic satellite data. Conclusions and the open issues needing closure regarding SREM2D are provided in Section Seven.

2 Why SREM2D?

Although existing rainfall error metrics and error models have undoubtedly advanced the application in terrestrial hydrology (Huffman, 1997; Gebremichael and Krajewski, 2004; Steiner et al., 2003; Ebert, 2008), some issues continue to remain open. Firstly, most error models treat error as a uni-dimensional (i.e., a single quantity) measure without an explicit recognition that rainfall is an intermittent process that can also affect the measurement accuracy. These models use the power law type relationships for estimating this aggregate error as a function of spatial and temporal sampling parameters. Such models may be acceptable for estimating the average error over large areal nd temporal domains (e.g 512 X 512 km^2, monthly or daily accumulations). However, there is no clear indication at this stage about the implication of using such coarse-grained error models for hydrologic error propagation experiments where the space-time covariance structure of the estimation error may not be preserved. For example, a satellite rainfall product with an error standard deviation of X mm/h can be generated from a multiplicity of distinct space-time patterns of rainfall. Each pattern, however, will have a different response in surface hydrology at fine space-time scales (see for example, Lee and Anagnostou, 2004).

Thus, there is a need to transition current error models to a level that recognizes at a minimum the need for preservation of covariance structure of the measured rainfall and the associated measurement accuracy as a function of space and time. With this need comes the recognition for a change in paradigm that single aggregate error metrics (such as error variance) are not sufficient metrics for error models that aim to simulate the hydrologically-relevant features of satellite rainfall uncertainty. SREM2D is one such error model developed for space-time generation of satellite rainfall fields in response to the limitations of current error models that tend to simplify the uncertainty.

3 General Modeling Structure Of SREM2D

SREM2D is designed as a collection of concepts, each having flexibility in modification or replacement with an alternative concept. The logical thought process behind the collection of concepts has already been outlined in a step by step manner by Hossain and Huffman (2008). For the convenience of our readers, we reiterate in this section the pertinent steps (Fig. 1) "as is" to highlight the general modeling structure of SREM2D. Hereafter, we use the term "reference" rainfall to refer to ground validation (GV) rainfall data that is corrupted by the error model to simulate less confident satellite-like observations of the rainfall process.

Recognizing that it is the intermittency of the rainfall process in space and time that dictates the variability of a hydrologic process overland, the SREM2D conceptualizes that the error metrics in three general dimensions. These are: (1) temporal dimension (*How does the error vary in time?*); (2) spatial dimension (*How does the error vary in space?*), and (3) retrieval dimension (*How "off" is the rainfall estimate from the true value over rainy areas?*). A given satellite grid-box can be rainy or non-rainy. When compared to the corresponding reference rainfall data, a satellite estimate may fall into one of four possible outcomes:

1) Satellite successfully detects rain (successful rain detection, or "hit").
2) Satellite fails to detect rain (unsuccessful rain detection, or "miss").
3) Satellite successfully detects the no-rain case (successful no-rain detection).
4) Satellite fails to detect the no-rain case (unsuccessful no-rain detection, or "false alarm").

The grid-boxes that are successfully detected as rainy may exhibit three additional properties or dimensions listed above (in space, time and scalar difference, see Figure 1). Each of these properties may be considered fully or partially representative of the three general dimensions outlined earlier. At this stage, it is not clear how adequately these properties represent a given dimension. For example, the temporal variation of error probably results from a mixture of the true spatial and temporal correlations of the rain system in its Lagrangian (system-following) frame of reference, and the advection speed of that frame of reference. In SREM2D, the temporal dimension (*how does error vary in time?*) is modeled with a simple representation – assuming that only the mean field bias (systematic error) is correlated in time in an Eulerian (surface-based) frame of reference.

The successful rain or no-rain detection capability may exhibit a strong covariance structure (i.e., the probability of successful detection of a grid-box as rainy or non-rainy may be a function of the proximity to a successfully detected grid-box).

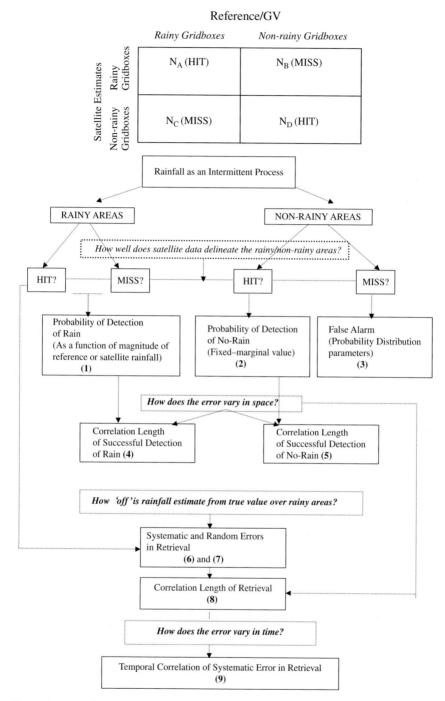

Fig. 1 Generalized framework for building error metrics and error models, (taken from – Hossain and Huffman(2008), "Investigating Error Metrics for Satellite Rainfall at Hydrologically Relevant Scales, Journal of Hydrometeorology vol. 9(3), pp. 563–575")

For grid-boxes that are detected as non-rainy, satellite rainfall data can be characterized by a marginal probability of no-rain. However, for grid-boxes that are detected as rainy, the probability of successful detection may depend on the magnitude of the rainfall rate. The functional dependency of probability of detection of rain may be tagged with reference (GV) or the estimated rain rate. For surface hydrology, users would likely be interested in the probability of rain detection benchmarked with respect to GV. On the other hand, according to Hossain and Huffman (2008), the data producers may find it almost impossible to tag the probability of detection of the satellite estimates in a likewise manner for the hydrologist on an operational basis due to lack of global scale GV data and hence, choose to use satellite estimates instead.

Collecting all these components, and by following the logical modeling steps outlined in Fig. 1, the SREM2D set of error metrics (e.g. in lieu of a single error metric concept) is: (1) Probability of rain detection (and as a function of rainfall magnitude) – POD_{RAIN}; (2) Probability of no-rain detection – POD_{NORAIN}; (3) First and second order moments of the probability distribution during false alarms; (4) Correlation lengths for the detection of rain-CL_{RAIN}, and (5) no rain–CL_{NORAIN}; (6) Conditional systematic retrieval error or mean field bias (when reference rain > 0); (7) Conditional random retrieval error or error variance; (8) Correlation length for the retrieval error (conditional, when rain >0.0) – CL_{RET}; and finally, (9) Lag-one autocorrelation of the mean field bias. In the following section, we dwell on the mathematical formulation of each of these nine error metrics. For more details, the reader can refer to Hossain and Huffman (2008) or Hossain and Anagnostou (2006).

4 Formulation of SREM2D Error Metrics

4.1 Probabilities of Detection (Rain and No-Rain) (Metrics 1 and 2)

Consider first, the following contingency matrix for hits and misses associated with satellite rainfall estimates:

The probabilities of detection for rain and no-rain are defined as follows,

$$\text{Probability of Detection for Rain (PODRAIN)}: \frac{N_A}{N_A + N_C} \quad (1)$$

$$\text{Probability of Detection for No Rain (PODNORAIN)}: \frac{N_D}{N_B + N_D} \quad (2)$$

We also define the (successful) rain detection probability, POD_{RAIN}, as a function of rainfall magnitude of either the reference rainfall or satellite estimate. The functional form is usually identified through calibration with actual data (see Hossain and Anagnostou, 2006). Based on observations with actual satellite data, SREM2D

SREM2D Guide

models the dependency of the probability of rain detection in the form of a logistic regression model as follows:

$$\text{PODRAIN (RREF)} = \frac{1}{A + \exp(-BR_{REF})} \qquad (3)$$

Subscript "REF" refers to reference rainfall (A and B are logistic parameters). The use of an idealized rain detection efficiency function may have its demerits when the empirical detection property deviates significantly from the logistic form. Users are therefore encouraged to verify the form and consider modeling POD$_{RAIN}$ from an empirical look-up table (discussed in detail in Section Five).

The POD$_{NORAIN}$, is the unitary probability that satellite retrieval is zero when reference rainfall is zero, which is also determined on the basis of actual satellite and reference rainfall data (Eq. 2).

4.2 False Alarm Rain Rate Distribution (Metric 3)

A probability density function (D_{false}) is defined to characterize the probability distribution of the satellite estimates when there are misses over non-rainy areas. This function is also identified through calibration on the basis of actual sensor data. Hossain and Anagnostou (2006) have reported that this D_{false} probability density function typically tends to appear exponential. Hence, both the moments (first and second) can be defined using only one parameter (a SREM2D metric) of the distribution, λ. This can be computed using the chi-squared or maximum likelihood method. We must however stress that it is up to the user to verify the assumption of exponential distribution and use the appropriate probability distribution for sampling these false alarm rain rates.

4.3 Correlation Lengths (Metrics 4, 5 and 8)

To identify the correlation lengths of error (i.e., *how does the error vary in space*) a simple exponential type auto-covariance function is assumed in SREM2D (users may opt for more sophisticated approaches if necessary). The correlation length (the separation distance at which correlation $=\frac{1}{e}= 0.3678$) is thus determined on the basis of calibration with actual data over a large domain. For identifying the spatial correlation length of rain detection, CL_{RAIN} (or, no-rain detection – CL_{NORAIN}) from data, all successfully detected rainy (non-rainy) pixels are assigned a value of 1.0 while the rest has a value of 0.0. The empirical semi-variogram is then computed as follows:

$$\gamma(h) = \frac{1}{2n(h)} \sum_{i=1}^{n(h)} (z(x_i) - z(x_i + h))^2 \qquad (4)$$

where $z(x_i)$ and $z(x_i+h)$ are the binary pixel values (0 or 1) at distance x_i and x_i+h, respectively and h is the lag in km. n represents the number of data points at a separation distance of h. The term $\gamma(h)$ is the semi-variance at separation distance h. Assuming that the empirical variogram is best represented by an exponential model, the functional parameters describing the spatial variability can be fitted as follows,

$$\gamma(h) = c_0 + c(1 - e^{-h/CL}) \tag{5}$$

where c_0 represents the nugget variance, c is the sill variance and CL is the distance parameter known as "correlation length" (a SREM2D metric). Conversely, the correlation function is modeled as, $C = EXP(-h/CL)$, where C is the correlation.

For identifying the correlation length for retrieval error (i.e., when both satellite and reference rainfall simultaneously register HITs), CL_{RET}, a similar set of steps are adopted as above for rain/no rain detection, with the exception that the binary values (0–1) are no longer pertinent. Instead, one computes the correlation length in terms of retrieval error defined as the logarithmic difference between reference and satellite estimate.

4.4 Conditional Rain Rate Distribution (Metrics 6 and 7)

The conditional (i.e., reference rainfall > threshold unit) non-zero satellite rain rates, R_{SAT}, are statistically related in SREM2D to corresponding conditional reference rain rates, R_{REF}, as,

$$R_{SAT} = R_{REF} \cdot \varepsilon_S \tag{6}$$

where the satellite retrieval error parameter, ε_s, is assumed to be log-normally distributed. This assumption has its pros and cons. The advantage of such an assumption is that a log transformation $[\log(R_{SAT}) - \log(R_{REF})]$ of Eq. 6 allows the ϵ_s to be mapped to a Gaussian $N(\mu, \sigma)$ deviate, ε (hereafter referred to as "log-error"), where μ and σ are the mean and standard deviation, respectively. On the other hand, the assumption of log-normality implies that data on log-error is homoscedastic (i.e., the variance remains the same regardless of the magnitude of the log-error). Hence, it is the user's responsibility to verify the assumption of log-normality and homoscedasticity and assess if log-normality is sufficient to model the skewness expected from non-zero and positive rainfall rates. Skewness of rainfall is known to diminish at longer space-time accumulations (from hourly to monthly). Thus, for a particular application, such as optimizing satellite rainfall-based irrigation schedule at weekly timescales, there may not be any need to account for skewness in the satellite rainfall. Vice-versa, skewness will be important for assessing the use of half-hourly real-time satellite rainfall data for flash-floods forecasting.

Another aspect to highlight is the definition of the threshold rainfall rate to distinguish rainy events from non-rainy (dry) events. This is particularly

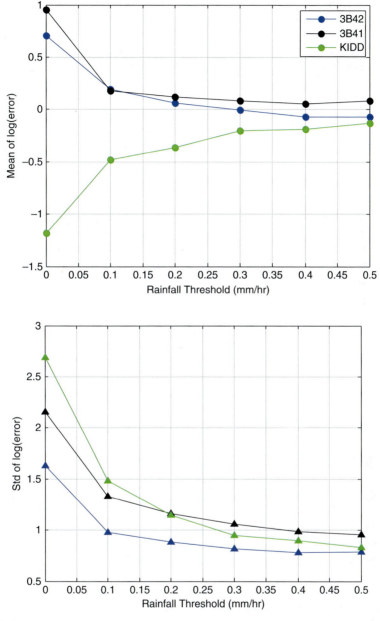

Fig. 2 Impact of reference rainfall threshold on the derivation of the mean and standard deviation of log-error for SREM2D for three high resolution satellite rainfall products (3B41RT, 3B42V6 and KIDD) over Northern Italy. Here, KIDD is a IR-based satellite rainfall product by Kidd et al. (2003)

important because of the multiplicative and log-transformed nature of the error model. A zero threshold can result in unrealistically high Gaussian standard deviation and bias because of exceedingly high multiplicative ratios that are obtained at near-zero reference rain rates. Figure 2 shows how the μ and σ of log-error varies as a function threshold for three existing satellite rainfall products remapped at 0.25° and 3 hourly timescales over Northern Italy. The reference GV data was derived from a dense gauge network. Our general recommendation is that the threshold be constrained to 0.1 mm/h or be subjectively decided after checking for reproducibility of SREM2D error statistics (discussed later in Section Six).

4.5 Lag-One Temporal Correlation (Metric 9)

The retrieval error parameter ε is both spatially and temporally auto-correlated and this space-time structure is accounted for in SREM2D. The spatial aspect has already been discussed earlier in Section 4.3. For temporal correlation, an autoregressive function is used to identify the temporal variability of μ (i.e., conditional satellite rainfall bias), with the pertinent metric being the lag-one correlation. This makes the treatment of temporal dependence of error in SREM2D somewhat subjective as the lag-one correlation will be dictated by the temporal resolution of data. A more robust treatment may be to incorporate the correlation length in time (i.e., the e-folding time of the temporal correlogram) in modeling of the temporal correlation of error. Again, this issue is for the user to verify depending on how adequately SREM2D captures the full spectrum of error at hydrologically relevant scales. More details on the temporal aspect is provided in the next section (Section Five).

5 Data QA/QC and Calibration of Metrics for SREM2D

5.1 Quality Assessment and Quality Control

SREM2D uses as input, a time-series of reference rainfall fields. This time-series is then corrupted in space and time according to the nine error metrics outlined in Section Four. The user needs to calibrate these nine SREM2D error metrics for a specific satellite rainfall product that he/she plans to assess. Collectively, these nine metrics represent the multi-dimensional error structure of the satellite data product under investigation. For calibration of SREM2D metrics, a sufficiently long period of synchronized rainfall fields (from a sufficiently large areal domain) of reference and satellite sources is required. The definition of "sufficiently long" is subjective. For example, 5 year of hourly reference and satellite rainfall data over the Upper Mississippi basin may yield a "climatologic" average SREM2D metrics for a specific satellite rainfall product that has matured in algorithmic formulation (such as Global Precipitation Climatology Project product available at 1° -Daily resolution).

On the other hand, 3 month-long hourly data during summer may be more informative of metrics a user should employ for simulation of satellite observation of thunder storms and other shorter-duration convective rain systems.

An important aspect of QA/QC during SREM2D calibration is that there should not be any missing data in space and time and that both sets (satellite and reference) must be synchronized very accurately. Users should resolve this QA/QC issue because most real-time HRPPs available today at sub-daily time scales are produced on a best-effort basis with a non-negligible portion of data often reported missing. We recommend the following two strategies for replacement of missing data: (1) if the percentage missing is small (< 5%), then reference rainfall may be substituted with minimal effect on the computation of error metrics; (2) if percentage of missing is considerably larger (~5–15%), persistence of preceding satellite data over missing periods may be considered. The argument for #2 is that in a real-world scenario, the user would have to continue using the last available satellite observation over ungauged regions until the next satellite overpass or data downlink.

A major problem arises when both satellite and reference data are missing in significant portions. For such cases, we recommend that the period of data not be included in SREM2D error metric calibration. As an example, Table 1 shows missing data statistic for one particular data set of Stage IV NEXRAD radar rainfall data over the United States spanning six years (2002–2007). The Northwestern region appears to have a significant amount missing data (mainly east of the Cascade Mountains) that can result in spurious error calibration of SREM2D if attempted.

Table 1 Missing data statistics for Stage IV NEXRAD data over different regions of the United States spanning 6 years (2002–2007) at 4 km and 1 hourly scale

	ALL	Northwest	Southwest	Midwest	Northeast	Southeast
% Missing	11	32%	9.1%	0.8%	1.3%	12.7%

Because the primary motivation of an error modeling technique is to understand how erroneous a satellite rainfall product is compared to a reference GV dataset both in rainfall and in hydrologic simulation, SREM2D does not account for the possible effects of errors in the "reference" rainfall estimates. However, users must also recognize that the SREM2D estimation technique of the nine error metrics will incorporate the uncertainties arising from both the satellite and reference rainfall.

5.2 Error Metric Calibration

After proper QA/QC of calibration data, the user needs to calibrate the nine metrics that serves as input to the SREM2D error model. In this section, we show examples of calibration for four global satellite HRPPs at 0.25° 3 hourly scales over the United States spanning two regions (Florida and Oklahoma; Fig. 3) and four seasons in 2004 (Winter, Spring Summer, and Fall). These four satellite products are: (1) 3B41RT; (2) 3B42RT; (3) CMORPH and (4) PERSIANN. Literature on

Fig. 3 Two regions (Oklahoma and Florida) in the United States selected for SREM2D calibration of error metrics for four global satellite rainfall products (*shown in boxes*)

the first two products (hereafter referred to as 3B41RT and 3B42RT) are available from Huffman et al. (2007), while readers can refer to details on CMORPH and PERSIANN from Joyce et al. (2004) and Hong et al. (2005), respectively. The reference GV data pertained to NEXRAD (Stage III) rainfall product. The regions are bounded, for Oklahoma, by 32.0°N to 39.0°N and –92.0°W to –102.0°W; and, for Florida, by 20.0°N to 26.0°N and –84.0°W and –80.0°W (Fig. 3).

Table 2 summarizes the missing data statistic at that native scale as part of QA/QC of calibration data. All data were then remapped to the consistent scale of 0.25° and 3 hourly to allow inter-comparisons among products. Figure 4 demonstrates the POD$_{NORAIN}$ for various products across the two regions and seasons. The nuances across products and seasons (particularly for CMORPH) are apparent in this figure. Figure 5 shows the POD$_{RAIN}$ as a function of NEXRAD rain rate. As mentioned earlier in Section

Table 2 Missing data statistic for four global satellite rainfall products at native scale over the United States for 2004 (the two regions – Oklahoma and Florida are combined)

Products	Native scale Temporal (h)	Spatial (°)	Percentage of missing data Winter (JF)	Spring (AM)	Summer (JJA)	Fall (SON)
3B41RT	1	0.25	0.97	2.18	1.18	1.00
3B42RT	3	0.25	1.46	2.10	1.45	1.00
PERSIANN	1	0.04	2.30	1.43	1.22	1.10
CMORPH	3	0.25	0.00	0.00	0.00	0.00

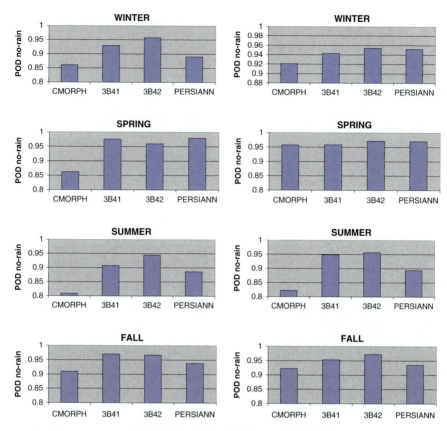

Fig. 4 POD$_{NORAIN}$ for CMORPH, 3B41RT, 3B42RT and PERSIANN across four seasons in 2004. Left panels – Oklahoma; Right panels – Florida

Four, the functional form of POD$_{RAIN}$ is almost invariably found to obey the logistic pattern. Users need to fit appropriate parameter values for A and B of Equation 3 to model the POD$_{RAIN}$ as a function of NEXRAD rain rate. There are several non-linear optimization routines that can be used to robustly derive A and B values. However, we recommend that the user also applies some human judgment to check for the closeness of the idealized logistic curve with empirical one derived (Fig. 5) at low rain rates (∼1–5 mm/h).

Figure 6 shows the probability distribution of false alarm rain rates of satellite products. The distribution appears exponential like. The mean (expected value) of this distribution comprises another SREM2D metric ($1/\lambda$). Care must be applied in the derivation of the false alarm distribution as it is sensitive to the choice of bin size. Users can apply more rigorous statistical tests and the maximum likelihood method to derivemore robust estimates of the false alarm metric. Figure 7 shows the spatial covariance structure of rain retrieval (conditional), rain detection and

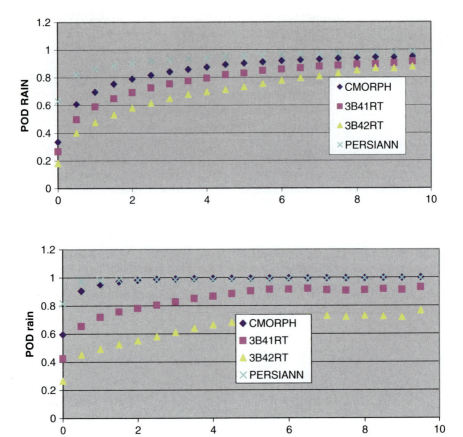

Fig. 5 POD$_{RAIN}$ as a function of NEXRAD rain rate. *Upper panel* – Florida for Winter 2004; *Lower panel* – Oklahoma for Fall 2004. X-axis represents NEXRAD rain rates at 0.25° 3 hourly resolution

no-rain detection for Florida (Summer 2004). Assuming that an exponential correlation model is representative, the separation distances where the correlation drops to 1/e (=0.368) comprise the correlation length (CL) error metrics for SREM2D for generation of correlated random fields. Certain instances may result in the correlation never (at least over the domain of the study region) dropping to 1/e. For example, in arid and clear-sky climates, the correlation length CL$_{NORAIN}$ for an Infra-red satellite rainfall product will probably be associated with large values. For such cases, we recommend that the user constrain the spatial structure by applying correlation length values compatible with the domain size of interest. A downside of large correlation lengths in error modeling, particularly for rain retrieval, is that the conditional error standard deviation may be under-simulated due to spatial similarity of the generated random values. This aspect is discussed in more detail in the next section.

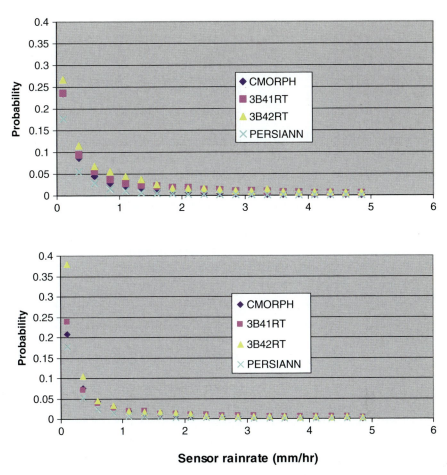

Fig. 6 False alarm rain rate distribution for satellite rainfall products. *Upper panel* – Florida-Summer; – Oklahoma-Spring. Sensor rainrate is the satellite rain estimate

6 SREM2D Simulation And Reproducibility Of Error Statistics

6.1 Simulation Issues

As model developers, we initially coded the first SREM2D error model using Fortran 77. However, we believe that the general modeling structure (Section 3) is tangible enough for any user to develop his/her own custom-built code. We therefore encourage users to rather understand the SREM2D philosophy first, assess if the complexity of the error modeling is compatible with the intended application and then apply/modify or simplify the error model accordingly using the preferred computing platform.

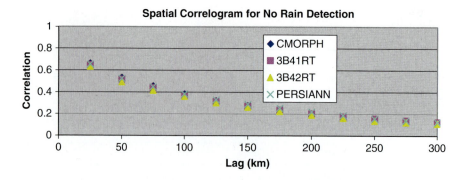
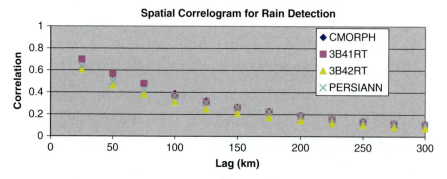
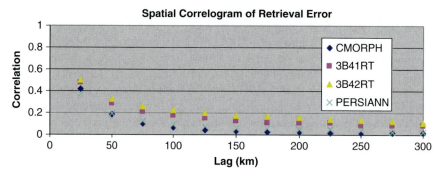

Fig. 7 Spatial covariance structure of rain retrieval, rain detection (*middle panel*) and no-rain detection (*upper panel*) for Summer 2004 in Florida

An aspect that adds to the computational burden of SREM2D is the need for generation of correlated Gaussian random fields. First, the spatial structure of rain and no-rain joint detection probabilities is modeled using Bernoulli trials of the uniform distribution with a correlated structure that is generated from Gaussian random fields. These two Gaussian random fields (one each for rain detection and no-rain detection) are transformed to the uniform distribution random variables via an error function transformation. Spatially correlated field of Gaussian $N(0,1)$ random deviates is generated in 2-D space based on Turning Bands (Mantoglou and

Wilson, 1982). The N(0,1) spatially correlated random field is then transformed to uniform U[0,1] field as follows:

$$x_j = \frac{1}{2} + \frac{1}{2} erf(\varepsilon_j/\sqrt{2}) \tag{7}$$

where x_j, is a U[0,1] random deviate for pixel j generated from the corresponding N(0,1) deviate, ε_j. The $erf(\varepsilon_j)$ is the error function defined by the following integral,

$$\mathrm{erf}(\varepsilon_j) = \frac{2}{\sqrt{\pi}} \int_0^x e^{-w^2} dw \tag{8}$$

The uniform random fields are then scaled by its standard deviation to yield a unitary variance (this ensures the maximum covariance of 1.0 at lag 0). Numerical consistency checks have revealed that correlation length is altered significantly by this non-linearity only at lags (grid spaces) beyond 10 and should be accordingly accounted for modeling the join probability of detection if necessary. Execution of this procedure yields a spatially correlated uniform field of U [0,1] random deviates that are now amenable for Bernoulli trials for rain and no-rain detection with *a priori* spatial structures. A third Gaussian random field is generated next for the simulation of correlated retrieval error field pertaining to N (μ,σ).

Hossain and Anagnostou (2006) provide the simulation algorithm for SREM2D that outlines each simulation step for the error model in the form of a programming flow-chart. We recommend that users refer to that algorithm flow-chart to clarify the individual process calculations that SREM2D computes in space and time.

6.2 Reproducibility of SREM2D Error Statistics

Before the assessment of satellite rainfall products for decision-making can begin, users must verify that the ensembles of satellite rainfall data simulated by SREM2D are adequately realistic. In other words, the reproducibility of error statistics (metrics) by SREM2D needs to be verified. Like any other mathematical model, SREM2D does not perfectly mimic the uncertainty as expected from the calibrated metrics. Nevertheless, the user must set some minimum standards on reproducibility based on the intended application. We recommend two particular ways by which SREM2D can be verified of this "reproducibility" property. These are as follows:

1) Checking the consistency of ensemble of cumulative rainfall hyeotograph against actual satellite rainfall data.
2) Checking the accuracy of error metrics computed from simulated satellite rainfall data against actual reference rainfall data.

The first method checks if the actual satellite cumulative rainfall hyetograph is enveloped reasonably realistically by the ensemble of SREM2D generated synthetic satellite hyetographs. Because actual satellite rainfall data is not used in the generation of SREM2D synthetic data, this test can considered an independent check. Users are recommended to perform this test over the whole domain and a few random smaller sub-domains within the study region. An additional aspect to check is to verify if the simulated hyetographs exhibit a pattern of jumps and plateaus similar to the actual data. The second method computes the nine SREM2D error metrics from synthetic satellite data against actual reference rainfall data to check the closeness of the values with calibrated measures. This check may be done on individual realizations or over a set of ensembles. The latter is likely to yield more accurate results due to the larger space-time sample size that minimizes the randomization effects per each realization.

In the following, we provide an example of the two error reproducibility tests over an alpine basin in Northern Italy.

6.2.1 Checking the Consistency of Ensemble of Cumulative Hyetograph Against Actual Satellite Rainfall Data

Figure 8 shows the alpine region of Northern Italy over which SREM2D error metrics were calibrated for three satellite rainfall products. The three shaded grid boxes represent the location of actual satellite pixels at $0.25°$ scale for three satellite products

3B41RT, 3B42V6 and KIDD. Herein, KIDD represents a high resolution ($0.04°$) Infrared (IR)-based satellite rainfall product produced by Kidd et al. (2003). Six months of satellite data spanning June–November 2002 were used for calibration of SREM2D metrics. Reference data comprised gage rainfall from a dense network represented by the black circles shown in the figure. Table 3 shows the SREM2D metrics calibrated for the satellite products at the $0.25°$ 3 hourly scale. A threshold of 0.1 mm/h was assigned to separate the rainy events from non-rainy events. Figure 9 demonstrates the cumulative hyetographs generated from 100 SREM2D realizations (mean and $\pm\sigma$) and actual satellite rainfall data for 3B41RT and 3B42V6. We observe that 3B41RT is relatively more accurately enveloped than 3B42V6. Overall, the simulation of both products appear reasonably realistic for the domain of interest in Northern Italy.

6.2.2 Checking Reproducibility of Error Metrics

In Table 4, the reproducibility of the mean and standard deviation of log-error for retrieval is demonstrated for a few random SREM2D realizations against the calibrated values (that served as input to the error model) for the KIDD satellite product. While the POD_{NORAIN} and bias of log-error is reasonably well reproduced for each selected realization, the standard deviation of log-error is found to be consistently underestimated by margins of 10–15%. A recently-identified limitation of the SREM2D model is that the generation of correlated random fields with long

Fig. 8 Alpine region of Northern Italy. Shaded grey boxes represent the actual location of the 0.25° satellite pixels for 3B41RT and 3B41V6 data used in the calibration of SREM2D error metrics. Black circles represent the location of tipping bucket gages that comprised reference rainfall data

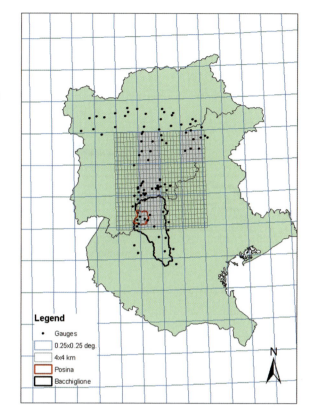

Table 3 SREM2D error metrics calibrated for 3B41 and 3B42 for the region of Northern Italy

Metrics	3B41	3B42	KIDD
A	1.05	1.1	1.1
B	1.85	1.08	1.2
Mean (mu-Gaussian of log-error)	0.026	−0.1102	−0.226
Sigma (std.dev Gaussian of log-error	0.942	0.764	0.733
False Alarm mean rain rate (mm/hr)	0.433	0.760	0.680
Lag-one correlation	0.41	0.13	0.41
POD no-rain	0.81	0.97	0.99
*CL_{ret} km	50	50	50
*$CL_{rain\ det}$ km	0	0	0
*$CL_{no\ rain\ det}$ km	75	75	75

correlation lengths for retrieval error tend to conflict with the standard deviation of retrieval error and result in under-simulation (i.e. underestimation). This underestimation appears to magnify as the domain size increases. We do not know yet how

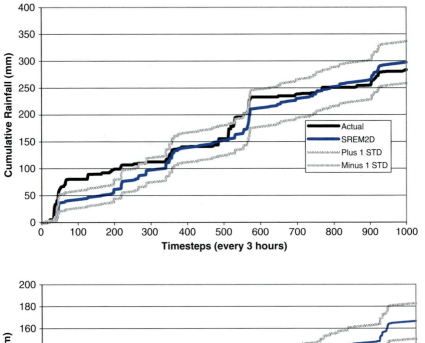

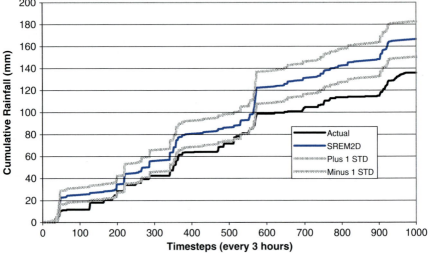

Fig. 9 Cumulative rainfall hyetographs over Northern Italy. Blue line represents the mean of 100 SREM2D realizations. Solid black line represents the actual satellite hyetograph. Upper panel – 3B41RT; Lower panel – 3B42V6

to address this problem at this stage, but it is certainly an aspect that users should be cognizant of and strive for rectification in future improvements of the SREM2D model. Users should also perform similar consistency checks for all other SREM2D metrics and not just of conditional bias and standard deviation.

Table 4 Reproducibility of some SREM2D error metrics for a few random realizations over Northern Italy for KIDD (KIDD is the IR-based satellite rainfall product by Kidd et al. 2003)

	POD $_{\text{NORAIN}}$	Bias (log-error)	Std Dev (log error)
Empirical	0.986	0.727	1.19
Realization 1	0.983	0.672	0.98
Realization 2	0.983	0.496	1.04
Realization 3	0.990	0.545	1.05
Realization 4	0.990	0.747	1.01

7 Conclusions

For continual refinement of error models and their promotion in prototyping satellite-based hydrologic monitoring systems, a practical user guide that readers can refer to is useful for potential users of HRPPs. In this chapter, we have provided our readers with one such practical guide on a space-time stochastic error model called SREM2D (A Two Dimensional Satellite Rainfall Error Model) developed by Hossain and Anagnostou (*IEEE Transactions on Remote Sensing and Geosciences*, 44(6), pp. 1511–1522, 2006). This practical guide overviewed the philosophy behind SREM2D and emphasized the need to flexibly interpret the error model as a collection of modifiable concepts always under refinement. We stressed at various stages of the guide the importance of verifying that the complexity and assumptions of error modeling were compatible with the intended application. Our motivation behind the compilation of this practical guide was that readers should learn to apply SREM2D recognizing the strengths and limitations simultaneously and thereby minimize any black-box or unrealistic applications for surface hydrology. We also hope that developers of other error models will produce similar "guides" to make the pros and cons of the error modeling philosophy open for the user.

Like any other model, SREM2D is not without limitations. The requirement of continuous data (reference and satellite) in space and time may be considered a short coming for calibration of SREM2D error metrics. For advancing the application of satellite HRPPs, the associated uncertainty information is critical for users to understand the realistic limits to which these HRPPs can be applied over an ungauged region. However, this represents a paradox. Satellite rainfall uncertainty estimation requires reference (ground validation-GV) data. On the other hand, satellite data will be most useful over ungauged regions in the developing world that are lacking in GV data. Consequently, we need to ask ourselves several questions for SREM2D. *Can the model parameters/metrics be transferred from one region to another? Can they be regionalized?* At this stage, there is no clear answer, although there is work on-going by the authors to resolve this paradox and understand how reliable is the "transfer" of error from a gauged location to an ungauged one.

On the computational side, the need to generate three independent and correlated random fields increases simulation runtime for SREM2D. The need to convert Gaussian random fields to uniform random fields by the non-linear error transformation also results in an unknown change of spatial structure that is not yet completely constrained at large space lags (> 10). The spatial correlation also has the effect of imparting negative bias to the standard deviation of retrieval error.

Despite these limitations, SREM2D represents a unique hydrological transition from current error models because it explicitly recognizes the need for preservation of covariance structure of rainfall and the associated measurement accuracy as a function of space and time. It also provides greater versatility in error modeling by moving away from the single aggregate error metric models to a multi-dimensional one comprising nine metrics. We believe that subject of space-time error modeling of high resolution satellite rainfall products can reach closure with the systematic evolution of the philosophy and concepts embedded in the SREM2D model.

Acknowledgements Support for this work was provided by the NASA New Investigator Program Award (NNX08AR32G) to the first author and NASA Precipitation Measurement Mission to authors Anagnostou and Hossain. Authors Nikolopoulos and Tang were supported by NASA Earth System Science Fellowship.

References

Anagnostou, E. N. and W. F. Krajewski. (1997). Simulation of radar reflectivity fields: Algorithm formulation and evaluation. *Water Resources Research*, vol. 33, 6, pp. 1419–1428.

Bras, R. L. and I. Rodriguez-Iturbe. (1976). Rainfall generation: A non-stationary time-varying multidimensional model. *Water Resources Research*, vol. 12, 3, pp. 450–456.

Ciach, G. J., W. F. Krajewski, and G. Villarini. (2007). Product-error-driven uncertainty model for probabilistic quantitative precipitation estimation with NEXRAD data. *J. Hydrometeorology.*, vol. 8, pp. 1325–1347.

Eagleson, P. S.. (1972). Dynamics of flood frequency. *Water Resources Research*, 8, 878–898.

Ebert, E. E.. (2008). Fuzzy verification of high resolution gridded forecasts: A review and proposed framework. *Meteorological Applications*, In press.

Fowler, H. J., C. G. Kilsby, P. E. O'Connell, and A. Burton. (2005). A weather-type conditioned multi-site stochastic rainfall model for the generation of scenarios of climatic variability and change. *Journal of Hydrology*, vol.8, 1–4, pp. 50–66.

Gebremichael, M. and W. F. Krajewski. (2004). Characterization of the temporal sampling error in space-time-averaged rainfall estimates from satellites. Journal of Geophysical Research, vol. 109, D11, doi: D11110 0.1029/2004JD004509.

Hong, Y., K. -L. Hsu, S. Sorooshian, and X. Gao. (2005). Self-organizing nonlinear output (SONO): A neural network suitable for cloud patch–based rainfall estimation from satellite imagery at small scales. Water Resources Research, vol. 41, W03008, doi:10.1029/2004WR003142.

Hossain, F. and E. N. Anagnostou. (2006). A two-dimensional satellite rainfall error model. *IEEE Transactions on Geosciences and Remote Sensing*, vol. 44, pp. 1511–1522, DOI: 10.1109/TGRS.2005.863866.

Hossain, F. and G. J. Huffman. (2008). Investigating error metrics for satellite rainfall at hydrologically relevant scales. *Journal of Hydrometeorology*, vol. 9, 3, pp. 563–575.

Huffman, G. J.. (1997). Estimates of root mean square random error for finite samples of estimated precipitation. *Journal of Applied Meteorology*, vol. 36, pp. 1191–1201.

Huffman, G. J.. (2005). Hydrological applications of remote sensing: Atmospheric states and fluxes – precipitation (from satellites) (active and passive techniques). *Encyclopedia for Hydrologic Sciences*, eds. M. G. Anderson and J. McDonnell. ISBN: 978–0–471–49103–3.

Huffman, G. J., R. F. Adler, D. T. Bolvin, G. Gu, E. J. Nelkin, K. P. Bowman, Y. Hong, E. F. Stocker, and D. B. Wolff. (2007). The TRMM Multi-satellite Precipitation Analysis: Quasi-Global, Multi-Year, Combined-Sensor Precipitation Estimates at Fine Scale. *Journal of Hydrometeorology*, vol. 8, pp. 38–55.

Jordan, P. W., A. W. Seed, and P. E. Weinmann. (2003). A stochastic model of radar measurement errors in rainfall accumulations at catchment scale. *J. Applied Meteorology*, vol. 4, pp. 841–855.

Joyce, R. L., J. E. Janowiak, P. A. Arkin, and P. Xie. (2004). CMORPH: A method that produces global precipitation estimates from passive microwave and infrared data at high spatial and temporal resolution. *Journal of Hydrometeorology*, vol. 5, pp. 487–503.

Kidd, C., D. R. Kniveton, M. C. Todd, and T. J. Bellerby. (2003). Satellite Rainfall Estimation Using Combined Passive Microwave and Infrared Algorithms. *Journal of Hydrometeorology*, vol. 4, 1088–1104.

Lee, K. H. and E. N. Anagnostou. (2004). Investigation of the nonlinear hydrologic response to precipitation forcing in physically based land surface modeling. *Canadian Journal of Remote Sensing*, 30, 5, pp.706–716.

Mantoglou, A. and J. L. Wilson. (1982). The Turning Bands method for simulation of random fields using line generation by a spectral method. *Water Resources Research*, vol.18, 5, pp. 1379–1394.

Sorooshian, S., K. -L. Hsu, X. Gao, H. V. Gupta, B. Imam, and D. Braithwaite. (2000). Evaluation of PERSIANN system satellite-based estimates of tropical rainfall. *Bulletin of American Meteorological Society*, vol. 81, pp. 2035–2046.

Steiner, M., T. L. Bell, Y. Zhang, and E. F. Wood. (2003). Comparison of two methods for estimating the sampling-related uncertainty of satellite rainfall averages based on a large radar dataset. *Journal of Climate*, vol. 16, pp. 3759–3778.

Stewart, R. E., J. D. Marwitz, and J. C. Pace. (1984). Characteristics through melting layer of stratiform clouds. *Journal of Atmospheric Science*, 41, 22, pp. 3227–3237.

Waymire, E. and V. K. Gupta. (1981). The mathematical structure of rainfall representations. 1. A review of the stochastic rainfall models. *Water Resources Research*, vol. 17, 5, pp. 1261–1272.

Regional Evaluation Through Independent Precipitation Measurements: USA

Mathew R.P. Sapiano, John E. Janowiak, Wei Shi,
R. Wayne Higgins, and Viviane B.S. Silva

Abstract This chapter concerns the validation of high resolution (mostly 0.25°, daily and three-hourly) precipitation products over the United States. A synthesis of relevant studies is followed by comparisons of high resolution estimates based on satellites and models, with in situ ground validation data over the US. All the comparisons use multiple satellite estimates as well as model data (from the NCEP GFS). First, daily results are shown from the ongoing, web-based, real-time International Precipitation Working Group validation activity over the US. Next, validation data from 15 sub-daily gauges over Kansas and Oklahoma are used to assess the performance of three-hourly precipitation estimates, with attention to the distribution of precipitation. Finally, results from the comparison of several products against data collected from the North American Monsoon Experiment are given. Results show that existing high resolution products have great skill over many areas of the US, even at the three-hourly time-scale. Significant issues still exist over orography and results are seasonally dependent.

Keywords Precipitation · Validation · Satellite · Model forecast · US

1 Introduction

The usefulness of validation studies is reliant on the quality and quantity of comparison data. The validation of satellite estimates of precipitation is traditionally achieved by comparison to in situ observations (including gauges and radar) since these present the most realistic picture of the true state. Generally speaking, such observations are concentrated over the populous areas, usually over the most developed or prosperous countries. As might therefore be expected, the observing system

M.R.P. Sapiano (✉)
Cooperative Institute of Climate Studies (CICS), Rm 3004, Earth System Science Interdisciplinary Center (ESSIC), University of Maryland, College Park, MD, 20742, USA
e-mail: msapiano@essic.umd.edu

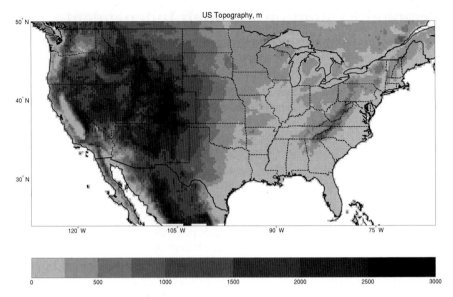

Fig. 1 Topography of the US, in m

over the US is amongst the best in the world with comprehensive gauge and radar networks which are freely available. In addition, the US has diverse meteorological conditions due to its large area and varied topography (Fig. 1), with the Rockies being the major orographic feature. Figure 2 shows the mean precipitation for each season from the Higgins et al. (2000) 0.25° gauge analysis which distinctly shows

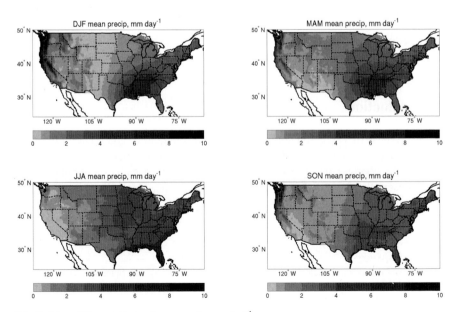

Fig. 2 Mean US precipitation by season in mm day^{-1}

some of the climatically different zones over the US such as the wet north-east coast, the relatively dry central plains and the moderately wet east coast. A strong annual cycle is also apparent with, for example, higher rainfall from convective systems in the warm season over the south-east coast and lower precipitation (with significant orographic effects) over the west coast during the cold season. Thus, the US observing system is an excellent place to conduct validation activities.

Table 1 Summary of satellite datasets used in this chapter

Product name	Key reference	Native resolution	Notes
TRMM multi-satellite precipitation analysis (TMPA)	Huffman et al. (2007)	0.25°; three-hourly	Combined passive microwave (PMW) and microwave-calibrated infrared (IR). Corrected to match gauge analysis over land
CPC morphing technique (CMORPH)	Joyce et al. (2004)	8 km; half-hourly	IR data used to obtain motion vectors to advect PMW rain rates
Hydro-estimator	Scofield and Kuligowski (2003)	0.25°; three-hourly globally; 4 km, quarter-hourly over US	Based on IR, but some model output also used
NRL-blended	Turk and Miller (2005)	0.25°; hourly	Weighted combination of PMW and PMW-calibrated IR
Precipitation estimation from remotely sensed information using artificial neural networks (PERSIANN)	Hsu et al. (1997); Sorooshian et al. (2000)	0.25°; half-hourly; 4 km version available over US	Adaptive neural network calibrated with PMW to get estimates from IR

Several studies have compared the currently available high resolution precipitation estimates over the US. Gottschalck et al. (2005) compared the precipitation from the TMPA, PERSIANN (see Table 1 for expansion of acronyms) and Stage IV radar (Baldwin and Mitchell 1998; Lin and Mitchell 2005), along with several model datasets, with the Higgins et al. (2000) gauge analysis over one year from March 2002 to February 2003. Their comparison showed that the 3B42RT (a version of TMPA without the gauge adjustment) and PERSIANN overestimated spring and summer precipitation over much of the Central US, a phenomenon they attributed to cold cirrus clouds which lead to higher rain rates (since colder infrared temperatures are assigned higher rain-rates). The TMPA (without the gauge adjustment) also overestimated precipitation during the winter and spring seasons over the plains and the Northern Rockies due to snow and ice contamination in the Passive Microwave (PMW) record. Gottschalck et al. (2005) also presented correlations between the Higgins et al. (2000) gauges and the precipitation estimates studied. The TMPA and

PERSIANN showed surprisingly similar skill with high correlations over the eastern half of the US (from Texas to the Great Lakes) but with generally low correlations over the western half throughout most of the year. Correlations were lower in many areas over the winter season with correlations greater than 0.5 over the south-east only.

Ebert et al. (2007) described results from the validation activities of three of the validation sites associated with the International Precipitation Working Group (IPWG) over Australia, the UK and the US (new results from the latter are shown in this chapter). They assessed a large number of satellite and model precipitation estimates against gauge and radar networks in these three territories, and continue to produce results in real-time on the internet. Over the US, they found that the satellite estimates performed better for warm-season convective precipitation, whilst Numerical Weather Prediction (NWP) estimates performed best for cold season synoptic-scale precipitation. They also showed that datasets based on PMW data tended to have higher correlations with validation data than infrared (IR) estimates, but that a blend of the two was superior to either PMW or IR estimates.

Tian et al. (2007) validated CMORPH and the TMPA against the Stage IV radar/gauge estimates and the Higgins et al. (2000) gauge combination over the US. They found that the gauge-adjustment of the TMPA was highly effective at removing the bias, especially when compared to CMORPH which heavily overestimated precipitation over the central US during the warm season. However, they also showed that both datasets tended to underestimate cold season precipitation over the north-east coast (although CMORPH had a more severe underestimate). Correlations against the validation data were similar to those found by Gottschalck et al. (2005) with high correlations over the Eastern US in the warm season and the south-east in the cold season (with low correlations elsewhere). They reported a higher Probability of Detection (POD) and False Alarm Ratio (FAR) over the south-east for CMORPH than for TMPA, but they showed that both datasets faithfully reproduced the diurnal cycle.

Ruane and Roads (2007) described the variance at different timescales for the TMPA, CMORPH and PERSIANN. They reported a higher variance at all frequencies in CMORPH and PERSIANN, and more high frequency noise in TMPA, particularly over land. PERSIANN had higher variance at short, synoptic timescales (3–6 days) which they linked to better tracking of storm systems and CMORPH had higher annual variance, particularly at high latitudes, which might indicate more signal at these time-scales.

Curtis et al. (2007) assessed the performance of the TMPA in estimating the total precipitation associated with Hurricane Floyd over three river basins in North Carolina. When compared to a gauge network and radar estimates, they found that the TMPA over-estimated the total precipitation of extreme events, although the overestimation was much smaller for very heavy events. Villarini and Krajewski (2007) compared a single $0.25°$ gridbox of the gauge-adjusted TMPA with a dense rain gauge network over Oklahoma. They found that the TMPA had higher correlations with the gauge data during the warm season but tended to underestimate low precipitation values.

The rest of this chapter extends on these results with new comparisons between ground observations and some of the high resolution satellite (and model) estimates. Section 2 shows synthesized results from the IPWG daily US validation website (Ebert et al. 2007; http://cics.umd.edu/~johnj/us_web.html). Section 3 shows extended results based on the validation work of Sapiano and Arkin (2008) over the Great Plains of Kansas and Oklahoma at three hourly resolution. Finally, Section 4 shows an evaluation of the warm season precipitation estimated by several different types of data during the North American Monsoon Experiment (NAME)

2 Results From IPWG Daily US Validation Site

For several years, the IPWG US validation site has been producing standard, daily validation metrics for several different precipitation products (based on satellites and models) in near real-time. A summary of the results from the archive is given in this section. One important caveat for the interpretation of these results is that the archive is based on the product that was available at the time of the comparison, thus the results are not updated to reflect enhancements and reprocessing of any of the datasets. The potential effect of this is that some datasets may appear worse than they would currently be expected to be.

Time series of spatial correlation (computed daily over the entire conterminous U.S.) present several interesting features (Fig. 3). First, the performance of the satellite products and model forecasts are out of phase, with the former performing best during the warm season (May-September) and the latter performing best in the cool season (November-March). This behavior has been noted previously in Ebert et al. (2007), and the explanation, in general, is that the satellite algorithms are challenged by cold surface features (snow and ice) that occur during winter. Specifically, the passive microwave scattering algorithms have difficulty discriminating the scattering from frozen water at the surface from that due to hydrometeors in the atmosphere that result in precipitation. Conversely, the models are quite skillful in predicting the large-scale stratiform precipitation that dominates during winter but are challenged with predicting convective precipitation that is much smaller in scale and which occurs most frequently during the warm season.

Another interesting feature is the amplitude of the seasonal differences in spatial correlation with the largest annual fluctuations observed in the model forecasts and the smallest in the Hydro-Estimator. In terms of mean correlation over the 2004–2008 period, the highest correlations with the validating rain gauge analyses are with CMORPH, the National Center for Environmental Prediction (NCEP) Global Forecast System (GFS; Kalnay et al. 1990) and Navy Operational Global Atmospheric Prediction System (NOGAPS; Hogan and Rosmond 1991) model forecasts, and the Hydro-Estimator (0.62, 0.61, 0.60, 0.60, respectively), while the 3B42RT and PERSIANN techniques have the lowest mean correlation over the period (0.51 and 0.55, respectively). Despite improvements in the algorithms (and models) with time, there is no discernable improvement in spatial correlation over the 5-year period.

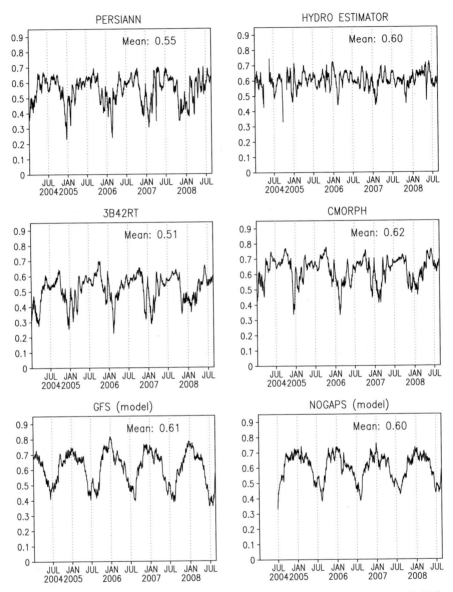

Fig. 3 Time series of spatial correlation (over the conterminous U. S.) between the CPC daily U. S. precipitation analyses and various satellite estimates and numerical model short-range predictions over the period January 1, 2004 to July 15, 2008. The time series is displayed as a 21-day running mean to improve clarity

Whilst these time series are informative, they are unable to portray geographical variations in the performance of the precipitation estimates. Hence, we present in Figs. 4 and 5 the spatial distribution of temporal correlation for the cool and warm seasons separately. The most striking observation is the large difference in performance of the model forecasts between the warm and cool seasons. Note the tremendous improvement during the cool season when correlation coefficients exceed 0.70 over most of the nation. In contrast, the satellite algorithms perform

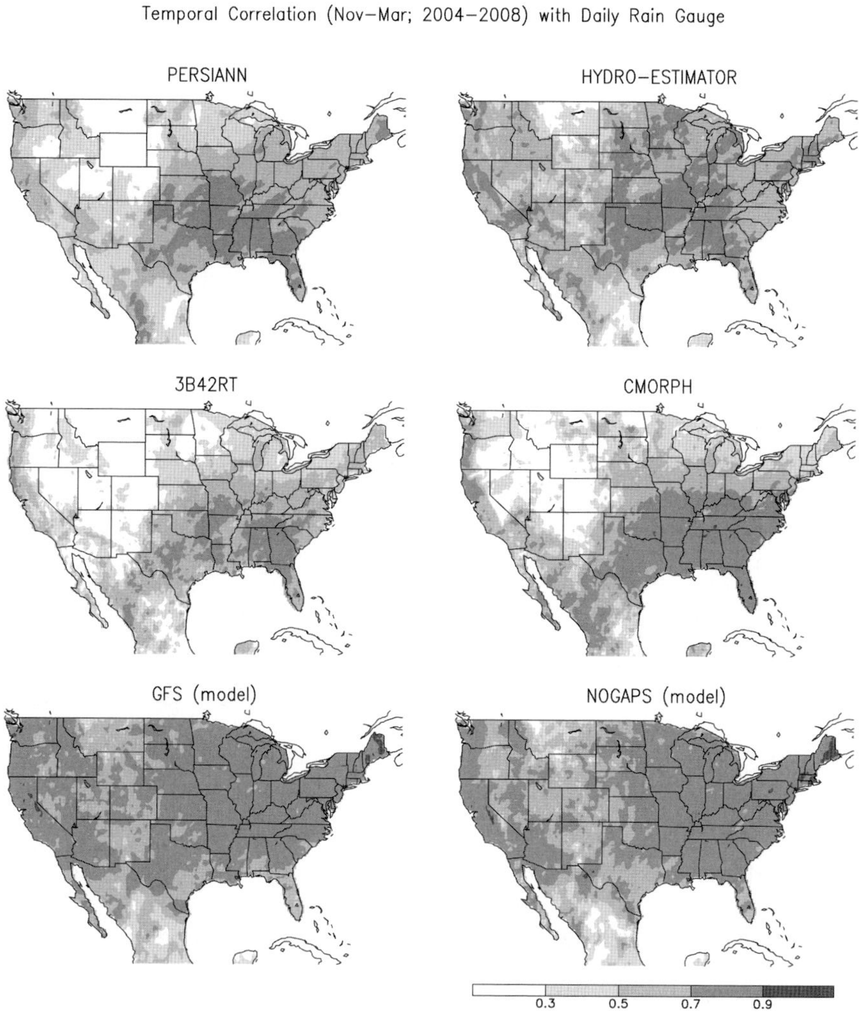

Fig. 4 Temporal correlation between the CPC daily U. S. precipitation analyses and various satellite estimates and numerical model short-range predictions for the November–March periods during 2004–2008. The 99% significance level is very conservatively estimated to be 0.30 to account for autocorrelation by reducing the degrees of freedom from approximately 900 to 70

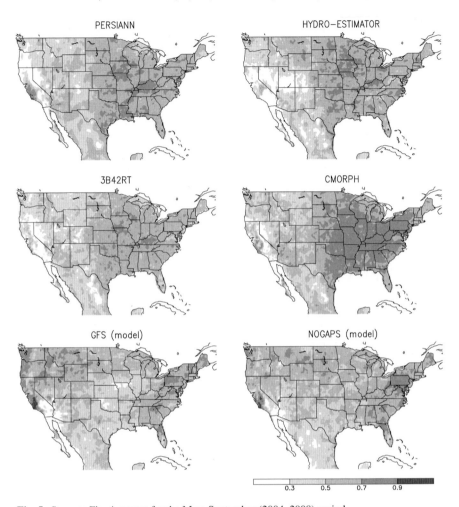

Fig. 5 Same as Fig. 4, except for the May–September (2004–2008) period

quite poorly over much of the West where snow cover is often observed during winter. Note the relatively strong correlations during winter (> 0.70) in the southeastern quadrant among all of the satellite analyses. These high correlations are likely due to the lack of snow and ice in that region and to the generally warmer and often convective nature of the precipitation there. The Hydro-Estimator also exhibits relatively strong correlation over portions of the north-central U.S., perhaps due to the inclusion of radar or model data in that technique.

During the warm season, the model performance is generally at or below that of the satellite estimates although they tend to exhibit higher correlation with the

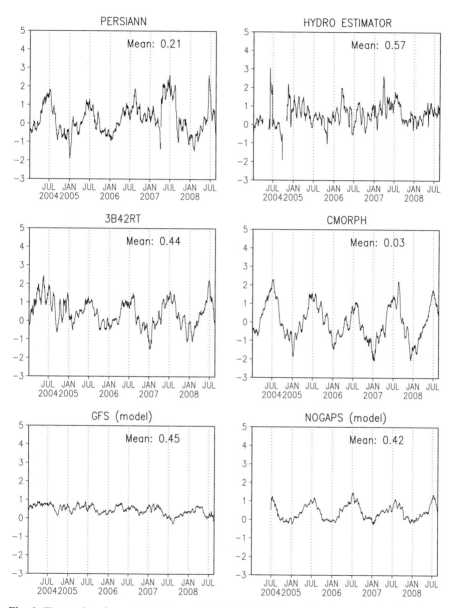

Fig. 6 Time series of mean bias between the CPC daily U.S. precipitation analyses and various satellite estimates and numerical model short-range predictions over the period January 1, 2004 to July 15, 2008. Mean bias is defined as the nationwide mean daily precipitation from a given technique or model *minus* the nationwide daily mean from the rain gauge analysis. Units are mm day^{-1}. The time series is displayed as a 21-day running mean to improve clarity

rain gauge data in the western half of the country compared to the satellite estimates. CMORPH exhibits correlation coefficients in excess of 0.70 over almost the entire eastern half of the U. S. during this season and the other satellite techniques have higher correlation with the rain gauge data over the eastern half of the nation compared to the western half.

Time series of mean bias, i.e. the difference in the daily spatial mean precipitation for the nation as a whole between the satellite/model estimates and the validating rain gauge analyses, are shown in Fig. 6. Most striking is the small bias in the model forecasts (about 0.5 mm day^{-1} in the GFS model) compared to the satellite estimates which range between (about +/− 2 mm day^{-1}) and that the bias in the models, although relatively low, is almost exclusively positive. Also interesting is the pronounced annual cycle in bias in all of the satellite estimates except the Hydro Estimator, and in the NOGAPS model forecasts but not in the GFS forecasts. That annual cycle is characterized by overestimates during the warm season and underestimates during winter. CMORPH and PERSIANN exhibit the most amplified annual cycle.

3 Sub-Daily Validation

Sapiano and Arkin (2008) performed an inter-comparison of five high resolution, satellite-derived precipitation products (CMORPH, Hydro-Estimator, NRL-Blended, PERSIANN and TMPA) as well as forecast precipitation from the NCEP GFS model. Global validation of sub-daily precipitation is not currently possible due to the lack of quality controlled, homogeneous, sub-daily in situ measurements, thus the precipitation estimates were compared with sub-daily gauge data from the Atmospheric Radiation Measurement (ARM; Ackerman and Stokes 2003) Program sites over the Southern Great Plains (SGP; over Oklahoma and Kansas). In addition, the NCEP Stage IV (Baldwin and Mitchell 1998; Lin and Mitchell, 2005) merged radar/gauge product was used for comparison.

Here, we show a similar but extended analysis featuring comparisons between three-hourly precipitation estimates from the SGP sites and CMORPH, the Hydro-Estimator, NRL-Blended, PERSIANN, TMPA (see Table 1). For each of these precipitation estimates, the four grid-points nearest to each SGP gauge were bi-linearly interpolated to create matched, three-hourly pairs. The satellite estimates were all evaluated on 0.25° grids (the finest common grid). The Stage IV data were included in the analysis using the same methodology, but incorporated at the native 4 km and 0.25° resolutions to allow for comparison of the effect of scale.

First, statistics based on the three-hourly resolution matched pairs for each SGP gauge are shown for the six-month warm and cold seasons (October–March and April–September). Boxplots are used in order to show the spread over all sites graphically. The box includes the 25th and 75th percentiles and the line in the middle of the box represents the median. The whiskers extend to the furthest outlying values that are no more than 1.5 times the inter-quartile range (difference between the 75th and 25th percentiles) away from the median. The "plus" symbols beyond the whiskers denote observations which are further than 1.5 times the inter-quartile range from the median. The crosses represent the mean of the three-hourly correlations and the circles show the mean correlation over all sites obtained using daily accumulations.

Figure 7 shows boxplots of correlations between each of the high resolution precipitation estimates and the SGP gauges as well as correlations between the gauges and the Stage IV data extracted from the 4 km and 0.25° resolutions. Of the satellite estimates, CMORPH has the highest mean three-hourly correlation (denoted by the cross) in both the warm and cold seasons with mean correlations of around 0.55. In the convective warm season, the TMPA has the next highest correlation, followed by PERSIANN and the Hydro-Estimator. In the cold season, the TMPA, PERSIANN and the Hydro-Estimator all have similar three-hourly correlations of around 0.45. NRL-Blended has consistently lower correlations in both seasons. Similar results are obtained for daily data (mean denoted by the circle) but with higher mean daily correlations of around 0.6–0.65 for CMORPH. In the warm season, TMPA, PERSIANN and the Hydro-Estimator are all perform about as well as CMORPH at the daily time-scale. As a benchmark, the 4 km Stage IV yields mean three-hourly correlations of around 0.85 and the 0.25° version yields lower correlations of around 0.7–0.75. This difference demonstrates the effect of the coarser 0.25° when compared to a point estimate. The results show that the satellite estimates have high correlations with daily and sub-daily gauge estimates, but that they are inferior to the radar/gauge estimate of Stage IV.

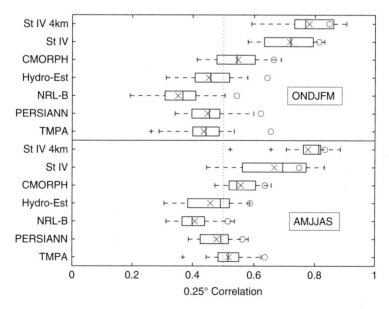

Fig. 7 Boxplots of three-hourly correlations at each SGP with the precipitation estimates extracted from 0.25° data

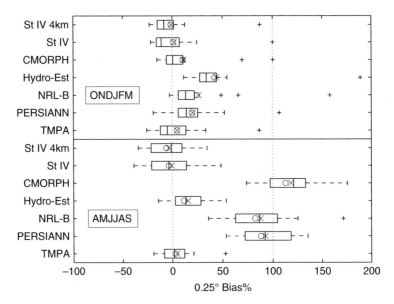

Fig. 8 Boxplots of three-hourly bias of the precipitation estimates extracted from 0.25° data, as a percentage of the mean precipitation at each SGP site

Of the datasets included in this comparison, the TMPA and the Stage IV include a gauge adjustment and are expected to a have a near zero bias compared to these gauges. Figure 8 shows the bias as a percentage of the mean for each SGP site and confirms that these datasets have near zero bias. Biases are generally lower in the cold season which is dominated by large-scale stratiform events than in the convective warm season. CMORPH, NRL-Blended and PERSIANN over-estimate precipitation by as much as 100% during the warm season. Biases are far smaller in the cold season, although there are more extreme outliers which are in excess of 100%; although the mean rainfall is lower during the cold season so these are relatively small in absolute terms. Interestingly, the Hydro-Estimator tends to more seriously overestimate cold season precipitation than warm season precipitation and has a far lower bias than the other satellite-only datasets. This dataset was designed to estimate extreme events which are more likely to occur in the warm season.

A crucially important property for high resolution products is how they represent the distribution of precipitation, particularly the upper tail of the distribution which is corresponds to the extreme wet events. Quantile–Quantile (Q–Q) plots are used here to show agreement (or lack thereof) between the Cumulative Distribution Function (CDF) of the SGP gauges and the CDF of each of the precipitation estimates. Since the number of data points is equivalent, these plots are constructed by separately ordering the estimates and the validation data into ascending values and plotting them against each other. Since an examination of the distributions at each site is not feasible, all the values have been put together and a single Q–Q plot is shown for all sites (Fig. 9). The Stage IV data tends to correctly capture the lower values of the total distribution, but underestimates values greater than 4 mm day^{-1}. Similarly, the Hydro-Estimator correctly captures most of the

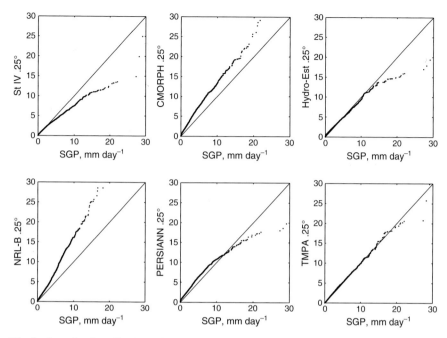

Fig. 9 Quantile–Quantile plots of each precipitation estimate against all SGP gauges for three-hourly, 0.25° data

distribution, but underestimates the most extreme values, which is surprising given that it was designed for the monitoring of extremes. CMORPH and NRL-Blended over estimate all values by 20–25%, which is consistent with their higher bias seen in Fig. 8, whereas the TMPA faithfully represents the overall distribution of the sites. These results suggest that a gauge correction applied to CMORPH and NRL-Blended would yield a similar agreement between distributions as seen for TMPA. PERSIANN overestimates values less than 10 mm day^{-1} and underestimates values greater than 10 mm day^{-1}.

These three hourly results show that CMORPH has the highest correlations but has relatively low HSS when compared to the other datasets, suggesting that it tends to overestimate the occurrence of an event, but gives more accurate estimates when it does rain. The bias tends to be larger in the convective warm season. Of the non-gauge adjusted datasets, the Hydro-Estimator has the lowest warm season bias, but the bias adjustment of the TMPA is a major advantage and such a correction should be incorporated into other datasets.

4 Evaluation of Warm Season US Precipitation Using Gauges From NAME

The North American monsoon system has substantial influences on the warm season climate over North America and it provides a useful framework for describing and diagnosing warm season climate controls and the nature of the year-to-year variability. The North American Monsoon Experiment (NAME) was an internationally coordinated process study aimed at determining the sources and limits of predictability of warm season precipitation over North America (Higgins et al. 2006). The NAME community organized a Precipitation Assessment Project (PAP) in advance of the NAME 2004 Field Campaign, as an important prerequisite to improved understanding, simulations and ultimately predictions of warm season precipitation over North America. The NAME PAP is one of several NAME activities designed to ensure maximum coordination of NAME observational, model development and prediction activities (Higgins et al. 2006). The NAME PAP includes seven different satellite estimates (though only four are examined here) and model precipitation forecasts from the NCEP GFS. These estimates are inter-compared over the NAME domain with the Climate Prediction Center (CPC) Unified rain gauge analysis (Higgins et al. 2000).

We present a direct comparison of the different precipitation estimates with NAME gauges over nine zones in southwestern North America (Fig. 10) that were invoked during the NAME 2004 field campaign. It is well known, however, that semi-arid regions such as the NAME domain pose a difficult challenge for remote estimates of precipitation, including radar. This is partly due to the fact that satellites (and radars – depending on the distance from the site) retrieve information from hydrometeors and clouds, i.e. parameters in the free atmosphere not at the surface where rain gauges measure precipitation. Several studies have documented large evaporation over semi-arid regions (e.g. Scofield 1987). Rosenfeld and Mintz (1988) estimated conservatively that 30% of the rainfall evaporates in the first 0.6 km below the cloud base in semi-arid regions at rainfall intensities as high as 80 mm h^{-1}. The inter-comparison presented here also highlights daily statistics, including frequency, intensity and variability of rainfall over the region during the period of the NAME 2004 field campaign (June–September 2004).

Each group participating in the NAME PAP agreed to inter-compare their precipitation estimates for the period June-September 2004 (hereafter JJAS 2004), which corresponds to the NAME 2004 Enhanced Observing Period (EOP). Data was made available for the NAME Tier III domain (140°W–80°W, EQ-55°N) and a subset of this is analyzed here for the US. Daily averages were used on a 0.25° grid where the averaging was done over the 24-h period from 1200 UTC on day zero to 1200 UTC on day one. The GFS forecasts used in this study are the average of the 12–24 h and 24–36 h forecasts from the 0000 UTC run of the GFS. Two PMW/IR datasets were used: CMORPH and PERSIANN (see Table 1), as were the PMW-only AMSU-B estimates of Weng et al. (2003; note that this AMSU-B data is used in CMORPH and in the training of PERSIANN) and the IR-only Hydro-Estimator.

There is a known tendency for the CPC gauge analysis to overestimate the light precipitation amounts and to underestimate the heavy amounts when compared to station data. These are well known attributes of all objective analysis schemes, though some techniques minimize these aspects better than others (e.g. Shen et al. 2001). Figure 11 shows scatter plots of the data averaged over all 9 zones. Differences between the CPC gauge analysis and the raw (quality controlled) station observations are generally quite small with a very high correlation (0.98)

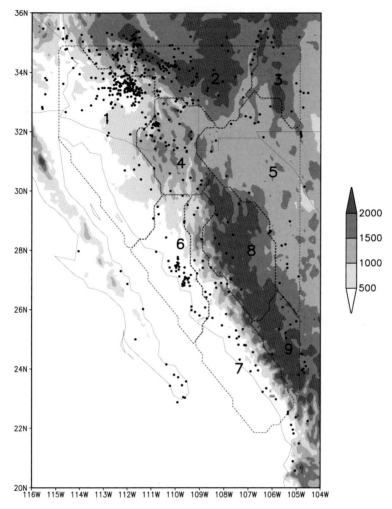

Fig. 10 A typical station distribution over the NAME domain for the CPC gridded gauge analysis and topography, shaded at intervals of 500 m. Also shown are the NAME zones used in this section

between the gauges and the CPC gauge analysis. The GFS forecasts are also in good agreement with the gauges with a relatively high correlation (0.81). All of the satellite products tend to overestimate precipitation amounts when the large area is considered as a whole. Correlations vary, with the AMSU data being the highest and PERSIANN being the lowest which shows the advantage of PMW estimates.

Some statistics computed for the region shown in Fig. 12 (125°W–95°W, 20°N–35°N) are given in Table 2. They were obtained after computing the mean daily precipitation for the period JJAS 2004. All of the products have mean daily precipitation (based on the period JJAS 04) exceeding 0.5 mm day^{-1} at most locations which is consistent with the influence of the objective analysis, which tends to

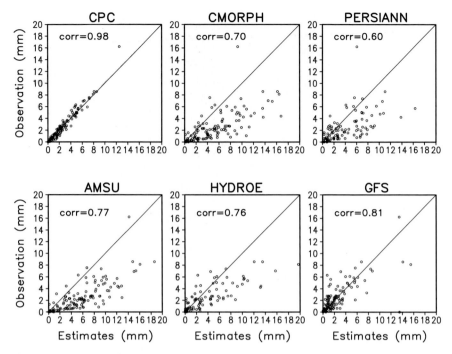

Fig. 11 Comparisons of daily precipitation amounts (mm) obtained by averaging over all 9 zones for the period JJAS 2004. Each panel shows an individual precipitation estimate product versus station observations, with each open circle representing one day during the period

spatially spread or extend light precipitation. There is a large difference in the area average of JJAS 2004 mean daily precipitation with the satellite estimates being generally higher (roughly 4–5 mm day^{-1}) than the gauge analysis and model forecasts (roughly 3 mm day^{-1}). In addition, the maximum values of the JJAS 2004 mean daily precipitation and the standard deviation (computed as the departure of the mean daily precipitation at each grid point from the area average mean daily precipitation) are also greater for most of the satellite estimates. Overall, these results are consistent with the results shown in Fig. 12, that indicate the tendency for the satellite estimates to overestimate precipitation during the NAME 2004 field campaign period. Spatial correlations of JJAS 2004 mean daily precipitation from the CPC gauge analysis with the satellite estimates (not shown) are generally quite good.

Higgins et al. (2006) identified two NAME sub-regions which are of particular interest for the North American monsoon: the core of the monsoon (CORE; 112°W–106°W, 24°N–30°N) and an area over Arizona and New Mexico (AZNM; 112.5°W–107.5°W, 32°N–36°N). Time series of daily precipitation amounts for the North American monsoon CORE and AZNM sub-regions show reasonable agreement amongst the estimates (Fig. 13). However, in the CORE region the satellite precipitation amounts almost always exceed the CPC and GFS amounts during the period shown. This is true whether the daily precipitation amounts are large or

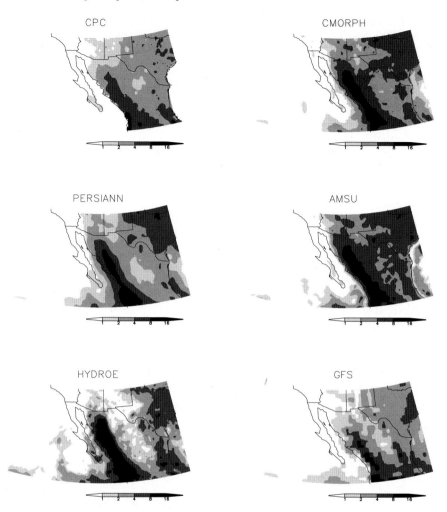

Fig. 12 Mean daily precipitation (mm) for JJAS 2004

small. In AZNM the satellite estimates exceed the CPC and CFS estimates when the precipitation amounts are large and compare more favorably when they are small. These results are consistent with those reported in Janowiak et al. (2004), who found that remotely-sensed precipitation estimates (including radar) exhibit a substantial positive bias compared to rain gauge estimates over semi-arid regions during the warm season due to evaporation of rain before it reaches the surface. Also shown in Fig. 13b, c are the mean daily precipitation amounts from the CPC (solid lines), GFS (dotted lines) and the average of the satellite estimates (dashed lines) for July–August 2004, for the corresponding sub-regions. The values are listed Table 3. In both sub-regions, the mean daily precipitation amounts from the GFS are very close to that from CPC's gauge analyses, while the averages of the satellite estimates at least double the amounts from either CPC or the GFS.

Table 2 Some statistics for the domain shown on Fig. 12 [125°W–95°W, 20°N–35°N] over land only for JJAS 2004. The total number of points in the domain is 3962 for all of the products

	Number of points with mean daily precipitation (> 0.5 mm)	Area average mean daily precipitation (mm)	Max. precipitation (mm)	Standard deviation (mm)	Spatial correlation with CPC-GAUGE
CPC-GAUGE	3513	3.03	11.15	1.57	NA
CMORPH	3604	4.78	17.59	2.64	0.79
PERSIANN	3631	3.96	13.93	2.33	0.77
AMSU	3602	5.29	16.98	2.71	0.79
Hydro-estimator	3407	3.99	28.73	3.49	0.70
GFS	3480	3.07	19.49	2.24	0.73

It is important to emphasize that satellite estimates do not always overestimate precipitation. In fact, during the cool season the satellite estimates tend to be lower than the CPC and GFS estimates (including the semi-arid CORE and AZNM regions). In addition, numerical model forecasts generally outperform the satellite estimates when validated against the CPC estimates during the cool season (Janowiak et al. 2004).

The NAME 2004 field campaign featured 10 Intensive Observing Periods (IOPs) embedded within the longer Enhanced Observing Period (EOP) which occurred during JJAS 2004. NAME IOP 2 focused on the influence of a tropical cyclone (Tropical Storm Blas) on a Gulf of California moisture surge event and the associated precipitation patterns. Johnson et al. (2004) documented the surge characteristics associated with this event (including the increases in rainfall over northwestern Mexico and southeastern Arizona from gauge precipitation analyses). In a more general context, Higgins and Shi (2005) documented relationships between tropical cyclones, moisture surges and the associated precipitation patterns using composites and historical data for the period 1979–2001.

The purpose of this case study is to examine the evolution of the observed precipitation associated with NAME IOP-2 for the various precipitation estimates. In particular, time-latitude diagrams are used to illustrate the northward progression of the precipitation associated with this event (Fig. 14). The northward evolution of the precipitation pattern is well captured in all of the estimates. The CPC and GFS estimates are considerably lighter throughout the event. The satellite estimates also show heavy precipitation in advance of the surge. This is likely due to the presence

Fig. 13 (**a**) CORE (112°W–106°W, 24°N–30°N) and AZNM (112.5°W–107.5°W, 32°N–36°N) subregions. Time series of daily precipitation (mm) for (**b**) CORE and (**c**) AZNM subregions for July–August 2004. The CPC analysis and GFS day 1 forecasts are indicated by the *solid and dotted lines*, respectively. The average of the satellite estimates is indicated by the dashed line. The shading surrounding the dashed line shows the spread in the satellite estimates

Evaluation Through Independent Precipitation Measurements: USA

(a)
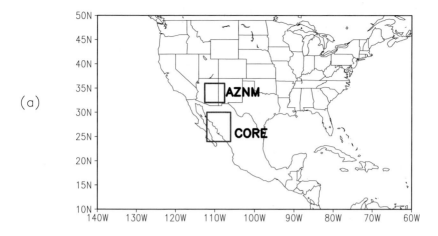

(b)
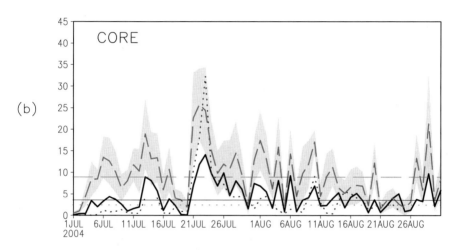

(c)
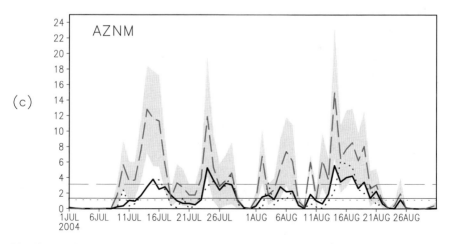

Fig. 13 (continued)

Table 3 Mean daily precipitation (mm day^{-1}) from the CPC, GFS and the average of the satellite estimates for July–August 2004 for the two sub-regions

	CPC	GFS	Mean-satellite
CORE	4.23	3.56	8.76
AZNM	1.43	1.41	3.27

of high cloudiness in advance of Tropical Storm Blas, which could contribute to overestimated precipitation in the IR satellite estimates.

Generally, the satellite, gauge and model precipitation estimates agree quite well with each other in terms of frequency, intensity and variability of precipitation in this region. The most substantial differences are found over the high elevation areas. Among these products, CPC's gauge-only gridded analysis is closest to the gauge observations themselves, both at individual gauge locations and for larger areas (including the NAME zones and NAME Tier 1). As for individual zones, the CPC analysis performs best in zones 4 and 7, where the stations are more uniformly distributed compared to the other regions. However, it underestimates the heavier precipitation amounts in several of the eastern zones (e.g. zones 5, 8 and 9) where the station density is particularly low. It is in these zones that the satellite estimates are particularly useful.

5 Discussion

Like other studies, our results show that high resolution satellite precipitation estimates are well-correlated with validation data over the US at daily and sub-daily time-scales at the 0.25° resolution. The diurnal cycle is generally well represented, although the three-hourly time resolution used is too coarse to resolve the 1–2 h lags that might be expected from PMW estimates. Like many other studies (e.g. Ebert et al. 1996; Arkin and Xie 1994; Ebert and Manton 1998), our results show that PMW estimates are superior to IR estimates, but that blends of PMW and IR information yield the highest skill. It is clear that such blends should be used in studies where precipitation data is required, and that gauge adjustment of blended PMW/IR data is an important step that removes much of the bias.

Many activities have largely focused on the 0.25° resolution which is common to most of the estimates currently used. Far higher resolutions are obviously desirable for hydrological purposes, but limits of the current sensors (most notably the footprint size of the PMW estimates) somewhat preclude this. Despite this limitation, many of the datasets have higher resolution estimates available (down to half-hourly at 4–8 km) and the validation community should make greater attempts to conduct inter-comparisons at higher resolutions. At the same time, users should take caution when using higher resolution estimates since many satellite estimates are very noisy on fine scales and spatial and/or temporal averaging is a necessary step to obtain

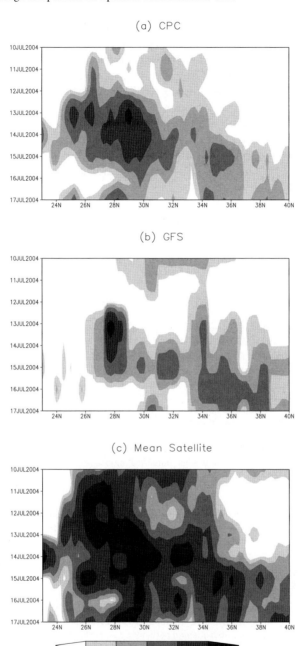

Fig. 14 Time-latitude cross section of daily precipitation (mm) from (**a**) CPC, (**b**) GFS and (**c**) satellite during NAME IOP-2. Precipitation data is averaged over the longitude band 112.5° W–107.5°W. The mean of the satellite estimates is shown in (**c**)

useful estimates. In this regard, we advise the user to exercise extreme caution in the interpretation of finer scale precipitation estimates.

Precipitation forecasts from numerical weather prediction models do well, although this skill might not be expected to be repeated elsewhere over the globe. In particular, we found higher skill in the cold season when synoptic-scale precipitation is more common than in the convective warm season. The GFS model shows very good skill over the US and tends to outperform the satellite estimates during the cold season. This result suggests that a combination of satellites and models, with the latter used during the cold season, would increase the skill of precipitation estimates. Such combinations are now becoming available (Sapiano et al. 2008) and will provide a valuable resource for many users, including hydrologists.

References

Ackerman, T. and G. M. Stokes (2003) The atmospheric radiation measurement program. Phys Today, 56: 38–45.
Arkin, P. A. and P. P. Xie (1994) The Global Precipitation Climatology Project: First Algorithm Intercomparison Project. Bull Amer Meteor Soc, 75: 401–419.
Baldwin, M. E. and K. E. Mitchell (1998) Progress on the NCEP hourly multi-sensor U.S. precipitation analysis for operations and GCIP research. Preprints, 2nd Symposium on Integrated Observing Systems, 78th AMS Annual Meeting 10–11.
Curtis, S., T. W. Crawford, and S. A. Lecce (2007) A comparison of TRMM to other basin-scale estimates of rainfall during the 1999 Hurricane Floyd flood. Natural Hazards, 43: 187–198.
Ebert, E. E. and M. J. Manton (1998) Performance of satellite rainfall estimation algorithms during TOGA COARE. J Atmos Sci, 55: 1537–1557.
Ebert, E. E., M. J. Manton, P. A. Arkin, J. R. Allam, C. E. Holpin, and A. Gruber (1996) Results from the GPCP Algorithm Intercomparison Programme. Bull Amer Meteor Soc, 77: 2875–2887.
Ebert, E. E., J. E. Janowiak, and C. Kidd (2007) Comparison of near-real-time precipitation estimates from satellite observations and numerical models. Bull Amer Meteor Soc, 88: 47–64.
Gottschalck, J., J. Meng, M. Rodell, and P. Houser (2005) Analysis of multiple precipitation products and preliminary assessment of their impact on global land data assimilation system land surface states. J Hydrometeor, 6: 573–598.
Higgins, R. W. and W. Shi (2005) Relationships between Gulf of California moisture surges and tropical cyclones in the Eastern Pacific basin. J Climate, 18: 4601–4620.
Higgins, R. W., W. Shi, E. Yarosh, and R. Joyce (2000) Improved United States Precipitation Quality Control System and Analysis. NCEP/Climate Prediction Center ATLAS No. 7
Higgins, R. W., et al (2006) The North American Monsoon Experiment (NAME) 2004 Field Campaign and Modeling Strategy. Bull Amer Meteor Soc, 87: 79–94.
Hogan, T. and T. Rosmond (1991) The description of the Navy Operational Global Atmospheric Prediction System's spectral forecast model. Mon Wea Rev, 119: 786–1815.
Hsu, K., X. Gao, S. Sorooshian, and H. V. Gupta (1997) Precipitation estimation from remotely sensed information using artificial neural networks. J App Meteorol, 36: 1176–1190.
Huffman, G. J., R. F. Adler, D. T. Bolvin, G. Gu, E. J. Nelkin, K. P. Bowman, Y. Hong, E. F. Stocker, and D. B. Wolff (2007) The TRMM multisatellite precipitation analysis (TMPA): Quasi-global, multiyear, combined-sensor precipitation estimates at fine scales. J Hydromet, 8: 38–55.
Janowiak, J. E., P. P. Xie, R. J. Joyce, M. Chen, and Y. Yarosh (2004) Validation of daily satellite precipitation estimates over the U.S. Proceedings of the 29th Climate Diagnostics and Prediction Workshop, Madison WI
Johnson, R., P. Ciesielski, and P. Rogers (2004) Preliminary results of the NCAR ISS deployment in NAME. Proceedings of the 29th Climate Diagnostics and Prediction Workshop, Madison WI

Joyce, R. J., J. E. Janowiak, P. A. Arkin, and P. Xie (2004) CMORPH: A method that produces global precipitation estimates from passive microwave and infrared data at high spatial and temporal resolution. J Hydrometeor, 5: 487–503.

Kalnay, E., M. Kanamitsu, and W. E. Baker (1990) Global numerical weather prediction at the National Meteorological Center. Bull Amer Meteor Soc, 71: 1410–1428.

Lin, Y. and K. E. Mitchell (2005) The NCEP Stage II/IV hourly precipitation analyses: development and applications. Preprints 19th Conf on Hydrology, San Diego CA, Amer Meteor Soc P1.2

Rosenfeld, D. and Y. Mintz (1988) Evaporation of rain falling from convective clouds as derived from radar measurements. J Appl Meteor, 27: 209–215.

Ruane, A. C. and J. O. Roads (2007) 6-hour to 1-year variance of five global precipitation sets. Earth Interactions, 11: 1.

Sapiano, M. R. P. and P. A. Arkin (2008) An inter-comparison and validation of high resolution satellite precipitation estimates with three-hourly gauge data. J Hydrometeor, 10: 149–166.

Sapiano, M. R. P., T. M. Smith, and P. A. Arkin (2008) A New Merged analysis of precipitation utilizing satellite and reanalysis data. J Geophys Res, 113:D22103. doi:10.1029/2008JD010310.

Shen, S. S. P., P. Dzikowski, G. Li, and D. Griffith (2001) Interpolation of 1961–1997 daily temperature and precipitation data onto Alberta Polygons of Ecodistrict and Soil Landscapes of Canada. J App Meteor, 40: 2162–2177.

Scofield, R. A. (1987) The NESDIS operational convective precipitation estimation technique. Mon Wea Rev, 115: 1773–1792.

Scofield, R. A. and R. J. Kuligowski (2003) Status and outlook of operational satellite precipitation algorithms for extreme-precipitation events. Weather Forecast, 18: 1037–1051.

Sorooshian, S., K. Hsu, X. Gao, H. V. Gupta, B. Imam, and D. Braithwaite (2000) Evaluation of PERSIANN system satellite-based estimates of tropical rainfall. Bull Am Meteor Soc, 81: 2035–2046.

Tian, Y., C. D. Peters-Lidard, B. J. Choudhury, and M. Garcia (2007) Multitemporal analysis of TRMM-based satellite precipitation products for land data assimilation applications. J Hydrometeor, 8: 1165–1183.

Turk, F. J. and S. D. Miller (2005) Toward improved characterization of remotely sensed precipitation regimes with MODIS/AMSR-E blended data techniques. IEEE Trans Geosci Rem Sens, 43: 1059–1069.

Villarini, G. and W. F. Krajewski (2007) Evaluation of the research version TMPA three-hourly $0.25° \times 0.25°$ rainfall estimates over Oklahoma. Geophys Res Lett, doi:10.1029/2006GL029147.

Weng, F., L. Zhao, R. R. Ferraro, G. Poe, X. Li, and N. C. Grody (2003) Advanced microwave sounding unit cloud and precipitation algorithms. Radio Sci, 38: 8068.

Comparison of CMORPH and TRMM-3B42 over Mountainous Regions of Africa and South America

Tufa Dinku, Stephen J. Connor, and Pietro Ceccato

Abstract Two satellite rainfall estimation algorithms, CMORPH and TMPA, are evaluated over two mountainous regions at daily accumulation and spatial resolution 0.25°. The evaluated TMPA products are TRMM-3B42 and TRMM-3B42RT. The first of the two validations region is located over the Ethiopian highlands in the Horn of Africa. The second is located over the highlands of Columbia in South America. Both sites are characterized by a very complex terrain. Relatively dense station networks over the two sites are used to validate the satellite products. The correlation coefficients between the reference gauge data and the satellite products were found to be low. Besides, the products underestimate both the occurrence and amount of rainfall over both validation sites. These were attributed, at least partly, to orographic warm rain process over the two regions. The performance over Colombia was better compared to that for Ethiopia. And CMORPH has exhibited better performance as compared to the two TRMM products.

Keywords Satellite · Rainfall estimation · Passive microwave · Infrared · Validation

1 Introduction

Thermal infrared (TIR) and passive microwave (PM) sensors are the most widely used for satellite rainfall estimation. TIR provides useful information mainly about storm clouds based on the low temperatures of the top of these clouds. PM sensors are more attractive mainly because their measurements have better physical relationships to rainfall as compared to TIR observation. Thus, PM observations

T. Dinku (✉)
International Research Institute for Climate and Society (IRI), The Earth Institute at Columbia University, Palisades, NY 10964, USA
e-mail: tufa@iri.columbia.edu

contain information about rain areas rather than clouds. The problem is that PM sensors are not yet available on geosynchronous satellites, and this makes the observation frequency just a couple of times per day. As a result, most current rainfall estimation techniques are based on TIR observations, or combination of TIR and PM observations. Both TIR and PM rainfall retrievals have some serious limitations. The limitations for TIR include inability to provide information beneath the top of the cloud, underestimation of rainfall from warm clouds particularly over coastal and mountainous regions, and the cold cirrus clouds that could be confused with deep convective. The main limitations of PM rain retrieval are background emission from the land surface which varies significantly depending on vegetation type and soil characteristics, stratiform and warm rain that does not produce much ice aloft, low repetition rate (typically twice per day) that makes aggregation over daily time periods impossible, and coarser spatial resolution compared to TIR sensors. For more detail on the different aspects of satellite precipitation estimation the reader may refer to, among many others, Levizzani et al., 2002, 2007; and Gruber and Levizzani, 2008.

Mountainous regions pose unique challenges to satellite rainfall retrieval from both TIR and PM observations. The challenge to TIR rainfall retrieval comes mainly from warm orographic rains. Most TIR algorithms use cloud-top temperature thresholds to discriminate between raining and non-raining clouds. And these thresholds are usually too low for the relatively warm orographic clouds, resulting in underestimation. The rainfall signal for over-land PM rainfall retrieval comes mainly from ice scattering aloft. However, orographic clouds may produce heavy rainfall without much ice aloft, and this may result in underestimation of surface rainfall. The other challenge to PM rain retrieval algorithms comes from cold surfaces and ice cover over mountain-tops. These cold surfaces could be misidentified as raining clouds.

Validation of satellite rainfall products over mountainous region will offer an insight into how the different algorithms perform over such regions. However, there has not been much validation, particularly at daily time scale, over mountainous regions. The main problem is lack of raingauge observations over mountainous regions, or lack of access to available observations. Here we take advantage of the availability of raingauge observation over two mountainous regions to compare the performances two satellite rainfall algorithms over these regions. A relatively dense station network over Ethiopia (Africa) and Colombia (South America) are used to validate the products of two algorithms. The first algorithm is the NOAA-CPC morphing technique (CMORPH, Joyce et al., 2004). The second is produced by the Tropical Rainfall Measuring Mission (TRMM) and is named "TRMM Multi-satellite Precipitation Analysis" product (TMPA, Huffman et al., 2007). One of these products is named TRMM-3B42. These algorithms are selected for comparison because (i) they are currently the state of the art algorithms; (ii) have been available for some time and are being used widely; and (iii) are available over both Africa and South America.

2 Study Regions and Data

2.1 Study Region

Two validation sites were selected to explore the performance of the two rainfall retrieval algorithms over mountainous regions. The first site is located over Ethiopia in the Horn Africa (Fig. 1), while the second site is part of Colombia in South America (Fig. 3). The rectangular boxes over the validation regions represent the specific validation sites. These sites are selected mainly because of three factors: (i) both have a very complex orography; (ii) they are located within the tropics; and (iii) there is good quality data from a relatively dense station network.

The site over Ethiopian is characterized by the Rift Valley, which is part of East African Rift Valley, and plateaus and mountain ranges on either side of the Rift Valley (Fig. 1). The main rainy season is during June to September (Fig. 2). Figure 2 represent the area average of the box outlined in the figure, but there are some variations within the box. The area-averaged monthly mean is calculated from gridded data for the years 1971–2000. The gridded data is produced by the Global Precipitation Climatology Center (GPCC) at a spatial resolution of $0.5°$ (Schneider et al., 2008). The full data product, which uses relatively more gauges than the other GPCC products, is used here. The main synoptic feature during the main rainy season over Ethiopia is the Inter Tropical Convergence Zone (ITCZ). However, the effect of ITCZ is modulated by the topography.

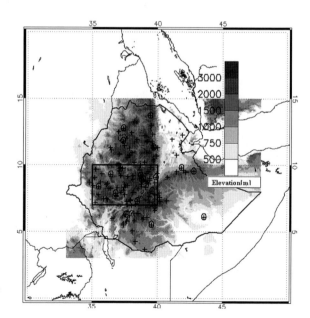

Fig. 1 Topography and raingauge distribution (+) over Ethiopia. *Circles* (O) indicate stations whose data is available through GTS. Only stations in the specified box are used for evaluation of the satellite products

Fig. 2 Mean (1971–2000) monthly rainfall for the validation site over Ethiopia

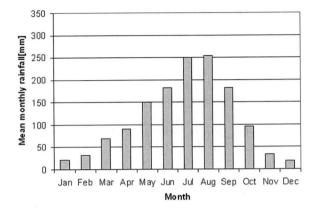

Fig. 3 Topography and distribution of raingauge stations (+) used for the validation over Colombia. *Circles* (O) indicate stations whose data is available through GTS. Only stations in the specified box are used for evaluation of the satellite products

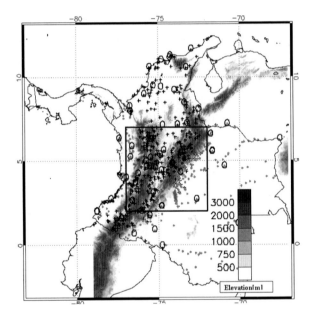

The main topographic features for the Colombian site are the three mountain ranges divided by two valleys (Fig. 3), which is the northern tip of the Andes Mountains. As opposed to the site over Ethiopia, the two valleys in Colombia are deeper (<500 m above mean sea level), and the valley on the eastern side is very prominent. Both valleys are associated with major rivers. This site has rainfall throughout the year with two relative peaks in May and October (Fig. 4). Figure 4 represents an area average over the specified box, and it is also computed from GPCC full data product (Schneider et al., 2008). Columbia has one of the wettest regions in the world with mean annual rainfall exceeding 11 000 mm over the south-western part of the country (Hurtado, 2005). The ITCZ, which stays over this region

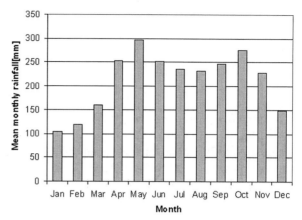

Fig. 4 Mean (1971–2000) monthly rainfall for the validation site over Colombia

throughout the year, is the main synoptic feature for Colombia. Topography plays a significant role in modifying the effect of ITCZ.

There are some significant differences between the two validation sites. The first difference is that the site over Colombia is very close to the Pacific Ocean, while the site over Ethiopia is relatively far away from any ocean. The other feature is that the ITCZ remains in the vicinity of Colombia throughout the year, while its stay over Ethiopia is limited to the summer months. Due to these factors the Colombian site has heaver rainfall than the site over Ethiopia. The rainy season is also longer over Colombia with two rainfall peaks in May and October (Fig. 4). The peak rainfall months for the site over Ethiopia are July and August (Fig. 2). The other difference, which might affect the validation results, is that the raingauge station network is denser over Colombia.

2.2 Gauge Data

Daily rain gauge data for about 145 stations was obtained from the National Meteorology Agency (NMA) of Ethiopia. After quality control 137 stations were retained, and 75 of these are located within the validation box in Fig. 1. Only data for the main rainy season (June to September) from 2003 and 2004 are used for the current work. The data for Columbia was provided by Columbian Meteorological Agency (IDEAM). Daily data from about 400 stations from April to June of 2003 to 2005 was used. From the 400 stations 246 fall within the validation box in Fig. 3. The distribution of the stations particularly over the mountain ranges (Fig. 3) is very impressive. Both Figs. 1 and 3 also show the distribution of raingauges whose data may be available through the Global Telecommunication System (GTS). Data from these stations may have been used for adjusting some of the satellite products evaluated here. Thus, these stations were excluded from the analysis. The rest of the data from both sites were subjected to rigorous quality check. The quality-controlled gauge measurements were then interpolated into regular grids, which are then averaged at 0.25° spatial resolution for comparison with the satellite products. All available stations, except the GTS gauges, were used for interpolation; however, only 0.25° grid boxes falling within the validation boxes and with at least one gauge were used for comparison with the satellite products.

2.3 Satellite Data

Three satellite rainfall products are compared. The first product is form the Climate Prediction Center (CPC) at the National Oceanic and Atmospheric Administrations (NOAA) named NOAA-CPC morphing technique (CMORPH, Joyce et al., 2004). The other two products are from the TRMM project at the National Aeronautics and Space Administration (NASA). The TRMM products compared here are TRMM-3B42 and its real time version TRMM-3B42RT (Huffman et al., 2007). Table 1 gives a summary of the main characteristics of the three satellite rainfall products.

Table 1 Summary of the satellite products evaluated here; the PM and Gauge columns indicate whether the product includes passive microwave or gauge observations

	Time res.	Space res.	Existence	PM	Gauge
TRMM-3B42RT	3-hourly	0.25°	2002-present	Y	N
TRMM-3B42	3-hourly	0.25°	1998-present	Y	Y
CMORPH	3-hourly	0.25°	2003-pressent	Y	N

CMORPH is a relatively new algorithm, which combines different PM rainfall estimates with information derived from TIR observations. As such CMORPH is not a rainfall estimation algorithm. It is a technique where by PM rainfall estimates produced by different algorithms from different sensors are propagated in space and time using motion vectors derived from half-hourly TIR observations (Joyce et al., 2004). The algorithm starts with detecting the time sequence of feature motions from the TIR data, and then uses this information to compute the displacement vector for morphing from one instantaneous PM estimate to the next. In additions, a time-weighted interpolation is performed to modify the shape and intensity of precipitation features during the time between PM observations. This creates spatially and temporally complete PM estimates, which are independent of the TIR rainfall estimates. This way, CMORPH combines the superior retrieval accuracy of PM estimates and the higher temporal and special resolution of TIR data.

The TRMM-3B42 algorithm (Huffman et al., 2007) combines TIR data from geostationary satellites and PM data from different sources: the TRMM microwave imager (TMI), Special Sensor Microwave Imager (SSM/I), Advanced Microwave Sounding Unit (AMSU), and Advanced Microwave Sounding Radiometer-Earth Observing System (AMSR-E). The TRMM-3B42 estimates are produced in four steps: (i) the PM estimates are adjusted and combined, (ii) TIR precipitation estimates are created using the PM estimates for calibration, (iii) PM and TIR estimates are combined, and (iv) the data is rescaled to monthly totals where by gauge observations are used indirectly to adjust the satellite product. TRMM-3B42 is available a couple of days after the end of each month. The real time version (TRMM-3B42RT) is a product from the third step above. Thus it does not use gauge adjustment. TRMM-3B42RT is available with a lag time of few hours after the TIR and PM inputs are obtained. The PM estimates used in step (i) above are adjusted using probability matching method to a "best" estimate. In the case of 3B42, the best estimates used for adjustment is the TRMM combined instrument, which is based on TMI and the TRMM precipitation radar (PR). For 3B42RT, the best estimate is just TMI. The method used to combine the PM and TIR estimates, in step (iii) above, is a very simple one: the PM estimates are used as is and pixels that do not have PM observations are simply replaced by TIR estimates. This may result in a heterogeneous time series for a given location.

3 Comparison of the Satellite Rainfall Products

Standard validation statistics are used to compare the performance of the two algorithms over the two mountainous regions. These include linear correlation coefficient (CC), multiplicative bias (Bias), mean absolute error (MAE), Frequency Bias (FBS), probability of detection (POD), false alarm ratio (FAR), critical success index (CSI), and the Heidke Skill Score (HSS). Correlation coefficient and MAE represent pixel-by-pixel comparison, while FBS, POD, FAR, CSI, and HSS are categorical validation statistics. The pixel-by-pixel comparison statistics are used to evaluate the performance of the satellite products in estimating the amount of the rainfall, while the categorical statistics are used to assess rain detection capabilities. The expressions for these error statistics are given below.

$$MAE = \frac{\frac{1}{N}\sum |(S-G)|}{\overline{G}} \quad (1)$$

$$Bias = \frac{\sum S}{\sum G} \quad (2)$$

Where: G = gauge rainfall measurements, $\overline{G}$ = average of the gauge measurements, S = satellite rainfall estimate, and N=number of data pairs. The following validation statistics are based on contingency Table 2, where A, B, C and D represent hits, false alarms, misses, and correct negatives, respectively.

$$FBS = \frac{A+B}{A+C} \quad (3)$$

$$POD = \frac{A}{A+C} \quad (4)$$

Table 2 Contingency table comparing gauge area-averages and satellite rainfall estimates. The rainfall threshold used is 1.0 mm

	Gauge ≥ Threshold	Gauge < Threshold
Satellite ≥ Threshold	A	B
Satellite < Threshold	C	D

$$FAR = \frac{B}{A+B} \quad (5)$$

$$CSI = \frac{A}{A+B+C} \quad (6)$$

$$HSS = \frac{2 \cdot (A \cdot D - B \cdot C)}{(A+C) \cdot (C+D) + (A+B) \cdot (B+D)} \quad (7)$$

The validation statistics is presented in Table 3a for Ethiopia and Table 3b for Colombia. The first impression from these statistics is that the correlation coefficients between the satellite products and the reference raingauge data are very low, particularly over Ethiopia. This shows that the satellite products may not reliably give the amount of daily rainfall. There is also an overall underestimation of both the amount rainfall (Bias < 1) and the frequency of rainfall occurrences (FBS < 1) over Ethiopia. For Colombia, all the three products underestimate the occurrence of rainfall, but only TRMM-3B42 seems to underestimate the amount. The underestimation of the frequency of occurrence is also reflected in the POD for both regions, which is low particularly for the TRMM products. However, the false alarm ratio is small over both validation sites showing that the main problem is under detection of the occurrence of the rainfall. The HSS, which measures the accuracy of the estimates accounting for matches due to random chance, is lower than what has been obtained for a plain region over Zimbabwe (Dinku et al., 2008).

The overall performance of the products is better over Colombia with higher correlation coefficients, POD, CSI and HSS. Correlation coefficients and HSS values are particularly higher for Colombia compared to the values obtained for Ethiopia. One source of this difference could be the denser station network over Columbia.

The bold values in Table 3a and b show a relatively better statistics for the specific product. Comparing the three products, CMORPH has an overall better statistics over both validation sites. Particularly POD, CSI and HSS values for CMORPH are better over both validation regions compared to the values for the TRMM products. TRMM-3B42 uses an indirect gauge adjustment as described earlier. Of course the number of GTS gauges, which are used in the adjustment, from these mountainous regions is very limited (see Figs. 1 and 3). But the current version of CMORPH does not employ any gauge adjustment scheme. Thus, the better performance of CMORPH could be more significant than what could be inferred from the results in

Table 3 Comparison of error statistics of the three satellite rainfall products over Ethiopia (**a**) and Columbia (**b**). The bold values highlight a relatively better value for the specific product. N is the number of data values used to compute the statistics. The threshold used for rain detection is 1 mm

(a) N=161348	3B42RT	3B42	CMORPH
CC	0.46	0.57	**0.60**
Bias	1.13	0.86	**1.11**
MAE	0.99	**0.73**	0.81
FBS	0.86	0.88	**0.95**
POD	0.73	0.76	**0.81**
FAR	0.14	0.13	0.14
CSI	0.65	0.68	**0.72**
HSS	0.40	0.45	**0.48**
(b) N = 15006	3B42RT	3B42	CMORPH
CC	0.37	**0.39**	0.32
Bias	0.83	0.84	**0.91**
MAE	0.93	**0.84**	0.86
FBS	0.88	0.93	**0.97**
POD	0.60	0.69	**0.81**
FAR	**0.11**	**0.11**	0.14
CSI	0.56	0.64	**0.72**
HSS	0.24	0.30	**0.33**

Table 3a. On the other hand, the gauge input in TRMM-3B42 has resulted in a slight improvement over TRMM-3B42RT.

As discussed above, the major problems with satellite rainfall estimates over the two mountainous regions are poor correlations coefficients and underestimation of rainfall occurrences and amount. This is attributed mainly to the complex orography of the regions and the associated warm rain process. Dinku et al. (2007) has shown the effect of orography on satellite rainfall estimates over Ethiopia by comparing two versions of NOAA-CPC Africa Rainfall Estimation (RFE). The previous version, V.1.0 (RFE1, Herman et al., 1997), was operational from 1995 to 2000. The current version, V.2.0 (RFE2, Xie et al., 2002), has been in operation since 2001. The RFE2 algorithm produces daily rainfall estimates at a spatial resolution of 0.1° by combining PM retrieval, TIR estimates and raingauge data obtained through GTS. RFE1 has a provision for taking into account the orographic warm rain process, which is absent from the current version. As a result, the previous version was found to perform better than the current version although the previous version did not use PM inputs. The performance of the different satellite rainfall products over Ethiopia was also compared with results for a relatively flat region over Zimbabwe (Dinku et al., 2008). The performance of the products was much better over Zimbabwe.

The above results suggest that topography must be taken into account to improve rainfall estimation over mountainous regions. One way of doing this is what Herman et al. (1997) implemented in RFE1, which combines relative humidity,

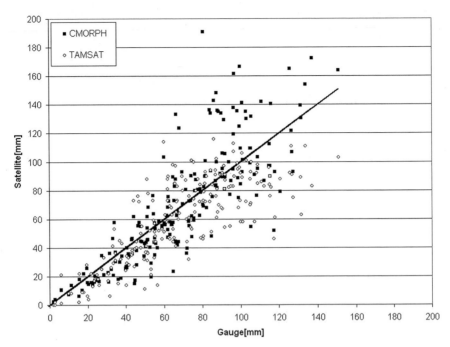

Fig. 5 Comparison of ten-day total satellite rainfall estimates over Ethiopia by the CMORPH and TAMSAT algorithms

wind direction and the terrain slope to estimate rainfall over regions where cloud-top temperatures are between 275 and 235K. The other approach could be local calibration using available raingauge data; of course this is possible only for areas where raingauge observation is available. A good example that demonstrates the advantage of local calibration is the TAMSAT (Tropical Applications of Meteorology using Satellite data) algorithm. This algorithm uses cold-cloud-duration (CCD), which is the length of time that a satellite pixel is colder than a given temperature threshold, then assumes a linear relationship between CCD and rain rate at the surface(Grimes et al., 1999; Thorne et al., 2001). The main advantage of the TAMSAT algorithm is that the temperature threshold and the parameters of the linear regression are determined through calibration using locally available gauges. As a result, the threshold temperature and regressions parameters will vary over different regions, and this may take the topographic effect into account. Figure 5 compares the performance of the TAMSAT algorithm with that of CMORPH over Ethiopia at ten-day accumulations. There is a big difference in the level of sophistication between the two algorithms; yet, TAMSAT performs as well as CMORPH. And this is attributed to the specific calibration of the TAMSAT algorithm over Ethiopia.

4 Conclusion

Two satellite rainfall algorithms, CMORPH and TMPA, have been evaluated over two mountainous regions in Africa and South America. The objective of this work has been to explore how the two algorithms would perform over the two validation sites, which are characterized by a very complex topography. The products had low correlation, and underestimated both the occurrence and the amount of surface rainfall over both validation regions. The performance of CMORPH was slightly better than that of the TRMM products over both validation sites. The better performance of CMORPH is significant when one considers the fact that TRMM-3B42 includes gauge adjustment, and CMORPH does not. The overall performance of both satellite products is better over the South American site.

Acknowledgements We are very grateful to the National Meteorological Agency (NMA) of Ethiopia, and the Meteorological Service of Colombia (IDEAM) for providing the rain gauge data free of any charge.

References

Dinku, T., P. Ceccato, E. Grover-Kopec, M. Lemma, S. J. Connor, and C. F. Ropelewski, 2007: Validation of satellite rainfall products over East Africa's complex topography. *International Journal of Remote Sensing*, **28**(7), 1503–1526.

Dinku, T., S. Chidzambwa, P. Ceccato, S. J. Connor, and C. F. Ropelewski, 2008: Validation of high-resolution satellite rainfall products over complex terrain in Africa. *International Journal of Remote Sensing*, **29**(14), 4097–4110.

Grimes, D. I. F., E. Pardo, and R. Bonifacio, 1999: Optimal areal rainfall estimation using raingauges and satellite date. *Journal of Hydrology*, **222**, 93–108.

Gruber, A. and V. Levizzani, 2008: Assessment of Global Precipitation Products, WMO/TD-NO. 1430, (Available online at: http://www.isao.bo.cnr.it/~meteosat/papers/AssessmentGlobalPrecipitation-2008.pdf)

Herman, A., V. B. Kumar, P. A. Arkin, and J. V. Kousky, 1997: Objectively determined 10-day African rainfall estimates created for famine early warning. *International Journal of Remote Sensing*, **18**(10), 2147–2159.

Huffman, G. J., R. F. Adler, D. T. Bolvin, G. Gu, E. J. Nelkin, K. P. Bowman, Y. Hong, E. F. Stocker, and D. B. Wolf, 2007: The TRMM multisatellite precipitation analysis (TMPA): Quasi-global, multiyear, combined-sensor precipitation estimates at fine scales. *Journal of Hydromet*eorology, **8**, 38–55.

Hurtado, G., 2005: Precipitation in Colombia. Climate Atlas of Colombia. IDEAM Bogotá, Colombia

Joyce, R. J., J. E. Janowiak, P. A. Arkin, and P. Xie, 2004: CMORPH: A method that produces global precipitation estimates from passive microwave and infrared data at high spatial and temporal resolution. *Journal of Hydromet*eorology, **5**, 487–503.

Levizzani, V., R. Amorati, and F. Meneguzzo, 2002: A Review of Satellite-based Rainfall Estimation Methods, Consiglio Nazionale delle Ricerche Istituto di Scienze dell'Atmosfera e del Clima, (Available online at http://www.isao.bo.cnr.it/~meteosat/papers/MUSIC-Rep-Sat-Precip-6.1.pdf)

Levizzani, V., Bauer, P., and Turk J. F. (Eds.),, 2007: Measuring Precipitation from space: EURAINSAT and the Future. Springer, New York, 724 pp..

Schneider, U., T. Fuchs, A. Meyer-Christoffer, and B. Rudolf 2008: Global precipitation analysis products of the GPCC. Available online at: (http://gpcc.dwd.de).

Thorne, V., P. Coakeley, D. Grimes, and G. Dugdale, 2001: Comparison of TAMSAT and CPC rainfall estimates with raingauges, for southern Africa. *Int. J. Remote Sensing*, **22**(10), 1951–1974.

Xie, P., Y. Yarosh, T. Love, J. Janowiak, and A. Arkin 2002: A real-time daily precipitation analysis over southern Asia. Preprints, *16th Conf. of the Hydro., Orlando, FL, Amer. Meteor. Soc.*

Evaluation Through Independent Measurements: Complex Terrain and Humid Tropical Region in Ethiopia

Menberu M. Bitew and Mekonnen Gebremichael

Abstract Evaluation of satellite rainfall products was conducted using ground-based daily rainfall measurements at 22 locations within a grid of 5×5 km collected during summer monsoon 2008 in a very complex terrain and humid tropical region in Ethiopia. Two high-resolution satellite rainfall products, namely, PERSIANN-CCS available at 1-h and 0.04° resolution, and CMORPH available at 30-min and 0.08° resolution. Both remotely-sensed products underestimated heavy events by about 50%, and so caution must be exercised when using CMORPH and PERSIANN-CCS as input for flood forecasting, as this could underestimate large flood events. The underestimation in monthly total rainfall was significant (32% for CMORPH, 49% for PERSIANN-CCS), and this error level needs to be acknowledged in applications that require monthly analyses. PERSIANN-CCS failed to detect half of the light events, and consistently those under 1.6 mm/day, indicating clearly that PERSIANN-CCS has difficulty detecting light rainfall events in complex terrain.

Keywords Rainfall · Remote Sensing · Validation · Complex terrain

1 Introduction

High-resolution satellite precipitation estimates are increasingly becoming available across the globe. However, examples of operational applications of these products are few, particularly in developing countries that do not have alternative reliable ground-based monitoring systems. One reason for this is the lack of quantitative information on the uncertainty level of these estimates. The recognition of this fact has led to the establishment of a number of ground validation sites in the developed countries. However, Africa in general does not have a ground validation site, and so

M.M. Bitew (✉)
Civil and Environmental Engineering Department, University of Connecticut, Storrs, CT, USA
e-mail: Menberu@engr.uconn.edu

only little is known about the error characteristics of the satellite estimates. In this study, we conducted an intensive 39-day field campaign to collect high-quality data on precipitation to evaluate the accuracy of satellite rainfall estimates in a complex terrain and humid tropical region in Ethiopia. In the following, we provide a discussion of the study region, types of satellite data tested, field campaign, results and discussion, and conclusions.

2 Data and Methods

2.1 Study Region

The Beressa Watershed is located in the central highlands of Ethiopia, within 9°33′43″N to 9°42′27″N and 39°28′34″E to 39°44′23″E (Fig. 1). It covers an area of 283 km^2. Geologically, it is part of a huge land mass that covers more than 300,000 km^2 in central Ethiopian high land. The high altitude and mountainous nature of the area is associated with the uplift during the rifting process and a series of volcanisms. The Beressa watershed is located on the western plateau edge of the rift system. It is characterized by diverse topographic conditions; elevation ranges from 1850 to 3700 m. Climate is considered humid, with 1100 mm of annual precipitation. Most of the annual rainfall comes from summer monsoon. Major rain-producing systems during summer monsoon are: the northward migration of the ITCZ; development and persistence of the Arabian and the Sudan thermal lows along 20°N latitude; development of quasi-permanent high-pressure systems over the south Atlantic and south Indian oceans; development of the tropical easterly jet and its persistence; and the generation of the low-level Somali jet, which enhances low level southwesterly flow (Seleshi and Zanke 2004).

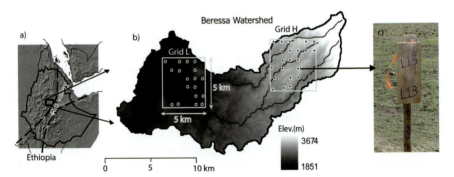

Fig. 1 (a) Map of Ethiopia showing the location of the Beressa watershed with respect to the rift valley. (b) The Beressa Watershed (283 km^2) delineated using a 90-m DEM. The inset contains two grids (5×5 km) located in two areas of the watershed, and the rain gauges within them. (c) A picture showing an example of one of the installed non-recording rain gauges

2.2 Types of High-Resolution Satellite Products Used

We evaluated the accuracy of two very high-resolution satellite rainfall products that have fundamental differences in their retrieval algorithms. These products are CMORPH (NOAA's Climate Prediction Center Morphing technique; Joyce et al. 2004) and PERSIANN-CCS (Precipitation Estimation from Remotely Sensing Imagery using Artificial Neural Network cloud classification system; Hong et al. 2004). CMORPH has 30-min and 0.08° latitude×0.08° longitude resolution, whereas PERSIANN-CCS has 1-h and 0.04° latitude×0.04° longitude resolution. While both products use geostationary infrared data as well as non-geostationary microwave in their retrieval algorithms, they use fundamentally different approaches to combine the two datasets. CMORPH obtains its precipitation estimates from the microwave data exclusively, however, it uses infrared data to construct the advection of the precipitation structure in between the microwave overpasses. On the other hand, PERSIANN-CCS obtains its precipitation estimates from the infrared data, however, it uses the microwave data for training the neural network that assigns precipitation estimate to the infrared temperature.

2.3 Rainfall Field Experiment

We conducted a summer field campaign in the Beressa watershed, where we installed two networks of non-recording rain gauges over an area of 5×5 km area (Fig. 1). The first network consisted of 22 non-recording rain gauges over a 5×5 km area, in the upstream mountainous region of the watershed we call "Grid H". The second network consisted of 18 non-recording rain gauges over a 5×5 km area, in the downstream hilly region of the watershed we call "Grid L". Initially, we put rain gauges every kilometer within each grid (i.e. a total of 36 rain gauges per grid), but we lost some of them during the experimental period. We believe that the remaining density of 18 or 22 rain gauges is adequate to estimate accurately the mean rainfall over each grid (see, for example Gebremichael et al. 2003). We focused our current study on Grid H, which has a relatively higher number of rain gauges and better geometric match to the satellite grids. Grid H's rain gauges have elevations ranging from 3100 to 3270 m.

The non-recording rain gauges used were the "Tru-Chek®" plastic rain gauges manufactured by the Forestry Suppliers. The gauges have scales permanently marked on the front sides. We mounted each gauge vertically on a wooden pole at a height of 2 m above the ground. A typical rain gauge installation is shown in Fig. 1c. Our research group and trained local research assistants took readings off each rain gauge, every morning from 7 am to 8 am, for the period of July 2 to August 9, 2008. This is the major rainy period in the region; there were 37 rainy days during the 39-day experimental period.

2.4 Method of Analysis

In Fig. 2, we show the CMORPH and PERSIANN-CCS pixels superimposed on the rain gauge network. To evaluate CMORPH, we compared the rainfall value of the CMORPH pixel which lies over the rain gauge network (i.e. C1 pixel) against the rainfall value obtained by averaging rainfall values recorded at the 22 rain gauges. To match the approximate size of a CMORPH pixel C1, we chose PERSIANN-CCS pixels P1, P2, P3, and P4. To evaluate PERSIANN-CCS, we compared the average of the rainfall values of the P1, P2, P3, and P4 against the rainfall value obtained by averaging rainfall values recorded at all 22 rain gauges. We aggregated the hourly or 30-min satellite precipitation estimates to daily values, so that we could compare them with the daily observations from the non-recording rain gauges.

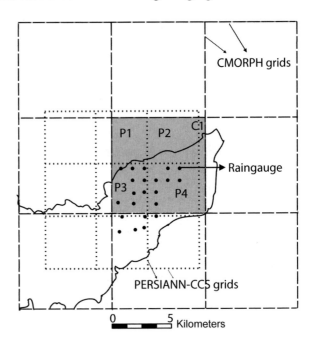

Fig. 2 CMORPH and PERSIANN-CCS grids over the study region with respect to the rain gauge network layout. *Dotted lines* show the grids for PERSIANN-CCS, and *dashed lines* show the grids for CMORPH. P1, P2, P3 and P4 PERSIANN grids, and C1 CMORPH grid are used in the current analysis

3 Results and Discussion

We begin with a description of the rainfall statistics using rain gauge data. In Fig. 3a, we present the cumulative spatial-average daily rainfall obtained by averaging data from the 22 rain gauges. The maximum daily rainfall was 48 mm. 50% of the total rainfall was contributed by daily events with rain rates greater than 25 mm/day, and 75% of the total rainfall from rain rates exceeding 14 mm/day.

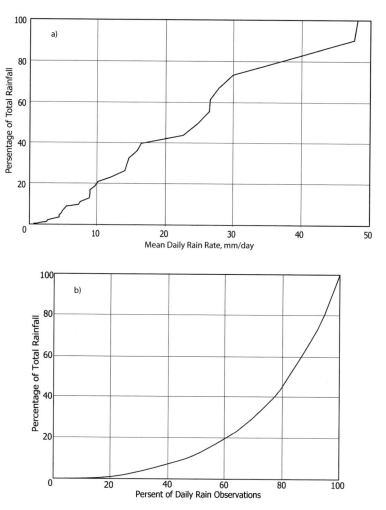

Fig. 3 (**a**) Percentage of total rainfall versus mean daily rain rate, and (**b**) percentage of total rainfall versus percentage of observations (no-rain observations were discarded), derived from the daily average of the rain gauge network

In Fig. 3b, we present a plot of the percentage of the running accumulation to the total accumulation against the percentage of observations to the total observations. The figure shows that the large contribution of the total rainfall came from infrequent yet heavy rains: 50% of the total rainfall was provided by the heaviest 18% of the observations, and 75% of the total rainfall by the heaviest 37% of the observations.

In Fig. 4, we present a time series of daily rainfall, as derived from the rain gauge network (i.e. average of 22 rain gauges), and CMORPH (i.e. C1) and PERSIANN-CCS (i.e. average of P1, P2, P3, and P4) satellite products. The rain gauge network showed large temporal variability in the daily rainfall, with a daily standard deviation (std) of 12.6 mm, mean of 12.4 mm, and coefficient of variation (cv) of

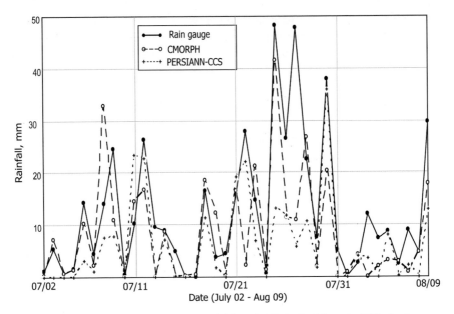

Fig. 4 Time series (July 2–August 9, 2008) of daily rainfall obtained from (*solid line*) rain gauge network; (*dotted line*) PERSIANN-CCS values averaged over four grids P1, P2, P3, and P4; and CMORPH value at grid C1

102%. CMORPH reported daily std of 10.0 mm, mean of 8.4 mm, and cv of 119%. PERSIANN-CCS reported daily std of 8.4 mm, mean of 6.3 mm, and cv of 134%. Overall, CMORPH tended to underestimate daily rainfall by 32%, while PERSIANN-CCS underestimated it by 49%. The correlation between the daily gauge-measured and remotely-sensed rainfall was moderate (0.68), and the lack of strong correlation likely results from the fact that the precipitable water concentration measured by remote sensors is not always highly correlated to rainfall reaching the Earth surface.

To examine in detail the performance of CMORPH and PERSIANN-CCS products for each daily event, we tabulate the daily rainfall values in Table 1, sorting them in descending order by the rain gauge network values. We also identify in the Table the heavy events (heaviest daily events which accounted for 50% of the total rainfall according to the rain gauge network data), moderate events (moderate events accounting for the subsequent 25% of the total rainfall), and light events (light events accounting for the remaining 25% of the total rainfall). We note the following observations:

– On heavy (> 25 mm) events:

- CMORPH underestimated all 7 heavy events; the underestimation ranged from 10% to 90%. PERSIANN-CCS underestimated all 7 heavy events; the underestimation ranged from 5 to 90%.

- On average, both CMORPH and PERSIANN-CCS underestimated heavy events by 50%.
- Both CMORPH and PERSIANN-CCS detected all heavy events.

– On moderate (14–25 mm) events:

- CMORPH overestimated 5 out of 7 medium events, while PERSIANN-CCS underestimated 6 out of 7 medium events.
- On average, CMORPH underestimated moderate events by 12%, whereas PERSIANN-CCS underestimated moderate events by 45%.
- The significant overestimation by CMORPH was accompanied by CMORPH significantly underestimating rainfall either the previous or the next day, indicating that the cloud tracking algorithm used in CMORPH may introduce errors in distributing rainfall temporally.
- Both CMORPH and PERSIANN-CCS detected all medium events.

– On light (< 14 mm) events:

- CMORPH missed 4 out of 24 lights events, while PERSIANN-CCS missed 12 light events.
- PERSIANN-CCS consistently failed to detect rain under 1.6 mm.

There are a couple of issues that might have influenced the accuracy of the above comparison. The first issue is that our network of 22 rain gauges did not adequately represent the four PERSIANN-CCS pixels (P1, P2, P3, and P4) and the one CMORPH pixel (C1), and this could cause differences between gauge-measured rainfall and remotely-sensed data, even if they all were perfect. To test the significance of this issue, we calculated the PERSIANN-CCS value averaged over P3 and P4, since the gauge network represented very well these two pixels (see also Fig. 2). Shown in the last column of Table 1 are the daily rainfall values obtained by averaging PERSIANN-CCS values at pixels P3 and P4. The PERSIANN-CCS values averaged over two pixels P3 and P4 are very similar to those averaged over four pixels P1, P2, P3, and P4. This suggests that either PERSIANN gives spatially smoother fields and/or the rain gauge network represents the rainfall field over the large C1 grid.

The second issue concerns the mismatch in pixels. Due to reasons, such as, oblique scattering of the microwave signals, CMORPH and PERSIANN-CCS may contain geolocation errors. To account for possible geolocation errors, we evaluated the performance of CMORPH and PERSIANN-CCS values at each adjacent grid with respect to the same rain gauge network comprised of 22 rain gauges. Figure 5 presents the correlation between the daily time series of gauge-measured rainfall and PERSIANN-CCS value at each adjacent grid. The differences among correlation values in adjacent boxes were not statistically significant.

It follows therefore that that the CMORPH and PERSIANN-CCS evaluation statistics reported in this study are fairly robust, and not significantly sensitive to the misalignment issues discussed above.

Table 1 Comparison of daily rainfall values obtained from the rain gauge network, CMORPH at grid C1, PERSIANN-CCS averaged over four grids (P1, P2, P3, and P4), and PERSIANN-CCS averaged over two grids (P3, and P4), sorted in descending order by the rain gauge network values

Date	Daily rainfall (mm) obtained from			
	Rain gauges	CMORPH	PERSIANN-CCS (P1+P2+P3+P4)/2	PERSIANN-CCS (P3+P4)/2
Heavy events (> 25 mm) contributing 50% of the total rainfall				
25-Jul	48.1	41.6	13.3	13.2
27-Jul	47.7	10.9	5.9	5.4
30-Jul	37.9	20.3	35.9	32.2
9-Aug	29.8	17.9	12.9	12.0
22-Jul	27.8	2.2	22.0	24.2
26-Jul	26.5	11.3	11.8	11.8
12-Jul	26.4	16.7	22.8	22.0
Medium events (14 – 25 mm) contributing 25% of the total rainfall				
9-Jul	24.7	10.9	7.7	8.0
28-Jul	22.5	26.8	10.6	10.8
18-Jul	16.5	18.6	11.3	11.3
21-Jul	15.9	16.5	19.1	18.2
23-Jul	14.7	21.2	6.9	6.9
6-Jul	14.3	10.3	3.2	2.1
8-Jul	14.1	32.8	7.6	6.7
Light events (< 14 mm) contributing 25% of the total rainfall				
3-Aug	12.0	0	3.5	1.7
11-Jul	10.2	14.4	23.3	21.4
13-Jul	9.7	0	0.0	0.0
7-Aug	9.0	0.8	2.2	2.5
14-Jul	9.0	8.8	7.2	6.9
5-Aug	8.8	3.1	8.0	7.7
29-Jul	7.5	2	1.7	1.7
4-Aug	7.3	2.1	0.0	0.0
3-Jul	5.5	7.2	0.0	0.0
31-Jul	5.1	0.1	0.0	0.0
15-Jul	4.9	0	0.2	0.2
8-Aug	4.7	4.7	0.0	0.0
7-Jul	4.4	2.2	1.2	1.2
20-Jul	4.4	0.3	0.1	0.1
19-Jul	3.9	12.3	1.9	2.0
2-Aug	2.7	4.2	3.9	4.4
6-Aug	2.5	2.8	0.2	0.1
5-Jul	1.6	1.3	0.0	0.0
2-Jul	1.1	0.2	0.0	0
4-Jul	0.7	0.7	0.0	0.0
24-Jul	0.6	1.3	0.0	0.0
10-Jul	0.4	0	0.0	0.0
1-Aug	0.1	0.9	0.0	0.0
16-Jul	0.0	0.3	0.0	0.0

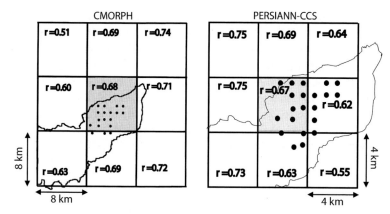

Fig. 5 Spatial map of daily correlation computed between the rain gauge average values and satellite values at each of the nine grids around the rain gauge network, for *(left)* CMORPH, and *(right)* PERSIANN-CCS. The standard error of the correlation estimate is 0.17 for all cases

Finally, let us compare our results to those reported for other regions. In a recent study, Zeweldi and Gebremichael (2009) evaluated the accuracy of CMORPH rainfall products over the Little Washita watershed of Oklahoma, US, and found that CMORPH overestimates rainfall in this region by a factor of 2 in July. A similar result has been reported for this region by Tian et al. (2007). Note that we reported an average 32% underestimation by CMORPH for Beressa. Differences in CMORPH performance between Little Washita and Beressa could be explained by regional differences in climate and topography. Compared to Beressa, Little Washita is hotter in July and therefore has more evaporation below the cloud base that typically leads to overestimation by the satellite rainfall product. Unlike Little Washita, Beressa is strongly characterized by orographic rainfall process.

Hong et al. (2007) evaluated the accuracy of PERSIANN-CCS products over the semiarid and highly mountainous region of Mexico, against the North American Monsoon Experiment (NAME) Event Rain Gauge Network (NERN) (see Gochis et al. 2003; Gebremichael et al. 2007). They reported that PERSIANN-CCS tends to miss light rainfall events at high elevations, which is consistent with our finding. This suggests that PERSIANN-CCS has difficulty detecting rainfall in comparatively shallow convective clouds over topographically complex regions. Hong et al. (2007) reported a positive bias for PERSIANN-CCS, which disagrees with our result in Ethiopia. This could be explained by the differences in climate between the semiarid Mexico and humid Beressa.

4 Conclusions

We evaluated the performance of high-resolution satellite products, particularly CMORPH and PERSIANN-CCS, in a complex terrain and humid tropical region in Ethiopia, using an event rain gauge network that we established in June 2008.

The comparison period (July 2–August 9, 2008) spans a major rainy period in the region. Our results reveal that both CMORPH and PERSIANN-CCS tend to underestimate heavy events by about 50%. Therefore, caution must be exercised when using CMORPH and PERSIANN-CCS products as inputs for flood forecasting, as this could underestimate large flood events. As far as monthly accumulations are concerned, CMORPH underestimated by 32% while PERSIANN-CCS underestimated by 49%. This error level needs to be acknowledged in analyses that require monthly data. PERSIANN-CCS missed half of the light events, and consistently those under 1.6 mm/day, clearly indicating that PERSIANN-CCS has difficulty detecting light rainfall in complex terrain. We note that these results are for the summer monsoon in complex terrain and tropical humid region. It is important to recognize the limitations on drawing the above conclusions based on a relatively short period of data (39 days), indicating that further study using a larger dataset is necessary.

Acknowledgment This research was supported by NSF – Office of International Science and Engineering (OISE-0651783) and NASA – New Investigator Program (NNX08AR31G). The authors thank Feyera Hirpa, Nathaniel Bergan, and Terrence McAuliffe for their help with data collection.

References

Bitew, M.M., M. Gebremichael, F.A. Hirpa, Y.M. Gebrewubet, Y. Seleshi, and Y. Girma, 2009: On the local-scale spatial variability of daily rainfall in the highlands of the Blue Nile: observational evidence. *Hydrol. Proc.*, (conditionally accepted in press).

Gebremichael, M., E.R. Vivoni, C.J. Watts, and J.C. Rodriguez, 2007: Sub-mesoscale spatiotemporal variability of North American monsoon rainfall over complex terrain. *J. Climate*, **20**(9), 1751–1773.

Gebremichael, M., W.F. Krajewski, M. Morrissey, D. Langerud, G. Huffman, and R. Adler, 2003: Error uncertainty analysis of GPCP monthly rainfall products: A data-based simulation study. *J. Appl. Meteor.*, **42**(12), 1837–1848.

Gochis, D.J., J.-C. Leal, W.J. Shuttleworth, C.J. Watts, and J. Garatuza-Payan, 2003: Preliminary Diagnostics from a new event-based precipitation monitoring system in support of the North American monsoon experiment. *J. Hydromet.*, **4**(5), 974–981.

Hong, Y., D. Gochis, J.-T. Cheng, K.-L. Hsu, and S. Sorooshian, 2007: Evaluation of PERSIANN-CCS rainfall measurement using the NAME event rain gauge network. *J. Hydromet.*, **8**, 469–482, DOI: 10.1175/JHM574.1.

Hong, Y., K.L. Hsu, S. Sorooshian, and X. Gao, 2004: Precipitation estimation from remotely sensed imagery using an artificial neural network cloud classification system. *J. Appl. Meteor.*, **43**, 1834–1852.

Joyce, R.J., J.E. Janowiak, P.A. Arkin, and P. Xie, 2004: CMORPH: A method that produces global precipitation estimates from passive microwave and infrared data at high spatial and temporal resolution. *J. Hydromet.*, **5**, 487–503.

Seleshi, Y. and U. Zanke, 2004: Recent changes in rainfall and rainy days in Ethiopia. *Int. J. Clim.*, **24**, 973–983.

Tian, Y., C.D. Peters-Lidard, B.J. Chaudhury, and M. Garcia, 2007: Multitemporal analysis of TRMM-based satellite precipitation products for land data assimilation applications. *J. Hydromet.*, 1165–1183. DOI: 10.1175/2007JHM859.1.

Zeweldi, D. and M. Gebremichael, 2009: Evaluation of CMORPH rainfall products at fine space-time scales, *J. Hydromet.*, **10**, 300–307, DOI: 10.1175/2008JHM1041.1.

Error Propagation of Satellite-Rainfall in Flood Prediction Applications over Complex Terrain: A Case Study in Northeastern Italy

Efthymios I. Nikolopoulos, Emmanouil N. Anagnostou, and Faisal Hossain

Abstract The study presented in this chapter evaluates the use of satellite rainfall for flood prediction applications in complex terrain basins. It focuses on a major flood event that occurred in October 1996 in a complex terrain basin of the northeastern region of Italy. A satellite rainfall error model is calibrated and used to generate rainfall ensembles based on two different satellite products and spatio-temporal resolutions. The generated ensembles are propagated through a distributed hydrologic model to simulate the hydrologic response. The resulted hydrographs are compared against the hydrograph obtained by using high-resolution radar-rainfall as the "reference" rainfall input. The error propagation of rainfall to stream runoff is evaluated for a number of basin scales that range from 100 to 1200 km^2. The results from this study show that (i) use of satellite-rainfall for flood prediction depends strongly on the scale of application (catchment area) and the satellite product resolution, (ii) different satellite products perform differently in terms of hydrologic error propagation and (iii) the propagation of error depends on the basin size; for example, this study shows that small watersheds (<400 km^2) exhibit a higher ability in dampening the error from rainfall-to-runoff than larger size watersheds.

Keywords Satellite rainfall · Flood prediction · Error propagation

1 Introduction

Recent advancements in space-based precipitation observations especially in terms of sampling frequency and sensor resolutions have opened new horizons in hydrological applications at global scale. Satellite sensors offer unique advantages because they provide (i) global coverage and (ii) observations in regions where in

---

E.I. Nikolopoulos (✉)
Department of Civil and Environmental Engineering, University of Connecticut, Storrs, CT, 06269 USA
e-mail: ein06002@engr.uconn.edu

situ data are inexistent or sparse. However, satellite observations are subject to a number of errors due to instrumental issues (calibration, measurement error) and the nonlinear and variable relationship between observables (i.e. brightness temperature) and precipitation. These errors introduce uncertainty in the satellite products, thus it is crucial to assess and quantify this uncertainty in order to make the use of satellite retrievals more meaningful in hydro-meteorological applications.

This necessity is recognized by the research community and several error studies have been reported dealing with the characterization of error structure for a number of global satellite rainfall products. However, as Hossain and Anagnostou (2006b) pointed out, these studies focus on the uncertainty of rain retrieval associated with large spatiotemporal scales, involving error statistics that are relevant to large scale meteorological phenomena, but do not provide insight towards the dynamic surface hydrologic processes such as floods (see for example, Griffith et al., 1978; Arkin and Meisner, 1987; Huffman et al., 2001; Steiner et al., 2003, among others). Highlighting this inherent complexity in the error structure of precipitation fields derived from space-based observations Hossain and Anagnostou (2006a) and Bellerby and Sun (2005) developed error models to provide explicit characterization of the complex stochastic nature of the satellite sensor derived rainfall fields. Both models use ensemble rainfall fields to represent scenarios of satellite sensor retrieval uncertainty.

As we now stand at the doorstep of a global scale mission, named Global Precipitation Measurement (GPM, http://gpm.gsfc.nasa.gov/), it is essential and most beneficial to investigate and evaluate the effective use of current satellite products in hydrologic studies. GPM is anticipated to create new potential on the use of satellite rainfall retrievals for the prediction of floods over ungauged basins worldwide, which is mainly due to the enhanced revisit frequency and global coverage (Smith et al., 2007; Hossain et al., 2004). Thus, a comprehensive investigation on the flood prediction uncertainties associated with current and upcoming (GPM era) satellite rain retrieval appears mandatory and can also serve as an important reference for highlighting the usefulness of GPM mission to the society.

Evaluating the error propagation of satellite-rainfall through the prism of surface hydrology is a very challenging task because it relates to many factors which include, among others, (i) specifications of the satellite-rainfall product, (ii) scale of the basin, (iii) spatio-temporal scale of the hydrologic variable of interest, (iv) the level of complexity and physical processes represented by the hydrologic model used and (v) regional characteristics. Researchers have identified and addressed several of those issues. Geuter et al. (1996) found that the propagation of satellite retrieval errors through a hydrologic model affected differently the streamflow error for different basin scales. Hossain et al. (2004) showed that the combined retrieval and sampling error from current satellite passive microwave (PM) sensors magnifies when it is transformed to flood prediction of a medium size (140 km^2) mountainous basin, and Hossain and Anagnostou (2004) quantified that flood prediction uncertainty to be up to 100% higher than the uncertainty associated with a hypothetical 3-hourly satellite sampling scenario (as planned for GPM) for a medium size watershed. Wilk et al. (2006) used passive microwave datasets and achieved reasonable estimates for the water balance over the Okavango River basin (165,000 km^2). Collischonn et al. (2008) used TRMM (Tropical Rainfall Measuring Mission) 3B42 dataset to estimate daily streamflow in the Amazon basin, using a distributed hydrologic model, and found that the results compared well with the raingauge driven streamflow simulations. Harris and Hossain (2008) showed using simple hydrological models that for a given satellite-rainfall product (3B41RT), the hydrologic model's representativeness of the rainfall-runoff processes can affect the uncertainty of flood prediction. A compelling evidence from summarizing findings from studies so far is that due to the multidimensionality of the problem, there is no specific answer to the question "how effective is the use of satellite-rainfall for hydrologic applications?" and we can only have specific answers for questions such as "what is the expected error propagation of a given satellite-rainfall product for a given basin scales and for a specific hydrologic application and model setup".

The scope of this chapter is to enlighten the reader as to how satellite-rainfall error propagates, for a specific hydrological setup and application. A distinct feature of this work is its focus on complex terrain environment and a range of basin scales from small to medium size. The reason is, as mentioned above, that space based observations are usually the only source of information for regions with complex terrain. The study will present results on precipitation error propagation in terms of flood prediction and its dependence on basin-size and satellite product resolution.

Section 2 describes the concept and methodology followed in this study and provides a brief description of the study area, data and models used. The results are presented and discussed in Section 3 and the conclusions are summarized in Section 4.

2 Methodology

Characterization of uncertainty in flood prediction forced with satellite-rainfall estimates is achieved in this study through an integrative data-modeling experiment. Ensemble satellite-rainfall fields generated from a satellite error model (hereafter named SREM2D out of Satellite Rainfall Error Model-2 Dimensional; Hossain and Anagnostou, 2006a), propagated through a distributed hydrologic model (hereafter named as tRIBS out of triangulated irregular network Integrated Basin Simulator, Ivanov et al., 2004) and the resulted simulated hydrographs are compared with the simulated hydrographs obtained by using radar data as the reference rainfall input. This comparison is carried out for a range of basin scales (100–1200 km^2) and for two satellite-rainfall products with different spatial resolutions and retrieval characteristics.

2.1 Study Area and Data

The basin chosen for this study (Bacchiglione basin) is located in the Veneto region, northeastern Italy (Fig. 1). With a drainage area of approximately 1200 km^2, the

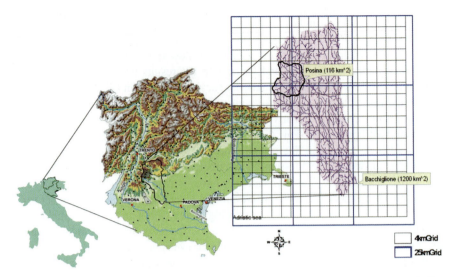

Fig. 1 Map showing the locations of the Posina and Bacchiglione basins in the northeastern Italian Alps. Note that the *thin* and *thick grid* corresponds to the 4 and 25 km scales

basin extends from the lowlands of the region (southern part), with an elevation of 30 m at the outlet, to the highlands (northern part) which are part of the eastern Italian Alps and reach a maximum elevation greater than 2000 m. The northern part is covered predominately by conifer and broadleaved forests while the southern part consists mainly of croplands. The high precipitation amounts in the area (>1000 mm annually), along with the very steep irregular terrain (slopes greater than 30° in the highlands) make the region prone to the generation of floods that makes it suitable for hydrologic investigations.

This study utilizes both in situ and remote sensing data. The in situ data include observations from rain and stream gauges while the remote sensing data include radar and satellite derived rainfall fields. The rain gauges, located in the region (see Fig. 1), provided half-hourly rainfall accumulations that were used for (i) bias adjustment of radar-rainfall fields and (ii) the calibration of the satellite-rainfall error model. Streamflow data were available at half-hourly scale and only for one sub-basin (called Posina, see Fig. 1), and were used to calibrate the hydrologic model parameters. The radar data were obtained from a C-band Doppler weather radar located at Mt. Grande, approximately 10 km southeast of the basin's outlet, and were available at 1 km spatial and 1 h temporal resolution. Two different satellite products were used, the TRMM 3B42 version 6 (see Huffman et al., 2007 for details) and another dataset obtained from the calibration of high resolution global IR data on the basis of available passive microwave satellite rainfall estimates (hereafter called KIDD) and is based on algorithm described in Kidd et al. (2003). Table 1 summarizes the nominal spatiotemporal resolution of each precipitation dataset used in this study.

Table 1 Nominal spatial and temporal resolution of the precipitation data used

Data type	Product name	Resolution Temporal (hr)	Spatial
Remote sensing	3B42	3	$0.25 \times 0.25°$
	KIDD	0.5	$\sim 4 \times 4$ km
	Radar	1	1 km
In situ	Gauge	0.5	Point

2.2 Satellite-Rainfall Ensembles

The generation of satellite-rainfall ensembles was based on the use of a two Dimensional Satellite Rainfall Error Model (SREM2D) developed by Hossain and Anagnostou (2006a). SREM2D uses as input "reference" rain fields of higher accuracy and resolution representing the "true" surface rainfall process, and stochastic space-time formulations to characterize the multi-dimensional error structure of satellite retrievals. The model parameters were calibrated for the region using six months (June–November, 2002) of gauge and satellite data for the 3B42 and KIDD satellite-rainfall products. For more details on the model's algorithmic structure as well as the calibration procedure and the evaluation of the generated ensembles, the reader is referred to Chapter 9 of this book.

The high resolution radar-rainfall fields were used as the "reference" to create the satellite-rainfall ensembles for the two different satellite products. Three sets of 100 realizations each were generated. The first set corresponds to the 3B42 product at its nominal scale (see Table 1). The second and third set is based on the KIDD product with corresponding resolution at high (4 km–1 h) and coarse (25 km–3 h) spatiotemporal scales, respectively. Due to computational limitations not all realizations from all sets were used to force the hydrologic model. Instead, ensembles for each set were ranked based on the total rainfall bias (compared to the reference field) and the realizations starting at the 5th–100th percentile were selected at step increments of 5 percentiles (i.e. 5th, 10th, 15th etc.). Thus, a total of 20 realizations plus the average of all 100 realizations from each set was used for the error propagation experiment.

2.3 Hydrologic Simulations

In October 1996 a major flood event occurred in the study area. The rainfall event that caused the flooding lasted for more than 60 h and resulted in mean areal rainfall accumulations (based on radar estimates) of 200 mm for the Bacchiglione basin ($\sim$1200 km^2) and approximately 350 mm (see Figs. 2 and 3) for the mountainous subbasin of Posina ($\sim$116 km^2). The error propagation experiment presented in this study is focused on this particular event. However, the reader should keep in mind

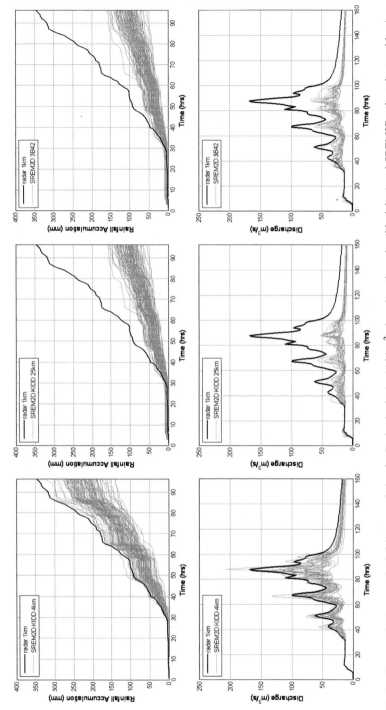

Fig. 2 *Top panel*: Mean areal rainfall accumulation for Posina basin (~ 116 km^2), based on the radar (*black line*) and SREM2D selected ensembles (*grey lines*) for KIDD 4 km (*left*), KIDD 25 km (*middle*) and 3B42 (*right*). *Bottom panel*: The respective simulated hydrographs for radar (*black line*) and SREM2D ensembles (*grey lines*) rainfall input. Data correspond to the flood event during October 1996

Evaluation Through Hydrologic Application: Europe 221

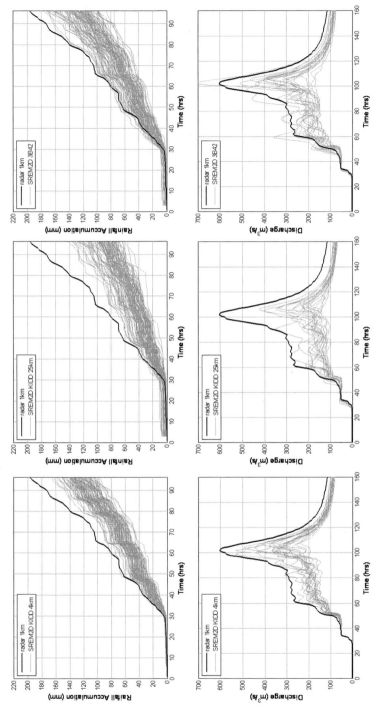

Fig. 3 Same as Fig. 2 but for the Bacchiglione basin (~1200 km^2)

that due to the highly nonlinear relationship of rainfall-to-runoff transformation, the propagation of error is expected to vary in different flood events.

The hydrologic model used in this study (tRIBS) is a fully distributed model that can simulate multiple storm events and account for the moisture losses during interstorm periods (see Ivanov et al., 2004 for more details). The distributed nature of the model allows retrieving the hydrograph response for several interior nodes thus providing the ability to compare the error propagation for different scales of drainage area. In this study we evaluated the error propagation for a number of basin scales that ranged between 100 and 1200 km^2.

For the calibration of the model, the SCE (shuffle complex evolution) optimization method (Duan et al., 1992) was used in order to minimize the error between the observed hydrograph and the simulated hydrograph based on the radar-rainfall input (reference hydrograph). As mentioned before, the only available discharge observations were for the Posina basin, thus were used as a reference to calibrate the parameters for the whole Bacchiglione basin. The total simulation time for the single flood event was 160 h and the total number of hydrologic simulations performed was 64 (1 for the reference rainfall input and 21 for each ensemble set).

3 Results

The results from the satellite-rainfall ensembles propagation through the distributed hydrologic model are presented and discussed in this section. The selected satellite-rainfall realizations that were used to force the hydrologic model along with the resulted simulated hydrographs, are shown in Figs. 2 and 3 for two basins of greatly different area sizes: Posina ~116 km^2 and Bacchiglione ~1200 km^2, which is about an order of magnitude scale difference. The differences shown between the KIDD-4 km and KIDD-25 km product demonstrate the effect of resolution in rainfall forcing while the comparison between KIDD-25 km and 3B42 can be used to assess how significant the differences between products can be in flood simulation. A point to note is that the satellite-rainfall ensembles consistently underestimate the total rainfall amount for both basin scales and all products. For the smaller basin, the high resolution product (KIDD-4 km) performs much better, in terms of bias, than the same aggregated product (KIDD-25 km) or the 3B42. However, in the case of the larger basin, the KIDD-4 km and the 3B42 behave similarly, and exhibit improved statistics compared to the KIDD-25 km.

The error in this study was quantified using the following metrics:

$$\text{Relative error} = \frac{\sum_{i=1}^{N} X_{\text{radar}}(i) - \sum_{i=1}^{N} X_{\text{SREM2D}}(i)}{\sum_{i=1}^{N} X_{\text{radar}}(i)} \qquad (1)$$

Evaluation Through Hydrologic Application: Europe 223

$$\text{Relative RMSE} = \frac{\sqrt{\frac{\sum_{i=1}^{N}(X_{\text{radar}}(i)-X_{\text{SREM2D}}(i))^2}{N}}}{\frac{\sum_{i=1}^{N}X_{\text{radar}}(i)}{N}} \quad (2)$$

where X_{radar} and X_{SREM2D} corresponds to rainfall (or discharge) obtained from radar and SREM2D ensembles respectively and N is the total number of data points in X_{radar} and X_{SREM2D}.

In order to facilitate the interpretation of the error propagation through the rainfall-runoff transformation, the error in rainfall versus error in runoff is presented in Fig. 4 based on the above metrics. A very distinct feature of the results is that the propagation of error exhibits a strong linear behavior in terms of its relative term. Especially in the case of Bacchiglione basin (1200 km^2), the points are aligned very

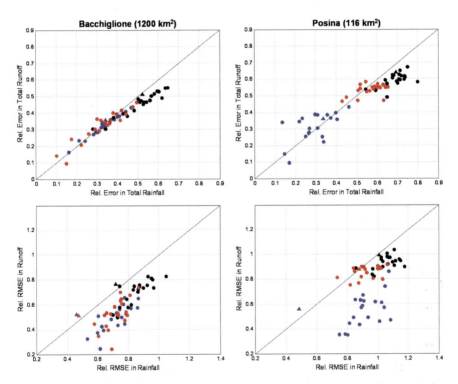

Fig. 4 Error propagation metrics: *Top panel* shows the relative error in total runoff versus relative error in total rainfall for Bacchiglione (*left*) and Posina (*right*) basin. *Bottom panel* shows relative RMSE in discharge versus relative RMSE in rainfall, respectively. Errors were calculated between SREM2D ensembles and the reference (radar). Note that blue black and red triangle corresponds to the ensemble average of KIDD 4 km, KIDD 25 km and 3B42 respectively

close to the 1–1 line, which indicates that the relative error in rainfall translates to an equal relative error in runoff. Another interesting feature apparent from Fig. 4 is that the performance of each satellite product manifests in distinct clusters of the rainfall-runoff error domain (those clusters are separated by different colors associated with different satellite products). In the case of Posina, the points of the high resolution satellite product (KIDD-4 km; blue color) cluster in a distinct (from the other products) region on the figure that is associated with lower relative error and higher dampening effect on the propagation of relative RMSE. For the largest size (Bacchiglione) basin, the two clusters of KIDD-4 km and 3B42 mix in the same domain since they perform equally as mentioned above.

To further investigate the dependency of error propagation with basin scale, the same analysis was carried out for a number of subbasins that range in size from 100 to 1200 km^2. In Fig. 5, results are presented in terms of the ratio of the error metric (mean relative error and relative RMSE) in runoff over the corresponding error metric in rainfall versus basin scale. Ratios equal to one indicate that statistics of the error in rainfall would translate to an equal statistical measure of the error in runoff, while ratios lower (higher) than one would indicate that the error dampens (magnifies) through the rainfall-runoff transformation process. The point to note from Fig. 5 is that while catchment area increases the average of the mean relative error ratios increases and approaches a plateau close to the value of one, which is more pronounced for the KIDD-4 km product. Another point to note is that the variability around the mean (i.e. error bars) tends to decrease as basin area increases. Similarly, the average of the relative RMSE ratios increases with catchment area, and after the size of 400 km^2 reaches a plateau around 0.7-0.8. This means that for smaller scale basins (<400 km^2), the dampening effect of the error is greater than in larger basins. The variability around the mean of the relative RMSE ratios does not show to vary much with catchment area.

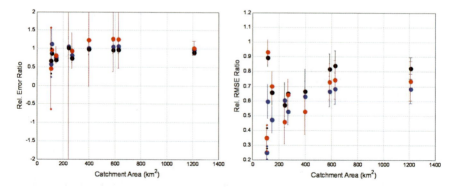

Fig. 5 Ratio of mean relative error (*left*) and relative RMSE (*right*) for different basin scales. The ratio is defined as error in discharge over error in rainfall. *Blue*, *black* and *red circles* correspond to the average of the 20 realizations for the KIDD 4 km, KIDD 25 km and 3B42 respectively. Error bars are equal to ±1 standard deviation

4 Conclusions

This chapter described a numerical investigation devised to evaluate the use of satellite-rainfall for flood prediction applications. Ensembles of satellite-rainfall were generated using the satellite error model described in Chapter 9. Those ensembles were used to force a distributed hydrologic model in order to assess the error propagation of satellite-rainfall in stream flow simulations for a range of basin sizes. The study was focused on a single flood event and two satellite products of different resolutions. Satellite rainfall estimates were compared against radar observations that were used here as the reference rainfall dataset.

The principal conclusions of the study are summarized in the following:

1. Satellite-rainfall consistently underestimates mean areal rainfall of the basin with relative errors ranging from 10 to 80% depending on the satellite product and the basin scale. This error in rainfall resulted in predicted runoff with a relative error of the same order.
2. The high resolution (KIDD-4 km) satellite product performed relatively well (~36% relative error of ensemble average) for both the small (Posina) and large size (Bacchiglione) basins, while the coarser resolution products (KIDD-25 km and 3B42) performed poorly (>55% relative error of ensemble average) in the case of the small size basin. In the case of the large size basin the two products (3B42 and KIDD-4 km) gave similar results. This indicates that the performance of a given product relates to both its resolution and scale of application.
3. Comparison between 3B42 and KIDD-25 km revealed that different satellite rainfall products (i.e., different retrieval algorithms) but at the same spatiotemporal resolution can give significantly different results in terms of streamflow simulations. This observation indicates the necessity that each satellite product must be evaluated separately in order to derive conclusions on its effective use for hydrologic applications (generalizations cannot apply)
4. The propagation of rainfall error was found to have a definitive dependence on basin scale. More specifically, catchment areas smaller than 400 km^2 exhibit a higher ability in dampening the error (from rainfall-to-runoff) than larger basins.

The study demonstrates that there is definitely potential in the use of satellite-rainfall in flood simulations. It is noted that to achieve better performance and define effective real-time flood prediction systems that are based on satellite observations, the satellite-retrieval community should focus on techniques that would advance the accuracy and quantify the uncertainty of high resolution rainfall products (at scales down to 4 km and 1 h). Overall, this study points out some key elements that can help us answer the question on "how effective is the use of satellites for flood prediction". To gain a holistic understanding on the error propagation issue, a series of similar studies must be carried out that will involve different satellite products, different satellite-rainfall error and hydrologic models and will be applied to different regions and for a number of flood events.

Acknowledgements Support for this work was provided by NASA Precipitation Measurement Mission (Anagnostou) and NASA Earth System Science Fellowship (Nikolopoulos). Prof. Chris Kidd of the University of Birmingham provided the Kidd et al. (2003) satellite rainfall product. The TRMM 3B42 data were obtained from Goddard Space Flight Center Distributed Active Archive Center's ftp site.

References

Arkin, P.A., and Meisner, B.N.: The relationship between large-scale convective rainfall and cold cloud cover Western Hemisphere during 1982–1984. Monthly Weather Review **115**, 51–74 (1987).

Bellerby, T.J., and Sun, J.: Probabilistic and ensemble representations of the uncertainty in an IR/MW satellite precipitation product. Journal of Hydrometeorology **6**, 1032–1044 (2005).

Collischonn, B., Collischonn W., and Tucci, C.E.M.: Daily hydrological modeling in the Amazon basin using TRMM rainfall estimates. Journal of Hydrology **360**, 207–216 (2008).

Duan, Q., Sorooshian, S., and Gupta, V.K.: Effective and efficient global optimization for conceptual rainfall-runoff models. Water Resources Research **28(4)**, 1015–1031 (1992).

Geuter, A.K., Georgakakos, K.P., and Tsonis, A.A.: Hydrologic applications of satellite data: 2. Flow simulation and soil water estimates. Journal of Geophysical Research **101**, 26527–26538 (1996).

Griffith, C.G., Woodley, W.L., and Grube, P.G.: Rain estimation from geosynchronous satellite imagery-visible and infrared studies. Monthly Weather Review **106**, 1153–1171 (1978).

Harris, A., and Hossain, F.: Investing the optimal configuration of conceptual hydrologic models for satellite-rainfall-based flood prediction. IEEE Geoscience and Remote Sensing Letters **5**, 532–536 (2008).

Hossain, F., and Anagnostou, E.N.: Assessment of current passive microwave and infrared based satellite rainfall remote sensing for flood prediction. Journal of Geophysical Research-Atmospheres **109** (2004), doi: 10.1029/2003JD003986.

Hossain, F., and Anagnostou, E.N.: A two-dimensional satellite rainfall error model. IEEE Transactions on Geosciences and Remote Sensing **44(4)**, 1511–1522 (2006a).

Hossain, F., and Anagnostou, E.N.: Assessment of a multidimensional satellite rainfall error model for ensemble generation of satellite rainfall data. IEEE Geoscience and Remote Sensing Letters **3**, 419–423 (2006b).

Hossain, F., Anagnostou, E.N., and Dinku, T.: Sensitivity analyses of satellite rainfall retrieval and sampling error on flood prediction uncertainty. IEEE Transactions on Geosciences and Remote Sensing **42(1)**, 130–139 (2004).

Huffman, G.J., Adler, R.F., Bolvin D.T., Gu, G., Nelkin, E.J., Bowman, K.P., Hong, Y., Stocker, E.F., and Wolf, D.B.: The TRMM multi-satellite precipitation analysis: quasi-global, multi-year, combined-sensor precipitation estimates at fine scale. Journal of Hydrometeorology **8**, 38–55 (2007).

Huffman, G.J., Adler, R.F., Morrissey, M.M. et al.: Global precipitation at one-degree daily resolution from multisatellite observations. Journal of Hydrometeorology **2**, 36–50 (2001).

Ivanov, V.I., Vivoni, E.R., Bras, R.L., and Entekhabi, D.: Catchment hydrologic response with a fully distributed triangulated irregular network model. Water Resources Research **40** (2004), doi: 10.1029/2004WR003218.

Kidd, C., Kniveton, D.R., Todd, M.C., and Bellerby, T.J.: Satellite rainfall estimation using combined passive microwave and infrared algorithms. Journal of Hydrometeor. **4**, 1088–1104 (2003).

Smith, E.A., Asrar, G., Furuhama, Y., Ginati, A., Kummerow, C., Levizzani, V., Mugnai, A., Nakamura, K., Adler, R., Casse, V., Cleave, M., Debois, M., Durning, J., Entin, J., Houser, P., Iguchi, T., Kakar, R., Kaye, J., Kojima, M., Lettenmaier, D., Luther, M., Mehta, A., Morel, P., Nakazawa, T., Neeck, S., Okamoto, K., Oki, R., Raju, G., Shepherd, M., Stocker, E., Testud, J., and Wood, E.: International global precipitation measurement (GPM) program and mission: An overview. In: Measuring precipitation from space – EURAINSAT and the future. V. Levizzani, P. Bauer,, and F.J. Turk, Eds., pp. 611–653. Springer (2007).

Steiner, M., Bell, T.L., Zhang, Y., and Wood, E.F: Comparison of two methods for estimating the sampling-related uncertainty of satellite rainfall averages based on a large radar dataset. Journal of Climate **16**, 3759–3778 (2003).

Wilk, J., Kniveton, D., Andersson, L., Layberry, R., Todd, M.C., Hughes, D., Ringrose, S., and Vanderpost, C.: Estimating rainfall and water balance over the Okavango River Basin For hydrological applications. Journal of Hydrology **331**, 18–29 (2006).

Probabilistic Assessment of the Satellite Rainfall Retrieval Error Translation to Hydrologic Response

Hamid Moradkhani and Tadesse T. Meskele

Abstract Satellite-based precipitation retrieval techniques and algorithms have been developed to estimate precipitation from satellite observation. The realistic characterization of uncertainty in satellite precipitation estimate and the corresponding uncertain hydrologic response can better aid water resources managers in their decision making. In this study, the standard error of satellite-based PERSIANN-CCS rainfall estimates conditioning on the assumed true field (i.e. radar rainfall) is obtained according to a multivariate function considering the spatial and temporal scales. Accepting the multiplicative nature of this error, the Monte Carlo simulation is used to generate the ensemble of precipitation and propagate them into a conceptual hydrologic model to investigate the impact of input error on streamflow simulation. The statistical assessment of the results through probabilistic measures explores the more in-depth quality and reliability of the hydrologic response resulted from input error characterization.

Keywords Radar · Streamflow ensemble · Uncertainty assessment · Probabilistic verification

1 Introduction

Precipitation is the key hydrologic variable which plays a dominant role in the climate system and links the atmosphere with land surface processes. Although considerable advances have been made in data collection and hydrologic models construction to improve simulation/forecasting accuracy, estimating forcing data

---

H. Moradkhani (✉)
Department of Civil and Environmental Engineering, Portland State University, Portland, OR 97201, USA
e-mail: hamidm@cecs.pdx.edu

(input) uncertainty is well-known to impose significant uncertainty into hydrologic modeling mainly due to its spatial and temporal variability. In deterministic hydrologic modeling the uncertainty in precipitation as the main driving force is neglected. The conventional approach in hydrologic model calibration relies on the assumption that the translation of uncertainty from input to output is specifically attributed to parameter uncertainty (Moradkhani, et al., 2005b; Kavetski, et al., 2006). Such assumption simply ignores the considerable errors in the input and instead relies on additive error term to attribute these errors to the model structural inadequacy. Several studies have demonstrated different approaches in explicit treatment of input uncertainty in a systematic framework particularly using rain gage data as the main forcing data (Kavetski, et al., 2006; Huard and Mailhot, 2006; 2006; Vrugt, et al., 2008). Rain gages are considered as the main sources of spatial precipitation estimates although it is acknowledged that they can contain significant measurement errors (Margulis, et al., 2006). Even though rain gages are the conventional techniques to measure rainfall directly, their very sparse network or no coverage over land or ocean/remote area limits their utility over these regions. The absence of gages at ocean and remote regions imposes using remotely sensed (radar and satellite) measurements by which rainfall estimation can be monitored as well. This provides an advantage of availability of data in real time and complete area coverage irrespective of terrain or climate. Owing to these reasons and the complex error structure as well as high uncertainty exhibited by intermittent measurement through sparse rain gages, identifying true rainfall fields has been recognized to be difficult. To overcome these issues, ground-base radar measurement adjusted with rain gages was found to be more representative of the true rainfall field. Also as an alternative, satellite based precipitation retrieval techniques and algorithms have been developed to estimate precipitation from satellite observation (Huffman, et al., 2001; Adler et al., 2003; Hsu, et al., 1997; Hong, et al., 2004). The wide spread availability and easy accessibility of satellite-based precipitation estimate has enhanced the hydrological modeling and forecasting at the watershed scale. The current procedure to estimate the precipitation from satellite observation is to combine infrared (IR) observation from geostationary satellite with passive microwave measurement from polar orbiting satellites. Satellite based precipitation measurements may overcome spatial sampling problem involve in measurement in rain gages (Gebremichael, et al., 2005), however they are indirect estimates of rainfall via an algorithm, depending on the properties of the cloud top (in the case of IR algorithms) and cloud liquid and ice content (in the case of passive microwave algorithms), with considerable error which may limit their applicability in operational settings.

Quantification of these errors and evaluating their impacts on hydrologic response is of major interest in the usage of satellite products in high resolution hydrologic applications. The common approach to assess the impact of remotely sensed satellite rainfall is to devise an error model and generate realistic ensemble of high resolution satellite rain fields from reference rain fields with higher accuracy such as dense rain gage network or ground radar data. Deriving the hydrologic model with erroneous forcing data will reveal the translation of such error into hydrologic model outputs (e.g. streamflow) which play a great importance in assessing the uncertainties related to hydrological applications such as flood risk analysis and data assimilation (Moradkhani, et al., 2005b).

In the past few years a number of studies have been undertaken assessing the error tied to satellite rainfall estimates. In particular the use of ensemble approaches have been found appealing to address the uncertainty in precipitation from remotely-sensed products (Hossain and Anagnostou, 2006; Hong, et al., 2006; Moradkhani, et al., 2006; Margulis, et al., 2006; Olson et al., 2006; Forman, et al., 2008). The main goal of the above studies has been to generate a realistic ensemble of precipitation realization that account for precipitation error that could be used in hydrologic data assimilation and uncertainty assessment of model states and fluxes. Hossain and Anagnostou (2005) also studied the uncertainty related to soil moisture prediction using Land Surface Model (LSM). The main sources of uncertainties considered in their study were the uncertainties tied to precipitation from the satellite and the model parametric uncertainties. In later studies, (Hossain and Anagnostou, 2006) presented a two dimensional satellite rainfall error model (SREM2D) with the intention of characterizing the multi-dimensional error structure of satellite rainfall retrieval. In similar effort (Hong, et al., 2006) analyzed the uncertainties linked to satellite based rainfall estimates from PERSIANN-CCS model (Hong, et al., 2004) and assessed the impact of these errors on hydrologic simulation using a parsimonious Hydrologic Model (HyMOD) (Boyle, et al., 2000). A parallel study by (Moradkhani, et al., 2006) demonstrated the application of the particle filter as a sequential data assimilation procedure to investigate the impact of individual and combined uncertainties connected to satellite rainfall estimates and hydrologic model states and parameters on streamflow forecasting. The findings from this study showed that the satellite precipitation error reflects a wide uncertainty range in streamflow forecasting as opposed to the narrow range ensuing from parameter uncertainty. It was also discovered that the ensemble filtering through combined state-parameter updating was capable of reducing the total uncertainty.

2 Methodology

In this study we devise a scheme to generate the satellite rainfall ensemble from remotely sensed precipitation and identify the impact of satellite rainfall retrieval error on hydrologic response. The ensemble of hydrologic response in terms of hydrological states and fluxes are simulated and the results are probabilistically evaluated. This is done utilizing a conceptual hydrologic model widely used operationally in the U.S. in all thirteen river forecast centers (RFCs).

3 Generating Satellite Precipitation Ensemble

The ensemble generation of satellite product in this study builds on previous studies conducted by Steiner, et al., (2003), Hong, et al., (2006) and Moradkhani, et al., (2006). As stated earlier, the ground radar rain field is considered as reference (true field) and the satellite-based precipitation error is estimated by conditioning the precipitation realizations of the CCS-PERSIANN satellite estimate on the National Weather Service WSR-88D stage IV data (see next section). The standard error associated with satellite precipitation can easily be calculated as follows:

$$\sigma_e \sqrt{\text{Var}\left(\hat{P}_{AT} - P_{AT}^{\text{ref}}\right)} \qquad (1)$$

Where $\hat{P}_{AT}$ is the satellite-based rainfall averaged over area A for the period of T, and P_{AT}^{ref} is the radar estimated rainfall over the same spatial and temporal domain. Following (Steiner, et al., 2003), the spatial coverage is substituted by spatial scale L which is the side length of A. After calculating the standard error represented in (1) the error model is devised by a power law multivariate function where the error is a function of spatial scale L, temporal coverage T and the precipitation intensity of $\hat{P}$ over the spatiotemporal scale of consideration. As illustrated in detail in Hong, et al., (2006), the error model is given as:

$$\sigma_e = f\left(\frac{1}{L}, \frac{\Delta t}{T}, P\right) = a \cdot \left(\frac{1}{L}\right)^b \cdot \left(\frac{\Delta t}{T}\right)^c (\hat{P})^d \qquad (2)$$

Where, $\Delta t = T/N$ is the satellite sampling frequency assuming that the satellite makes N visits over area A during the period of T.

Calibrating error model (2) for various spatial and temporal scales and also precipitation magnitudes yields the parameters a, b, c, and d. As depicted in Hong, et al., (2006), the spatial scale ranging from 0.04 to 0.96° and temporal scale ranging from hours to days with associated precipitation rate $\hat{P}$ were used in the calibration process for the central part of US where our study river basin is located in the southeast part of the domain.

Considering the multiplicative nature of error in precipitation data, the CCS satellite product is assumed to be lognormally distributed with mean $\mu_X = \hat{P}$ and standard deviation of $\sigma_X = \sigma_e$. In other words, the log transformation of CCS data transforms the data to Gaussian distribution $N(\mu_{LX}, \sigma_{LX})$ where μ_{LX} and σ_{LX} are the mean and standard deviation of the transformed data respectively and can be calculated as follows:

$$\mu_{LX} = \ln\left(\frac{\mu_X^2}{\sqrt{\mu_X^2 + \sigma_X^2}}\right) \qquad (3)$$

$$\sigma_{LX} = \sqrt{\ln\left(\frac{\sigma_X^2}{\mu_X^2} + 1\right)} \qquad (4)$$

Now, precipitation ensemble can be generated using the lognormal distribution for a desired ensemble size, n:

$$\hat{P}_{ens} = \log N(\mu_{LX}, \sigma_{LX}, n) \qquad (5)$$

Which is equal to:

$$\hat{P}_{ens} = \exp(\sigma_{LX} * \varepsilon^i + \mu_{LX}) \quad \text{and} \quad i = 1:n \qquad (6)$$

Where, ε is normally distributed random variable with zero mean and standard deviation of 1, i.e. $\varepsilon \sim N(0,1)$.

4 Study Basin and Datasets

The study area selected is the Leaf River Basin (1944 Km^2) located north of Collins, Mississippi, United States which has been widely used in various studies (see Moradkhani and Sorooshian, (2008) for list of references) (Fig. 1).

Fig. 1 The location of Leaf River basin

The satellite-based precipitation data was based on the PERSIANN-CCS (Precipitation Estimation from Remotely Sensed Information Using Artificial Neural Networks – Cloud Classification System) product (Hong, et al., 2004). The PERSIANN is a satellite infrared-based algorithm that produces global estimates of rainfall based on infrared brightness temperature image provided by geostationary satellites. These data covers 50°S–50°N globally and an improved version of PERSIANN algorithm estimates rainfall at spatial and temporal resolution of 0.04 × 0.04° and 30 min respectively. Five years of data from 2001 to 2005 was extracted and used for this study from (http://hydis8.eng.uci.edu/CCS/).

The other rainfall product used to estimate the error associated with the satellite product, is the National Weather Service WSR-88D Stage IV radar rainfall data. The Next generation weather radar (NEXRAD, formally known as the Weather Surveillance Radar-1988 Doppler -WSR-88D) of the United States provides precipitation data are obtained from 158 Weather radar deployed by the Weather Service

Agency across the country. The data was acquired from NCEP (National Center for Environmental prediction) for the same period and the spatial domain as that of PERSIANN-CCS and used here as reference (true) rainfall field.

The other hydrologic data sets used for the study were the rainfall from rain gages and temperature data extracted from National Climate Data Center (NCDC) and streamflow data from USGS for the period of 1948–2005. These data were utilized for calibrating and validating the hydrologic model (see next section).

5 Hydrologic Model and Ensemble Streamflow Simulation

The Sacramento Soil Moisture Accounting Model (SAC-SMA) is the test model utilized in this study. The model was originally developed at the California-Nevada River Forecast Center located in Sacramento, California as a "Generalized Stream flow Simulation System", (Burnash, et al., 1973; Burnash, 1995) to ameliorate the region's operational capability of stream flow prediction. The model passed through extensive development and testing on the other catchments in United States before its streamflow simulation capability got adopted as major component of the National Weather Service River Forecast System (NWSRFS).

The SAC-SMA model is a conceptual hydrologic model consisting of six sate variables that represents the soil water content of the upper and lower soil mantels in the form of free and tension water (Burnash, 1995). The soil moisture storages in

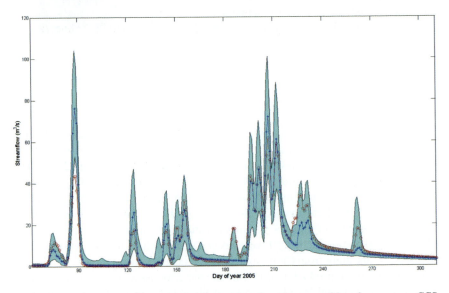

Fig. 2 Streamflow ensemble generated by propagating the ensemble of erroneous CCS-PERSIANN rainfall. The *shaded* area represents the 95 percentile of the streamflow ensemble. The *solid circle* and *solid asterisk lines* display the synthetic truth and deterministic simulation by deriving the SAC-SMA model with radar rainfall and CCE data respectively

both zones vary due to the non linear dynamics of precipitation, percolation, evaporation and lateral drainage. The working principle of the model designates that the precipitation which falls on the impervious area of the catchment produces immediately a direct runoff while the part that falls over pervious area of the catchment infiltrates to the upper zone of the soil mantel. If there is water deficiency in the upper zone of the soil, it gets absorbed till the tension water requirement is met and then starts to flow as free water that percolates to meet the demand of the lower zone. The model has 17 parameters of which 4 are considered fixed and the rest are to be calibrated using the Shuffled complex Evolution (SCE-UA) optimization algorithm developed at University of Arizona (Duan, et al., 1992).

As elaborated earlier, the ensemble of satellite-based precipitation data is generated over Leaf River Basin, Mississippi using Eq. (6) and the hydrologic model is derived with this ensemble for the period of 2001–2005. For clarity we display the ensemble streamflow generation for the portion of last year (2005) of analysis in Fig. 2.

6 Results with Statistical Ensemble Verification

In this section, we define the criteria used to assess the performance of ensemble model simulations/predictions. The common goal in hydrologic predictions is to maximize the accuracy and reliability. There exist various qualitative and quantitative methods to measure the model performance. Root Mean Square Error (RMSE), Absolute Bias (ABS) and Nash Sutcliffe Estimator (NSE) are commonly used for evaluating the accuracy (association of the expected value) of deterministic predictions with observation. In this study, we employed a number of criteria including Normalized Root Mean Square Error Ratio (NRR), Ranked Probability Skill Score (RPSS), Probability of Detection (POD) and False Alarm Ratio (FAR) to evaluate the model performance probabilistically. In probabilistic predictions, it is desired that the probability density function (PDF) of the predictions and observations are consistent.

Model error can be interpreted as model's failure to accurately fit the observations (here hydrologic model response where the model derived by radar rainfall). An ensemble with perfect reliability in an ensemble forecast system is the one that is statistically consistent with the verifying analysis (observations). In other words, in a perfectly reliable ensemble system, observations are expected to be statistically indistinguishable from the forecast members. A useful measure in assessing the effectiveness of the ensemble method in prediction is a comparison between the spread of an ensemble and the ensemble mean forecast error. The deficiency in spread is taken as a measure of the uncertainty associated with the ensemble mean. The ensemble spread can be diagnosed using rank histograms. Rank histograms describe the probability that an observation falls within the $n+1$ intervals defined by an ordered series of n ensemble members (Toth, et al., 2003). As explained in detail by Hamill, (2001), the desired situation is that the observation falls in each of the

$n+1$ intervals equally likely. A reliable or statistically consistent ensemble is the one that yields the rank histogram close to flat.

A simpler procedure to investigate the quality of ensemble is by means of NRR. (Moradkhani, et al., 2006, 2005a) used NRR according to Anderson (2001) to measure the ensemble dispersion for indicating how confidently the ensemble mean could be extracted from the ensemble spread. According to this method, the ratio of the time-averaged RMSE of the ensemble mean, R_1 to the mean RMSE of the ensemble members, R_2 is calculated and then the ratio is normalized by $E[R_a] = \sqrt{\frac{(n+1)}{2n}}$. This factor shows if the observation is statistically indistinguishable from n ensemble members (Anderson, 2001).

$$R_1 = \sqrt{\frac{1}{T}\sum_{t=1}^{T}\left[\left(\frac{1}{n}\sum_{t=1}^{n}y_t^i\right) - O_t\right]^2} \qquad (7)$$

$$R_2 = \frac{1}{n}\sum_{i=1}^{n}\sqrt{\frac{1}{T}\sum_{t=1}^{T}[y_t^i - O_t]^2} \qquad (8)$$

$$Ra = \frac{R_1}{R_2} \qquad (9)$$

$$\text{NRR} = \frac{Ra}{E[Ra]} \qquad (10)$$

Where, n and T are the ensemble size and length of analysis respectively.

It is desired that the ensemble yields NRR = 1. However, NRR > 1 indicates that the ensemble has too little spread, and NRR < 1 is an indication of an ensemble with too much spread. NRR of the streamflow ensemble was obtained annually for the 5 years of simulation from 2001 to 2005. As seen in Fig. 3 the NRR's for all years are above one meaning that the streamflow ensemble is under dispersed. This could be explained in part as a result of a large number of ensemble members having zero precipitation creating very narrow ensemble of precipitation resulting in narrow streamflow ensemble.

Several other forecast verification methods have been reported and are readily available to use (for example see WWRP/WGNE Joint Working Group on Verification http://www.bom.gov.au/bmrc/wefor/staff/eee/verif/verif_web_page.htm), however, in this study we put our focus into using few of these methods. Ranked Probability Score (RPS) is another widely used measure for evaluating the quality of probabilistic predictions (Wilks, 2006). By definition RPS is the sum of squared error of the cumulative probability forecasts averaged over multiple events. In streamflow prediction, the probability forecast is usually expressed using a non-exceedence probability forecast within pre-specified categories (i.e., 5, 10, 25, 50, 75, 90, 95 and 99% non-exceedence). The observed value for a given threshold (forecast category) takes on the value of 1 if the observed flow value is less than the threshold for that category. Otherwise, the observed value is 0. The discrete expression of RPS is given as:

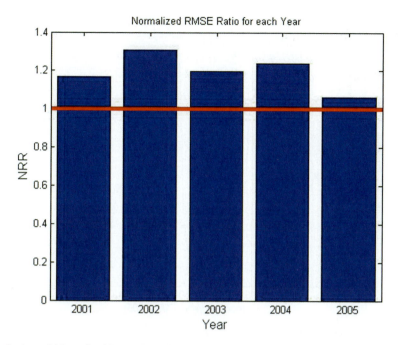

Fig. 3 Annual Normalized Root Mean Square Error Ratio (NRR) for the streamflow ensemble generated by CCS-PERSSIAN Satellite precipitation ensemble. The *solid line* shows the ideal NRR=1

$$\text{RPS}_t = \sum_{i=1}^{J} [F_i^t - O_i^t]^2 \quad (11)$$

Where F_i^t is the forecast probability at time t given by $P(\text{forecast}_i < \text{thresh}_i)$ and O_i^t is the observed probability given by $P(\text{observed} < \text{thresh}_i)$ where i is the probability category.

The average monthly RPS is shown in Fig. 4. A small value for RPS implies that the PDF of simulation is sharp and well calibrated. The RPS is a multi-category extension of the Brier score (*BS*), which is similar to the RMSE and measures the difference between a forecast probability of an event (F) and its occurrence (O), expressed as 0 or 1 depending on if the event has occurred or not.

The average RPS_t is calculated across all simulations over a verification period $t = 1: T$ and is written as:

$$\overline{\text{RPS}} = \frac{1}{T} \sum_{t=1}^{T} \text{RPS}_t \quad (12)$$

Verification statistics such as Root Mean Square Error and RPS are less meaningful when used in absolute terms. Therefore, forecasters prefer to calculate the relative scores and obtain skill scores which will range between 0 and 1 (Wilks, 2006; Jolliffe and Stephenson, 2003). Skill scores, like Rank Probability Skill Score (RPSS) are usually computed as the percentage improvement over a reference score (e.g. climatology) (Bradley, et al., 2004):

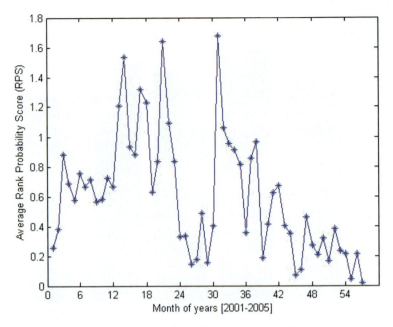

Fig. 4 Average monthly Rank Probability Score for 5 years of ensemble analysis

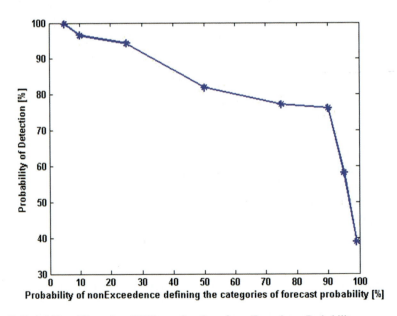

Fig. 5 Probability of Detection (POD) as a function of non-Exceedence Probability

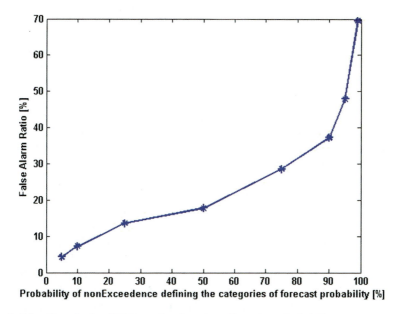

Fig. 6 False Alarm Ration (FAR) as a function of non-Exceedence Probability

$$\text{RPSS} = \left(1 - \frac{\overline{\text{RPS}}}{\text{RPS}_{\text{ref}}}\right) \times 100 = \left(1 - \frac{\overline{\text{RPS}}}{\text{RPS}_{\text{climatology}}}\right) \times 100 \quad (13)$$

Where $\text{RPS}_{\text{climatology}}$ is the rank probability score for the observation, i.e. the synthetic hydrologic response generated from radar rainfall.

Using the above definition, the RPSS was computed as 64.15% meaning that the forecast (here, ensemble simulation) was 64.15% better than using climatology (radar generated truth) when simulation is verified with multiple categories.

For categorical verification metrics, contingency tables were made (they are not shown here) relative to different nonexceedence probability thresholds (pre-specified multiple probability categories) explained earlier and the probability of detection (POD) and False Alarm Ratio (FAR) were calculated (Figs. 5 and 6). For example for the 95% nonexceedence probability which is associated with the flood level of 39 m^3/s, the POD is equal to 59% and the FAR is equal to 48%. Similar POD and FAR can be extracted from Figs. 5 and 6.

7 Summary and Conclusion

In this chapter, we presented a study on applicability of satellite-based precipitation estimation at watershed scale and demonstrated how the precipitation ensembles were generated. The Monte Carlo propagation of precipitation ensemble into a conceptual hydrologic model was made and the hydrologic response in terms of streamflow ensemble was obtained. To assess the usefulness of the product, the ensemble streamflow simulation was probabilistically evaluated.

We utilized rainfall estimates from PERSIANN-CCS (Hong, et al., 2004) in this study. PERSIANN-CCS is a high resolution precipitation estimation system currently operating at near real time mode (http://hydis8.eng.uci.edu/CCS/), and has been generating precipitation data at 30-min time intervals and $0.04 \times 0.04°$ spatial resolution since 2000. A nonlinear multivariate function (Eq. 2) was used according to previous study (Hong, et al., 2006) to estimate the error associated with satellite product. The error model was assumed to be dependent on spatial and temporal resolution of the estimates with the associated rain rate. Using the error model, an ensemble of satellite precipitation was generated assuming that the precipitation data error is multiplicative and therefore lognormally distributed.

To assess the impact of this precipitation uncertainty on hydrologic response in terms of streamflow, the precipitation ensemble was propagated into the SAC-SMA model (a conceptual model currently operated in National Weather Service in US). The resultant streamflow ensemble was statistically evaluated through various probabilistic measures including Normalized RMSE, Rank Probability Skill Score (RPSS), Probability of Detection (POD) and False Alarm Ratio (FAR).

This study aimed at demonstrating another application of satellite-based precipitation product at the watershed scale streamflow simulation. As discussed earlier and also seen through other chapters, similar efforts are underway for different satellite precipitation products and different error models. Considering that one of the major steps in hydrologic data assimilation efforts is to characterize the uncertainly in forcing data (here, precipitation), responsible for the model states and fluxes uncertainties, the importance of realistic characterization of forcing data uncertainty is perceived.

Acknowledgements The partial financial support for this work was provided by NOAA-CPPA grant NA070AR4310203.

References

Adler, F.R. et al. (2003), The version-2 Global Precipitation Climatology Project (GPCP) monthly precipitation analysis (1979–present), *Journal of Hydrometeorology*, *4*, 1147–1167.

Anderson, J.L. (2001), An ensemble adjustment kalman filter for data assimilation, *Monthly Weather Review*, *129*(12), 2884–2903.

Boyle, D.P., H.V. Gupta, and S. Sorooshian (2000), Toward improved calibration of hydrologic models: Combining the strengths of manual and automatic methods, *Water Resources Research*, *36*(12), 3663–3674.

Bradley, A.A., S.S. Schwartz, and T. Hashino (2004), Distributions-oriented verification of ensemble streamflow predictions, *Journal of Hydrometeorology*, *5*(3), 532–545.

Burnash, R.J.C. (1995), The NWS river forecast system- catchment model, in *Computer Models of Watershed Hydrology*, edited by V.P. Singh , pp. 311–365, Water Resources Publications, Littleton, CO.

Burnash, R.J., R.L. Ferral, and R.A. McGuire (1973), A generalized streamflow simulation system: Conceptual modeling for digital computers, Technical Report, Joint Federal-State River Forecast Center, US National Weather Service and California Department of Water Resources, Sacramento, CA, 2004 pp.

Duan, Q.S., S. Sorooshian, and V.K. Gupta (1992), Effective and efficient global optimization for conceptual rainfall- runoff models, *Water Resources Research*, *28*(4), 1015–1031.

Forman, B.A., E.R. Vivoni, and S.A. Margulis (2008), Evaluation of ensemble-based distributed hydrologic model response with disaggregated precipitation products, *Water Resources Research*, *44*(12), W12409.

Gebremichael, M., W.F. Krajewski, M.M. Morrissey, G.J. Huffman, and R.F. Adler (2005), A detailed evaluation of GPCP 1-degree daily rainfall estimates over the Mississippi River Basin, *Journal of Applied Meteorology*, *44*, 665–681.

Hamill, T.M. (2001), Notes and correspondence on "Interpretation of rank histograms for verifying ensemble forecasts", *Monthly Weather Review*, *129*, 550–560.

Hong, Y., K. Hsu, X. Gao, and S. Sorooshian (2004), Precipitation estimation from remotely sensed imagery using artificial neural network-cloud classification system, *Journal of Applied Meteorology*, *43*, 1834–1853.

Hong, Y., K. Hsu, H., Moradkhani, and S. Sorooshian. (2006), Uncertainty quantification of satellite precipitation estimation and monte carlo assessment of the error propagation into hydrologic response, *Water Resources Research*, *42*(8), W08421.

Hossain, F. and E.N. Anagnostou (2006), A two-dimensional satellite rainfall error model, *IEEE Transactions on Geoscience and Remote Sensing*, *44*, 1511–1522.

Hossain, F. and E.N. Anagnostou (2005), Numerical investigation of the impact of uncertainties in satellite rainfall estimation and land surface model parameters on simulation of soil moisture, *Advances in Water Resources*, *28*, 1336–1350.

Hsu, K., X. Gao, S. Sorooshian, and H.V. Gupta (1997), Precipitation estimation from remotely sensed information using artificial neural networks, *Journal of Applied Meteorology*, *36*, 1176–1190.

Huard, D. and A. Mailhot (2006), A Bayesian perspective on input uncertainty in model calibration: Application to hydrological model "abc", *Water Resources Research*, *42*(7), W07416.

Huffman, G.J., R.F. Adler, M. Morrissey, D. Bolvin, S. Curtis, S. Joyce, B. McGavock, and J. Susskind (2001), Global precipitation at one-degree daily resolution from multisatellite observations, *Journal of Hydrometeorology*, *2*, 36–50.

Jolliffe, I.T. and D.B. Stephenson (Eds.) (2003), *Introduction to Forecast Verification, A Practitioners Guide in Atmospheric Sciences*, Wiley Inc. Press, New York.

Kavetski, D., G. Kuczera, and S.W. Franks (2006), Bayesian analysis of input uncertainty in hydrological modeling: 1. Theory, *Water Resources Research*, *42*, W03407.

Margulis, S.A., D. Entekhabi, and D. McLaughlin (2006), Spatiotemporal disaggregation of remotely sensed precipitation for ensemble hydrologic modeling and data assimilation, *Journal of Hydrometeorology*, *7*, 511–533.

Moradkhani, H., K.L. Hsu, H. Gupta, and S. Sorooshian (2005b), Uncertainty assessment of hydrologic model states and parameters: Sequential data assimilation using the particle filter, *Water Resources Research*, *41*(5), W05012, doi: doi:10.1029/2004WR003604.

Moradkhani, H., K. Hsu, Y. Hong, and S. Sorooshian (2006), Investigating the impact of remotely sensed precipitation and hydrologic model uncertainties on the ensemble streamflow forecasting, *Geophysics Research Letters*, *33*, L12401.

Moradkhani, H. and S. Sorooshian (2008), General review of rainfall-runoff modeling: Model calibration, data assimilation, and uncertainty analysis, in *Hydrological Modeling and Water Cycle, Coupling of the Atmospheric and Hydrological Models*, vol. 63, Anonymous, pp. 1–23, Springer, Water Science and Technology Library, New York.

Moradkhani, H., S. Sorooshian, H.V. Gupta, and P.R. Houser (2005a), Dual state–parameter estimation of hydrological models using ensemble Kalman filter, *Advances in Water Resources*, *28*(2), 135–147.

Olson, W.S., et al., (2006), Precipitation and latent heating distributions from satellite passive microwave radiometry. Part I: Improved method and uncertainties, *Journal of Applied Meteorology and Climatology*, *45*, 702–720.

Steiner, M., T.L. Bell, Y. Zhang, and E.F. Wood (2003), Comparison of two methods for estimating the sampling-related uncertainty of satellite rainfall averages based on a large radar dataset, *Journal of Climate, 16*, 3759–3778.

Toth, Z., O. Talagrand, G. Candille, and Y. Zhu (2003), Probability and ensemble forecasts, in *Environmental Forecast Verification: A Practitioner's Guide in Atmospheric Science*, edited by I.T. Jollife and D.B. Stephenson. Wiley, New York.

Vrugt, J.A., C.J.F. Braak, M.P. Clark, and J.M. Hyman (2008), Treatment of input uncertainty in hydrologic modeling: Doing hydrology backward with Markov chain Monte Carlo simulation, *Water Resources Research, 44*, W00B09.

Wilks, D.S. (2006), *Statistical Methods in the Atmospheric Sciences*, Second Edition, Elsevier Academic Press, Amsterdam.

Part III
Real Time Operations for Decision Support Systems

Applications of TRMM-Based Multi-Satellite Precipitation Estimation for Global Runoff Prediction: Prototyping a Global Flood Modeling System

Yang Hong, Robert F. Adler, George J. Huffman and Harold Pierce

Abstract To offer a cost-effective solution to the ultimate challenge of building flood alert systems for the data-sparse regions of the world, this chapter describes a modular-structured Global Flood Monitoring (GFM) framework that incorporates satellite-based near real-time rainfall flux into a cost-effective hydrological model for flood modeling quasi-globally. This framework includes four major components: TRMM-based real-time precipitation, a global land surface database, a distributed hydrological model, and an open-access web interface. Retrospective simulations for 1998–2006 demonstrate that the GFM performs consistently at catchment levels. The interactive GFM website shows close-up maps of the flood risks overlaid on topography/population or integrated with the Google-Earth visualization tool. One additional capability, which extends forecast lead-time by assimilating QPF into the GFM, also will be implemented in the future.

Keywords Flood · Satellite Precipitation · Global flood modeling · TRMM

1 Introduction

Floods impact more people globally than many other type of natural disaster (World Disasters Report, 2003) and they usually return every year in flood-prone regions. It has been established by experience that the most effective means to reduce the property damage and loss of life caused by floods is the development of flood warning systems (Negri et al., 2004). However, progress in large scale flood warning has been constrained by the difficulty of measuring the primary causative factor, i.e. rainfall fluxes, continuously over space (catchment-, national-, continental-, or even global-scale areas) and time (hourly to daily), due largely to insufficient ground monitoring

Y. Hong (✉)
School of Civil Engineering and Environmental Sciences, University of Oklahoma, Norman, OK 73019, USA
e-mail: yanghong@ou.edu

networks, long delay in data transmission and absence of data sharing protocols among many geopolitically trans-boundary basins (Hossain, and Lettenmaier, 2006; Biemans et al., 2009). In addition, in-situ gauging stations are often washed away by the very floods they are designed to monitor, making reconstruction of gauges a common post-flood activity around the world (Asante et al., 2007), e.g., Hurricane Katrina in 2005 and Mozambique flood in 2000.

In contrast, space-borne sensors inherently estimate precipitation across international basin boundaries and safe against recurrent flooding. In reality, satellite-based precipitation estimates may be the only source of rainfall information available over much of the globe, particularly for developing countries in the tropics where abundant extreme rain storms and severe flooding events repeat every year. For instance, the Mekong River Commission, a partner in the Asia Flood Network, began downloading Tropical Rainfall Measuring Mission (TRMM) real-time data since 2003 to help calculate rainfall for the international Mekong River basins, located in China, Cambodia, Laos, Thailand, and Vietnam. These facts highlight the opportunity and need for researchers to develop alternative satellite-based flood warning systems that may supplement in-situ infrastructure for uninterrupted monitoring of extreme rainfall and dissemination of flood alerts when conventional data sources are denied due to natural or administrative causes (Asante et al., 2007; Hong et al., 2007a).

Today, multi-satellite imagery acquired and processed in real time can now provide near-real-time rainfall fluxes at relatively fine spatiotemporal scales (kilometers to tens of kilometers and 30-min to 3-h). These new suites of rainfall products have the potential for analyzing sub-daily variations and extreme flooding events. Shown in Fig. 1 is an example of quasi-global "heavy rain" event maps from TRMM website. The map shows the areas (in red) over land that has accumulated rain totals from a flow of TRMM-based real-time precipitation estimation above a pre-selected threshold. Using such a simple rainfall threshold-based rain map, researchers and decision-makers alike can look at the evolution of regional to global scale events on a daily basis.

While these global heavy rain outlooks on emerging flooding events are potentially useful, this rainfall threshold-based approach has limited implications from a terrestrial hydrologic perspective. First, they are independent of terrain, soil type, soil moisture and vegetation. Furthermore, they do not take into consideration the local/regional hydrologic regimes that determine the pertinent rainfall-runoff relationships. As a result, such a simple statistical approach of thresholds is inadequate in capturing the spatiotemporal variability of estimated runoff and consequently, does not optimally extract valuable hydrologic information contained in rainfall fluxes estimated by satellites. Improvement is particularly important in anticipating the planned Global Precipitation Measurement (GPM) mission that beckons hydrologists as an opportunity to improve flood prediction capability for medium to large river basins, especially in the underdeveloped world where ground instrumentation is absent. The GPM mission (http://gpm.gsfc.nasa.gov) is envisioned as a constellation of operational and dedicated research satellites to provide microwave-based precipitation estimates for the entire globe. Hossain and Lettenmaier (2006) have argued that before the potential of GPM can be realized, there are a number of hydrologic issues that must be addressed prior to the adoption of global satellite rainfall datasets in hydrologic models. Accuracy, in particular, will depend on the sensible use of the spatiotemporally varying rainfall fluxes as derived from satellites and not on the accumulated rainfall volumes that are currently adopted in thresholding techniques.

This chapter describes a module-structured framework for Global Flood Modeling (GFM) system that integrates the TRMM-based multi-satellite forcing data (Huffman et al., 2007) with a cost-effective hydrological model (Hong et al., 2007b), parameterized by a tailored geospatial database, in an effort to evolve toward a more hydrologically-relevant flood prediction with direct rainfall input from real-time satellite nowcasting or quantitative precipitation forecasting (QPF). The GFM is now running in real-time in an experimental mode with results being displayed on the TRMM website http://trmm.gsfc.nasa.gov/publications_dir/potential_flood_hydro.html. A major outcome of this framework is the availability of a global overview of flooding conditions that quickly disseminate through an open-access web-interface. We expect these developments in utilizing satellite remote sensing technology to offer a practical solution to the challenge of building a cost-effective early warning system for data-sparse and under-developed areas. Additionally, through the use of more hydrologically relevant approaches, we hope this framework will spur meteorologists engaged in satellite rainfall data production to communicate more effectively with hydrologic modelers for development of GPM satellite rainfall algorithms (Hong et al., 2006; Hossain and Lettenmaier, 2006).

Fig. 1 TRMM rainfall threshold-based flood potential *Map*: *red regions* show areas that have received a 24-hour rainfall accumulation (May 2, 2003) > 35 mm

2 A Quasi-Global Flood Modeling Framework

Shown in Fig. 2 is the conceptual framework for the GFM that puts forward a computationally simplified hydrological model to predict floods quasi-globally using a combination of data from the TRMM-based multi-satellite products, Shuttle Radar Topography Mission (SRTM), and other global geospatial data sets such as soil property and land cover types. This framework is modular in design and flexible that permits changes in the model structure and in the choice of components. It includes four major components: (1) multi-satellite high-resolution precipitation products; (2) characteristics of land surface including elevation, topography-derived hydrologic parameters such as flow direction, flow accumulation, basin, and river network; (3) spatially distributed hydrological models to infiltrate rainfall and route overland runoff; and (4) an implementation interface to relay the input data to the models and display the flood inundation results on website. The simulation results must be updated at regular intervals (~3 h), requiring computational efficiency in transferring space-borne observations into the operational web-based flood mapping interface, even at the expense of some loss of detail and accuracy. Another note is that the flexible module-structured framework also allows for optimal use of grid-based precipitation flux fields from multiple sources, including in situ, satellite, and numerical model forecasts.

2.1 Satellite-Based Precipitation Products

Precipitation displays high space-time variability that requires frequent observations for adequate representation. Such observations are not possible through surface-based measurements over much of the globe, particularly in oceanic, remote, or developing regions (Huffman et al., 2007). Continued development in the estimation of precipitation from space has culminated in sophisticated satellite instruments and techniques to combine information from multiple satellites to produce precipitation products at long-term coarse scale (Adler et al., 2003) and short-term finer time-space scales useful for hydrology including flood analysis (Sorooshian et al., 2000; Kidd et al., 2003; Joyce et al., 2004; Hong et al., 2004; Turk and Miller, 2005; Huffman et al., 2007). The key data set used in the framework is the TRMM Multi-satellite Precipitation Analysis (TMPA; Huffman et al., 2007), which provides a calibration-based sequential scheme for combining precipitation estimates from multiple satellites, as well as gauge analyses where feasible, at fine scales (0.25 × 0.25° and 3-hourly) over the latitude band 50°N-S (http://trmm.gsfc.nasa.gov).

The TMPA is a TRMM standard product computed for the entire TRMM period (January 1998–present) and is available both in real time and retrospectively. Although Huffman et al. (2007) verified that the TMPA is successful at approximately reproducing the surface-observation-based histogram of precipitation, as well as reasonably detecting large daily events, properly characterizing the impact of the rainfall estimation error structure on hydrological response uncertainty at small scales remains to be carried out (Voisin et al., 2008).

2.2 A Central Geospatial Database

A central geospatial database describing the land surface (topography, land cover, and soils etc.) is archived to derive comprehensive parameter sets for linking the precipitation input with hydrological flood simulation models. The basic topography data considered in this system include NASA SRTM (http://www2.jpl.>nasa.gov/srtm/) and US Geological Survey's GTOPO30 (http://edcdaac.usgs.gov/gtopo30/gtopo30.html). The 30 m horizontal resolution provided by

SRTM data is a major breakthrough in digital mapping of the world, particularly for large portions of the developing world. The digital elevation data are used to derive topographic factors (slope, aspect, curvature etc) and hydrological parameters (river network, flow direction, and flow path). Global soil property data sets are taken from Digital Soil of the World published in 2003 by the Food and Agriculture Organization of the United Nations (http://www.fao.org/AG/agl/agll/dsmw.htm) and the International Satellite Land Surface Climatology Project Initiative II Data Collection (http://www.gewex.org/islscp.html). The soil parameters used in this study are soil property information (including clay mineralogy and soil depth) and 12 soil texture classes, following the U.S. Department of Agriculture soil texture classification (including sands, loam, silt, clay, and their fractions). The Moderate Resolution Imaging Spectroradiometer (MODIS) land classification map is used as proxy of land cover/uses at its highest (250-m) resolution (Friedl et al. 2002). A large proportion of the supporting data for implementing the hydrological models is also available from the NASA Goddard Global Land Data Assimilation System (Rodell et al., 2004).

2.3 A Cost-Effective Hydrological Simulation Model

Many hydrological models have been introduced in the hydrological analysis to predict runoff (Singh, 1995) but few of these have become common planning or decision-making tools (Choi et al., 2002), either because the data requirements are substantial or because the modeling processes are too complicated for operational application. Our initial pool of candidate hydrological simulation models included a number of distributed hydrological models (Singh, 1995; Chen et al., 1996; DHI, 1999; Liang et al., 1994; Beven and Kirkby, 1979; Coe, 2000 etc.). However, many of these models are extremely time-consuming when used to update global flood simulation results at sub-daily scale. They require not only more parameterization of the hydrological processes, but additional climate variables such as air temperature or net radiation, vapor pressure and wind speed, which are not always available in real time and also introduce uncertainty due to their unknown error characteristics. Trade-offs between efficiency and complexity in terms of the quantity and quality of the data available to meet model input requirements and the modeling components to implement (Choi et al., 2002) argue for a more simplified model in the current works that:

(1) Treats only the most important processes related to the hydrological problem to be considered (i.e., flood);
(2) Allows the model to evolve as more data become available or as the modeler gains insight during the modeling process; and
(3) is ready to use, accepting a certain loss of detail and accuracy.

In fact, such simplification is already an accepted methodology among hydrologists to allow the hydrologic modeling effort evolve to the global solution through an iterative process (Klemes, 1983). Additionally, for rainfall-runoff simulation alone, parsimonious models often perform as well as sophisticated ones (Jian et al., 2003; Duan et al., 2003). Hence, we adopt the Hydrologic Engineering Center (HEC) Hydrologic Modeling System (HMS) developed by US Army Corps of Engineers (USACE, 2002) for prototyping the Global Flood Model. The HEC-HMS improves upon the capability of the predecessor HEC-1, providing additional capabilities for distributed modeling and continuous simulation (USACE, 2002). Its simplicity is especially critical for the vast un-gauged regions and geopoliticially trans-boundary basins of the world. The following section briefly describes the major modification of the Natural Resource Conservation Services-Curve Number (NRCS-CN) method in the HEC-HMS for incorporating TRMM-based real-time satellite rainfall for rainfall-runoff simulation.

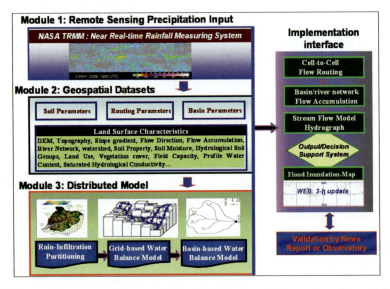

Fig. 2 A modular structured framework for global flood monitoring system: (1) precipitation input; (2) surface geospatial database; (3) hydrological models; and (4) an implementation interface

3 Modified NRCS-CN Method for Global Rainfall-Runoff Simulation

3.1 Mapping a Spatially Distributed Global NRCS-CN

In a literature review, Choi et al. (2002) concluded that NRCS-CN has useful skill because it responds to major runoff-generating properties including soil type, land/use/treatment, and soil moisture conditions. NRCS-CN has been successfully applied to situations that include simple runoff calculation (Heaney et al.,

2001), assessment of long-term hydrological impact on land use change (Harbor, 1994) for tens of years, stream-flow estimation for watersheds with no stream flow records (Bhaduri et al., 2000), and comprehensive hydrologic/water quality simulation (Srinivasan and Arnold, 1994; Engel, 1997; Burges et al., 1998; Rietz and Hawkins, 2000). Recently, Curtis et al. (2007) used satellite remote sensing rainfall and gauged runoff data to estimate CN for basins in eastern North Carolina. Harris and Hossain (2008) found that simpler approaches such as the NRCS-CN method to be more robust than more complicated schemes for the levels of uncertainty that exist in current satellite rainfall data products. As such, we adopt the NRCS-CN to estimate a first-cut global runoff by taking advantage of the multiple years of rainfall estimates from the TRMM Multi-satellite Precipitation Analysis (Huffman et al., 2007).

The NRCS-CN method generates runoff as a function of precipitation, soil property, land use/cover, and hydrological condition. The later three factors are empirically approximated by 1 parameter, CN. In this case, the set of Equations (1) and (2) is used to partition rainfall into runoff and infiltration:

$$Q = \frac{(P - IA)^2}{(P - IA + PR)} \tag{1}$$

$$PR = \frac{25{,}400}{CN} - 254 \tag{2}$$

where P is rainfall accumulation (mm/day); IA is initial abstraction; Q is runoff generated by P; PR is potential retention; CN is the runoff curve number, with higher CN associated with higher runoff potential; and IA was approximated by 0.2 PR.

CN values are approximated from the area's hydrologic soil group (HSG), land use/cover, and hydrologic condition, the two former factors being of greatest importance in determining its value (USDA, 1986). Following methodology adopted from the standard lookup tables in USDA (1986) and NEH-4 (1997), we derived a global CN map from infiltration characteristics of soils classified by USDA-NRCS (2005) and the MODIS land cover classification, at long-term averaged soil wetness conditions (Hong and Adler, 2008). Thus, for a watershed on a coarse grid, a composite CN can be calculated as:

$$CN_{com} = \frac{\sum A_i CN_i}{\sum A_i} \tag{3}$$

In which CN_{com} is the composite CN used for runoff volume computations; $i =$ the index of subgrids or watershed subdivisions. $A_i =$ the drainage area of area i.

3.2 Time-Variant NRCS-CN

Note that the CN values obtained from Equations (1), (2) and (3) are for the "fair" hydrologic condition from standard lookup tables, which are used primarily for design applications. However, for the same rainfall amount there will be more runoff under wet conditions than dry conditions. In practice, lower and upper enveloping curves can be computed to determine the range of CN according to the Antecedent Moisture Conditions (AMC):

$$CN_i^I = \frac{CN_i^{II}}{2.281 - 0.01281\, CN_i^{II}} \quad (4)$$

$$CN_i^{III} = \frac{CN_i^{II}}{0.427 + 0.00573\, CN_i^{II}} \quad (5)$$

Where upper subscripts indicates the AMC, *I* being dry, *II* normal (average), and *III* wet (Hawkins, 1993). The change of AMC is closely related to antecedent precipitation (NEH-4, 1997). We apply the concept of an Antecedent Precipitation Index (API) to provide guidance on how to estimate the variation of CN values under dry or wet antecedent precipitation conditions. Kohler and Linsley (1951) define API as:

$$API = \sum_{t=-1}^{-T} P_t k^{-t} \quad (6)$$

Where *T* is the number of antecedent days, *k* is the decay constant, and *P* is the precipitation during day *t*. The model is also known as "retained rainfall" (Singh 1989). Decay constant *k* is the antilog of the slope on a semi-log plot of soil moisture and time (Heggen, 2001). API practice suggests that *k* is generally between 0.80 and 0.98 (Viessman and Lewis 1996). Here we use decay constant *k* as 0.85 for demonstration purpose. API generally includes moisture conditions for the previous 5 days (or pentad; NEH-4, 1997). In order to obtain time-variant CN, the site-specified API is first normalized as:

$$NAPI = \frac{\sum_{t=-1}^{-T} P_t k^{-t}}{\overline{P} \sum_{t=-1}^{-T} k^{-t}} \quad (7)$$

Where *T*=5 for pentads, the numerator is API, and the denominator is a normalizing operator with two components: average daily precipitation $\overline{P}$ and the $\sum k^{-t}$ series. The "dry" condition is defined as NAPI < 0.33, the "wet" condition is defined as NAPI>3, and the intermediate range 0.33~3 is the "fair" hydrological condition. By definition, the surface moisture conditions are delineated as dry (or wet) if any pentad API is less than one third (or larger than three times) of the climatologically averaged pentad API, and fair conditions for all others. Summarizing, the CN can be converted to dry, fair, or wet condition using Equations

(4), (5), (6) and (7) according to the moisture conditions approximated by the pentad NAPI. Using the multi-year (1998–2006) satellite-based precipitation dataset from NASA TRMM, the 9-year climatological pentad API is shown in Fig. 3a. Thus, given any date, the pentad NAPI can be determined and thus CN can be updated with Equations (4), (5), (6) and (7). For example, on August 25, 2005, the pentad rainfall accumulation, pentad NAPI, resulting hydrological conditions (dry, fair, or wet), and the updated CN on the same date are shown in Figs. 3b–e, respectively. Note that part of text in Section 3.2 is from previous publication by Hong et al., 2007b.

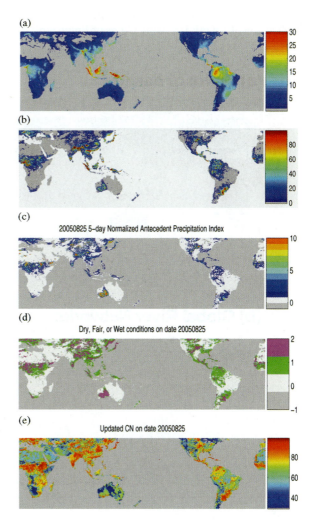

Fig. 3 (**a**) Climatological pentad antecedent precipitation index (API) averaged over 9 years (1998–2006). (**b**) Pentad antecedent rainfall accumulation (mm) ending on August 25, 2005. (**c**) Pentadnormalized API (NAPI) on August 25, 2005. (**d**) Hydrological condition, with −1, 0, 1, and 2 corresponding to no data, dry, fair, and wet conditions, respectively, determined by NAPI as of August 25, 2005. (**e**) Updated CN on August, 25 2005. (image source: Hong et al., 2007b)

4 Implementation of the GFM

4.1 Retrospective Simulation

The time-variant global CN map enables estimation of runoff by partitioning gridded satellite-based precipitation estimates, particularly useful at places lacking in-situ gauge data. Surface topography data from the geospatial database are used to define river networks and sub-catchments on a global basis such as shown in Fig. 4. Flow directions and flow speed are also calculated from the global DEM, soil, and land cover databases described in Section 2.2 driven by multiyear remote sensing rainfall, the NAPI and NRCS-CN methods are first used to compute the surface runoff for each grid independently. The first-order linear differential equation is used for routing the overland flow to the watershed outlet through downstream cells (Olivera et al., 2000; US Army Corps of Engineers, 2000). Quasi-global runoff data

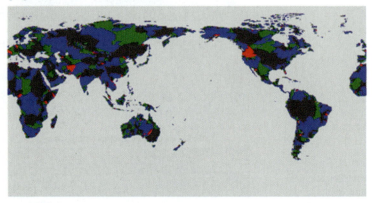

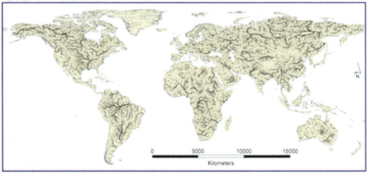

Fig. 4 Global Basins and River Networks

are thus simulated from this framework over the entire time span (1998–2006) of TMPA dataset at 3-h time scale. The simulated surface runoff is water depth with unit mm/3h. In general, observed river discharge can be used to evaluate the GFM simulation by dividing the contributing drainage area (Yilmaz et al., 2008).

The TRMM-simulated runoff (TRMM-CN) is compared with the three sets of Global Runoff Data Center (GRDC) runoff fields: observed (OBS), water balance model (WBM)-simulated, and composite (CMP) from the OBS and WBM (Fekete et al., Global Composite Runoff Data Set (v1.0), 2000). The WBM used the water balance model of Thornthwaite and Mather [1955] with a modified potential evaporation scheme from Vorosmarty et al. [1998], driven by input monthly air temperature and precipitation from Legates and Willmott [1990a, 1990b]. Table 1 shows the TRMM-CN runoff corresponds more closely with the WBM, having a relatively high correlation and low error. An intercomparison with the GRDC runoff observation demonstrates that the WBM has a moderate advantage over the TRMM-CN runoff: The correlation and root-mean-square difference (RMSD) between the GRDC OBS and WBM are 0.81 and 159.7 mm/year (or 0.44 mm/d), respectively, which is slightly better than the TRMM-CN case (Table 1).

Figure 5a shows the annual mean runoff (mm/year) driven by TRMM daily precipitation for the same 9-year period in comparison with the GRDC-observed runoff climatology (Fig. 5b). Note that the gray areas indicate no data or water surface in Figs. 5a and b. By averaging areas covered by both TRMM-CN and GRDC runoff data, Fig. 5c shows the TRMM-CN runoff zonal mean profile against the OBS, WBM, and CMP. In general, the TRMM-CN zonal mean runoff follows more closely with the three GRDC runoff profiles in the Northern Hemisphere than in the Southern Hemisphere. We believe that this difference is the result of having many more samples in the Northern Hemisphere as well as more accurate GRDC data. Considering the TRMM-CN runoff difference as a function of basin area shows the TRMM-CN performance deviates more for basins smaller than 10,000 km^2, with significantly better agreement for larger basins (Fig. 6).

Figure 7a shows the locations of GRDC gauge stations, which represents 72% catchment coverage of actively discharging land surface (excluding Antarctica, the glaciated portion of Greenland and the Canadian Arctic Archipelago). Figure 7b provides the scatterplot for all annual mean rainfall and runoff matchups of GRDC

Table 1 TRMM-CN runoff climatology in the latitude band 50S–50N compared to GRDC observed, water balance model, and the later two composite runoff[a]

Statistics	GRDC runoff climatology		
	OBS	WBM	CMP
Corr. Coef.	0.75	0.80	0.79
Bias ratio	1.28	1.12	1.12
RMSD	0.56 mm/day	0.48 mm/day	0.51 mm/day

[a] Abbreviations are OBS, observed; WBM; water balance model; CMP, composite; and RMSD, root-mean-square difference (Source: Hong et al., 2007b).

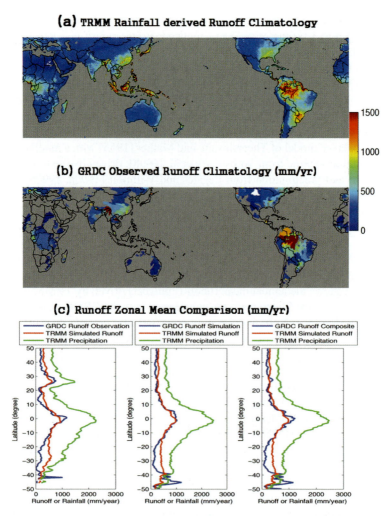

Fig. 5 (**a**) Annual mean runoff (mm/year) simulated using NRCS-CN methods from TRMM estimates for the period 1998–2006. (**b**) GRDC-observed runoff (mm/year). (**c**) Runoff zonal mean profiles comparing TRMM precipitation (*green*) and simulated runoff (*red*) to GRDC runoff (*blue*) from the (*left*) observed, (*middle*) WBGS, and (*right*) composite data sets. Note the *gray areas* in Figs. 3a and b indicate no data or water surface (image source: adapted from Hong et al., 2007b)

and TMPA. Both the rainfall and runoff scattergrams are quite consistent, with correlation coefficient above 0.8. However, it also indicates that TMPA overestimated at high range of rainfall and low range of runoff values, causing positive bias of 0.57 and 7.74% comparing to GRDC, respectively. The likely explanation is that a wind loss correction is applied to the gauge data used in the TMPA, but not to the GRDC. In contrast to the point comparison in Fig. 7, basin-averaged runoff scattergrams in Fig. 6 show that TRMM runoff significantly underestimated runoff over small basins

Flood and Landslide Modeling

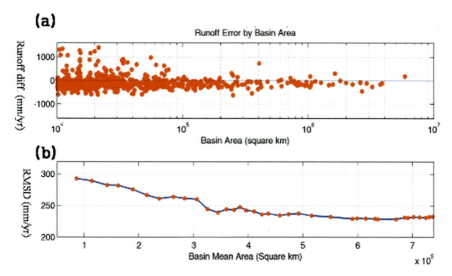

Fig. 6 TRMM-CN (**a**) runoff difference distribution and (**b**) root-mean-square difference (RMSD) as a function of basin area (image source: adapted from Hong et al., 2007b)

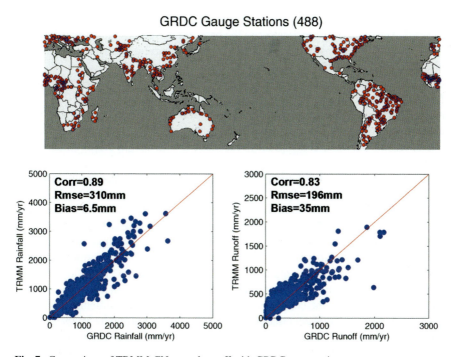

Fig. 7 Comparison of TRMM-CN annual runoff with GRDC gage stations

and increasingly performed better for larger basins. For more evaluation results of this simulation framework, please refer to Hong et al., 2007b.

4.2 Implementation Interface

Since 2007, the GFM trial version has been operating at near real-time on NASA TRMM website in an effort to offer a practical cost-effective solution to the ultimate challenge of building flood early warning systems for the data-sparse regions of the world. A hydrograph of rainfall-runoff time series is simulated for each grid and basin; and the hydrograph estimates are the basis for a web-based interface that displays color-coded maps of flood potentials by comparing current surface runoff with a predefined water depth or bankfull flow value. The interactive GFM website shows close-up maps of the flood risks overlaid on topography and integrated with the Google-Earth visualization tool (http://trmm.gsfc.nasa.gov/publications_dir/potential_flood_hydro.html).

Shown in Fig. 8 are examples of flood events detected and displayed by the framework. As shown in the example in Fig. 8, the GFM diagnosed flooding events of February 2007 in Mozambique as the result of tropical cyclone Favio. This event was verified by Reuters News reports: 69+ dead and more than 120,000 lives affected. The hydrograph estimates are the basis for a Web-based interface that displays color-coded maps of flood risk by comparing current surface runoff in each grid box with a predefined water depth. Figure 8c shows a map of excess water depth due to heavy rainfall from tropical cyclone George that caused widespread flooding in the northwestern Northern Territory, Australia, on March 5, 2007. This event is shown as an example of the current Web interface: a quasi-global interactive map with flood potential areas highlighted in red and close-up maps of the flood potential overlaid on topography or integrated with Google-Earth visualization tools.

Also shown in Fig. 8d is the 9-year rainfall-runoff simulation results over the Limpopo Basin, Southern Africa. The runoff spikes in February 2000 indicate the catastrophic Mozambique flood disaster within the TMPA's limited 9-year span. On February 27, floods inundated low farmlands in the worst flooding event in Mozambique for last 55 years (ARPAC, 2000). Two million people were affected by the floods, 50,000 lost homes, and about 800 were killed. The flood also had a tremendous effect on agriculture in Mozambique, destroying 1,400 km^2 of cultivated and grazing land, leaving 113,000 small farming households with nothing, and damaging 90% of the country's functioning irrigation infrastructure. One year later, another severe flood occurred in late February, 2001, caused by heavy seasonal rains (Fig. 7d), which killed 52 people and displaced almost 80,000 in the central Zambezi Valley (FAO, 2001). Figure 8b shows the 2007 flood in Mozambique, which worsened when Category 4 Cyclone Favio made landfall on February 22. The United Nations Office for the Coordination of Humanitarian Affairs reported approximately 121,000 people displaced. However, US Agency for International

Development (USAID) cited Mozambique's response to 2007 flood as a success: "Deaths from this year's disasters were kept to a minimum – less than 100 – due largely to the timely response and efficiency of Mozambican emergency operations", according to USAID regional director, Jay Knott in Mozambique. Overall, US humanitarian and development aid to Mozambique amounted to $150 million in 2006, with $200 million sought by the Bush administration for 2007. The TRMM website (http://trmm.gsfc.nasa.gov) provides more extreme rainfall and flooding examples.

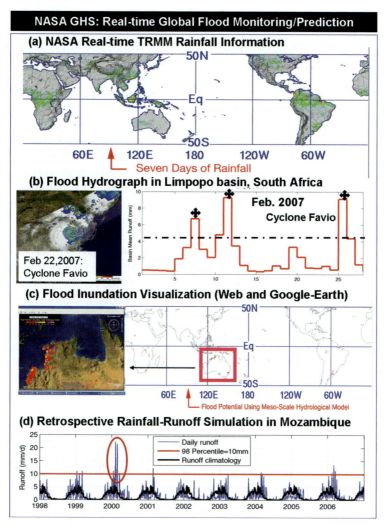

Fig. 8 Implementation of GHS-Flood: (**a**) Example of 7-day accumulation of real-time TMPA rainfall, and (**b–d**) examples of GHS-Flood detection/visualization results

5 Summary and Discussion

5.1 Summary

Satellite rainfall observations acquired in real time are valuable in improving our understanding of the occurrence of flood hazardous events and in lessening their impacts on the local economies and reducing injuries around the world. This chapter described a practical modular-structured framework, Global Flood Modeling (GFM) that predicts surface runoff by incorporating NASA TRMM-based multi-satellite real-time precipitation estimates into a cost-effective hydrological model that includes parameters from high-resolution topography and other geospatial data sets. As shown in Fig. 2, this framework includes four major components: (1) a real-time satellite-based precipitation measuring system; (2) a geospatial database containing global land surface characteristics; (3) a spatially distributed hydrological model; and (4) an open-access web interface.

Given the increasing availability of global geospatial data describing land surface characteristics, we adopt the HEC-HMS developed by US Army Corps of Engineers (USACE, 2002). The HEC-HMS improves upon the capability of the predecessor HEC-1, providing additional capabilities for distributed modeling and continuous simulation. The runoff generation method is the NRCS-CN that determines runoff as a function of precipitation, soil property, land use/cover, and hydrological conditions. The later three factors are empirically approximated by one parameter, the CN. First, this study estimated a global CN map primarily based on soil property and land use/cover information under the "fair" moisture condition. Then using Antecedent Precipitation Index (API) from TRMM rainfall as a proxy of initial moisture conditions, this study further estimated time-variant CN values bounded by dry and wet moisture conditions approximated by pentad normalized API (Hong et al., 2007b). Finally, driven by 3-h TMPA precipitation estimates, quasi-global runoff was simulated with this framework for period of 1998–2006. Compared to GRDC runoff, this framework provides consistent estimation of runoff at both stations and medium-to-large basins. Currently, the framework operationally runs in near-real-time and updates flood conditions every 3 hours with the most current satellite-based rainfall maps (http://trmm.gsfc.nasa.gov/publications_dir/potential_flood_hydro.html).

5.2 Discussion and Directions of Alternative Flood Modeling Work

Our effort is a first approach to understand a challenging problem that lies ahead in advancing satellite-based global runoff monitoring. Thus, the use of NRCS-CN should not be construed as a call for replacement of other more advanced methods for rainfall-runoff simulation. We expect that the successes and limitations revealed in this study will lay the basis for applying more advanced methods to capture the dynamic variability of the hydrologic processes for global runoff monitoring in real time. Although this study is able to demonstrate the potential of using this framework for quasi-global flood monitoring when driven by satellite-based rainfall estimates, there remain several unanswered questions: First, among many methods to estimate CN values, Hawkins (1993) recognized that remote sensing data may not be adequate to define the "true" value of a CN. Thus, field surveys of basin characteristics should be conducted wherever feasible in order to obtain "true" soil and land cover data. Second, while this study recognized the uncertainty of the estimates of actual CN values and assumed that they likely fall within the enveloping wet (upper) and dry (lower) conditions approximated by the 5-day Normalized API, it may be possible to adjust the CN more precisely to account for local or regional information given their availability in the future. Finally, one major unaddressed hydrological concern for rainfall-runoff applications of remotely sensed precipitation is the thorough evaluation of satellite-based rainfall estimation error and its nonlinear influence on rainfall-runoff modeling uncertainty in varying landscapes and climate regimes (Hong et al., 2006; Hossain and Anagnostou, 2006; Villarini and Krajewski, 2007). Thus, while we conclude that this cost-effective framework seems to provide a reliable tool when using state-of-the-art satellite precipitation data, we also urge similar studies using more sophisticated hydrological models, particularly seeking to serve the vast un-gauged regions and geopolitically trans-boundary basins of the world. An alternative approach as described by Brakenridge et al. (2005) is also encouraged to explore by the global flood monitoring community.

As an initial step to evolve toward a more hydrologically-relevant approach that can make better use of the valuable information contained in the state-of-the-art satellite precipitation data, the modular-structured framework allows the use of new components and the integration of locally existing flood management tools into a global flood alert system. One on-going activity is to continue calibrating and regionalizing this system with local in-situ data when they become available. Additionally, more complex hydrological models can be modified and implemented at regional or local scales by subsetting the TRMM rainfall input data. An important improvement to the GFM will be to link it to the NASA Land Information System (LIS; Peters-Lidard et al., 2004), allowing for use of LIS databases, local in-situ data, and various hydrological models. Particularly, we will utilize the LIS platform to improve/regionalize the land surface modeling capability (e.g. VIC) and the associated meteorological data. The LIS will provide three types of inputs for local regionalization: (1) Initial conditions, which describe the initial state of the land surface; (2) Boundary conditions, which describe both the upper (atmospheric) fluxes or states also known as "forcings" and the lower (soil) fluxes or states; and (3) Parameters, which are a function of soil, vegetation, topography, etc., for the selected land surface models in order to predict terrestrial rainfall-runoff processes. Early results of TMPA real-time rainfall application in VIC model show that the TMPA-driven VIC hydrological model simulations were able to capture the flooding events and to represent low flows, although peak flows tended to be biased upward (Su et al., 2008). After the successful launch and operation of the Regional Visualization and Monitoring System (SERVIR) for Mesoamerica (www.servir.net), the NASA Applied Science program has again partnered with

United States Agency for International Development and The Africa Regional Centre for Mapping of Resources for Development (RCMRD) to implement an operational flood warning system as part of the SERVIR-Africa project. The ultimate goal of the project is to build up disaster management capacity in East Africa by providing local governmental officials and international aid organizations a practical decision-support tool in order to better assess emerging flood impacts and to quantify spatial extent of flood risk, as well as to respond to such flood emergencies more expediently. Although the results (Li et al., 2009) suggest that TMPA real-time data can be acceptably used to drive hydrological models for flood prediction purpose in Nzoia basin at resolution of 1-km grid scale after downscaling, continuous progress in space-borne rainfall estimation technology toward higher accuracy and higher spatial resolution is highly appreciated.

The current GFM uses normalized 5-day antecedent precipitation index as a proxy of antecedent moisture conditions to provide initial soil wetness information. In future, we will explore use of daily AMSR-E soil moisture data as initial soil condition instead of the normalized antecedent precipitation index. One additional capability of the framework under consideration is to incorporate quantitative precipitation forecasts (QPF) from NOAA Global Forecast System (GFS) or other global forecasts to predict runoff within the short to medium ranges of lead time. The increased lead time provided by this framework will be a major improvement over in-situ monitoring infrastructures that may have to wait for delayed transmission of rainfall information from upstream countries and may be defeated by severe weather. We expect work reported here to become increasingly fruitful and practical with the advent of GPM-related products and research collaborations.

Acknowledgement This research is carried out with support from NASA's Precipitation Measurement Mission (PMM) and Applied Sciences program under Ramesh Kakar and Stephen Ambrose of NASA Headquarters. Partial research support is from the University of Oklahoma.

Refernces

Adler, R.F., G.J., Huffman, A. Chang, R. Ferraro, P.-P. Xie, J. Yanowiak, B. Rufolf, U. Scheider, S. Curtis, D. Bolvin, A. Bruber, J. Susskind, P. Arkin, and E. Nelkin, 2003: The Version-2 Global Precipitation Climatology Project (GPCP) monthly precipitation analysis (1979–Present). *Journal of Hydrometeorology*, **4**, 1147–1167.

Arquivo do Patrimonio Cultural (ARPAC), 2000: Cheias e Conhecimento Indigina na Bacia do Limpopo, Maputo, Mozambique: ARPAC.

Artan, G., J. Verdin, and K. Asante, 2001: A wide-area flood risk monitoring model. In *Fifth International Workshop on Application of Remote Sensing in Hydrology*. Montpellier, France: UNESCO/IAHS.

Asante, K.O., R.M. Dezanove, G. Artan, R. Lietzow, and J. Verdin, 2007: Developing a flood monitoring system from remotely sensed data for the Limpopo Basin, *IEEE Transactions on Geoscience and Remote Sensing*, **45**(6), 1709–1714.

Beven, K.J. and M.J. Kirkby, 1979: A physically-based variable contributing area model of basin hydrology, *Hydrological Sciences Journal*, **24**(1), 43–69.

Biemans H., R. Hutjes, P. Kabat, B. Strengers, and D. Gerten (2009) Impacts of precipitation uncertainty on discharge calculations for main river basins. *Journal of Hydrometeorology* (In Press).

Brakenridge, G.R., Nghiem, S.V., Anderson, E., and Chien, S., 2005, Space-based measurement of river runoff. *EOS Transactions of the American Geophysical Union*, **86**(19), 185–188.

Bhaduri, B., J. Harbor, B.A. Engel, and M. Grove, 2000: Assessing watershed-scale, long-term hydrologic impacts of land-use change using a GIS-NPS model, *Environmental Management*, **26**(6), 643–658.

Burgess, S.S.O., M.A. Adams, N.C. Turner, and C.K. Ong, 1998: Redistribution of water within plant root system, *Oecologia*, **115**, 306–311.

Chen, F.,K. Mitchell, J.C. Schaake, Y. Xue, H.-L. Pan, V. Koren, Q.Y. Duan, M. Ek, and A. Betts, 1996: Modeling of land-surface evaporation by four schemes and comparison with FIFE observations, *Journal of Geophysical Research*, **101**, 7251–7268.

Choi, J.Y., B.A. Engel, and H.W. Chung, 2002: Daily streamflow modeling and assessment based on the curve-number technique, *Hydrological Processes*, **16**, 313–3150

Coe, M.T., 2000: Modeling terrestrial hydrological systems at the continental scale: testing the accuracy of an atmospheric GCM, *Journal of Climate*, **13**, 686–704.

DHI (Danish Hydraulic Institute), 1999: MIKE 21 User Guide and Reference Manual, Hørsholm, Denmark.

Duan, Q., H. V. Gupta, S. Sorooshian, A. N. Rousseau, R. Turcotte, (Eds.), 2003: Calibration of watershed models, *AGU Water Science and Application*, Vol. **6**, American Geophysical Union, Washington, DC.

Engel, B.A., 1997: "GIS-based CN Runoff Estimation", Agricultural and Biological Engineering Departmental Report, Purdue University.

FAO, 2001: Food and Agriculture Organization Foodcrops and Shortages, *Report #1*

Fekete, B.M., C.J. Vorosmarty, and W. Grabs, 2000: Global Composite Runoff Data Set (v1.0), Complex Systems Research Center, University of New Hampshire, Durham, New Hampshire, USA. Available online at http://www.grdc.sr.unh.edu/

Friedl, M.A., D.K. McIver, J.C.F. Hodges, X.Y. Zhang, D. Muchoney, A.H. Strahler, C.E. Woodcock, S. Gopal, A. Schneider, A. Cooper, A. Baccini, F. Gao, and C. Schaaf, 2002: Global land cover mapping from MODIS: algorithms and early results, *Remote Sensing of Environment* **83**(1–2), 287–302.

Hawkins, R.H., 1993: Asymptotic determination of runoff curve numbers from data. *Journal of Irrigation and Drainage Engineering. American Society of Civil Engineers*, **119**(2), 334–345.

Harbor, J., 1994: A practical method for estimating the impact of land-use change on surface runoff, groundwater recharge, and wetland hydrology, *Journal of the American Planning Association*, **60**(1), 95–108.

Harris, A., F. Hossain, 2008: Investigating optimal configuration of conceptual hydrologic models for satellite rainfall-based flood prediction for a small watershed, *IEEE Geosciences and Remote Sensing Letters*, **5**(3).

Heaney, J.P., 2001: Long-term experimental watersheds and urban stormwater management, *Water Resources Impact*, **3**(6), 20–23.

Heggen, R.J., 2001: Normalized antecedent precipitation index, *Journal of Hydrologic Engineering*, **6**(5), 377–381.

Hong, Y., K.-L. Hsu, X. Gao, and S. Sorooshian, 2004: Precipitation estimation from remotely sensed imagery using artificial neural network – cloud classification system, *Journal of Applied Meteorology*, **43**(12), 1834–1853.

Hong, Y., K.-L. Hsu, H. Moradkhani, and S. Sorooshian, 2006: Uncertainty quantification of satellite precipitation estimation and Monte Carlo assessment of the error propagation into hydrologic response, *Water Resources Research*, **42**(8), W08421, 10.1029/2005WR004398

Hong, Y, R.F. Adler, A. Negri, and G.J. Huffman, 2007a: Flood and landslide applications of near real-time satellite rainfall estimation, *Journal of Natural Hazards*, **43**(2), DOI: 10.1007/s11069-006-9106-x.

Hong, Y., R.F. Adler, F. Hossain, S. Curtis, and G.J. Huffman 2007b, A first approach to global runoff simulation using satellite rainfall estimation, *Water Resources Research*, **43**(8), W08502, doi: 10.1029/2006WR005739.

Hong, Y and R.F. Adler, 2008, Estimation of Global NRCS-CN (Natural resource conservation service curve numbers), *International Journal of Remote Sensing*, **29**(2), 471–477, DOI: 10.1080/01431160701264292.

Hossain, F. and E.N. Anagnostou, 2006: Assessment of a multi-dimensional satellite rainfall error model for ensemble generation of satellite rainfall data, *IEEE Geosciences and Remote Sensing Letters*, **3**(3), 419–423, doi: 10.1109/LGRS.2006.873686.

Hossain, F. and D.P. Lettenmaier 2006. Flood prediction in the future: recognizing hydrologic issues in anticipation of the Global Precipitation Measurement Mission – Opinion Paper *Water Resources Researchvol*, **44**, doi:10.1029/2006WR005202.

Huffman, G.J., R.F. Adler, D.T. Bolvin, G. Gu, E.J. Nelkin, K.P. Bowman, Y. Hong, E.F. Stocker, and D.B. Wolff, 2007: The TRMM Multi-satellite Precipitation Analysis: Quasi-Global, Multi-Year, Combined-Sensor Precipitation Estimates at Fine Scale. *Journal of Hydrometeorology*, **8**(1), 38–55.

Jian, G.J., S.L. Shieh, and J. McGinley, 2003: Precipitation simulation associated with Typhoon Sinlaku (2002) in Taiwan area using the LAPS diabatic initialization for MM5. *Terrestrial Atmospheric and Oceanic Sciences*, **14**, 261–288.

Joyce, R.J., J.E. Janowiak, P.A. Arkin, and P. Xie, 2004: CMORPH: a method that produces global precipitation estimates from passive microwave and infrared data at high spatial and temporal resolution. *Journal of Hydrometeorology*, **5**, 487–503.

Kidd, C.K., D.R. Kniveton, M.C. Todd, and T.J. Bellerby, 2003: Satellite rainfall estimation using combined passive microwave and infrared algorithms. *Journal of Hydrometeorology*, **4**, 1088–1104.

Klemes V., 1983, Conceptualization and scale in hydrology. *Journal of Hydrology*. **5**, 1–23.

Kohler, M.A. and R.K. Linsley, 1951: Predicting Runoff From Storm Rainfall. US Weather Bureau, Research Paper 34.

Legates, D.R. and C.J. Willmott, 1990a: Mean seasonal and spatial variability global surface air temperature, *Theoretical and Applied Climatology*, **41**, 11–21.

Legates, D.R. and C.J. Willmott, 1990b: Mean seasonal and spatial variability in gauge-corrected, global precipitation. *International Journal of Climatology*, **10**, 111–127.

Li, L., Y. Hong, J. Wang, R. Adler, F. Policelli, S. Habib, D. Irwn, T. Korme, and L. Okello, 2009: Evaluation of the real-time TRMM-based multi-satellite precipitation analysis for an operational flood prediction system in Nzoia Basin, Lake Victoria, Africa, *Journal of Natural Hazards*: doi 10.1007/s11069-008-9324-5.

Liang, X., D.P. Lettenmaier, E.F. Wood, and S.J. Burges, 1994: A simple hydrologically based model of land surface water and energy fluxes for general circulation models, *Journal of Geophysical Research*, **99**(D7), 14415–14428.

Peters-Lidard, C.D., S.V. Kumar, Y. Tian, J.L. Eastman, and P.R. Houser, 2004: Global urban-scale land-atmosphere modeling with the land information system. Presented at the Symposium on Planning, Nowcasting, and Forecasting in the Urban Zone during the 84th Annual Meeting of the American Meteorological Society, 10–16 January 2004, Seattle, WA, USA

Negri, A, N. Burkardt, J.H. Golden, J.B. Halverson, G.J. Huffman, M.C. Larson, J.A. Mcginley, R.G. Updike, J.P. Verdin, and G.F. Wieczorek, 2004: The hurricane-flood-landslide continuum, *BAMS*, DOI:10.1175/BAMS-86-9-1241.

NEH-4 (National Engineering Handbook Section 4 Hydrology Part 630), 1997: U.S. Department of Agriculture Natural Resources Conservation Service, Washington, DC.

Olivera, F., J. Famiglietti, and K. Asante, 2000: Global-scale flow routing using a source-to-sink algorithm. *Water Resources Research*, **36**, 2197–2207.

Rietz, P.D. and R.H. Hawkins, 2000: Effects of land use on runoff curve numbers. Watershed Management, Am. Soc. Civil Engineers. Proceedings Watershed Management Symposium, Fort Collins CO (CD ROM).

Rodell, M., P.R. Houser, U. Jambor, J. Gottschalck, K. Mitchell, C.-J. Meng, K. Arsenault, B. Cosgrove, J. Radakovich, M. Bosilovich, J.K. Entin, J.P. Walker, D. Lohmann, and D. Toll, 2004: The global land data assimilation system, *Bulletin of the American Meteorological Society*, **85**(3), 381–394.

Scott, C., T.W. Crawford, and S.A. Lecce, 2007: A comparison of TRMM to other basin-scale estimates of rainfall during the 1999 Hurricane Floyd flood, *Journal of Natural Hazards*, **43**(2), doi:10.1007/s11069-006-9093-y.

Singh, A., 1989: Digital change detection techniques using remotely-sensed data, *International Journal of Remote Sensing*, **10**(6), 989–1003.

Srinivasan, R. and J.G. Arnold, 1994: Integration of a basin-scale water quality model with GIS, *Water Resources Bulletin. AWRA*, **30**(3), 453–462.

Singh, V.P., 1995: *Computer Models of Watershed Hydrology*, V.P. Singh, ed., Water Resources Publications, Littleton, Co, pp. 1–22.

Sorooshian, S., K-L. Hsu, X. Gao, H.V. Gupta, B. Imam, and D. Braithwaite, 2000: Evaluation of PERSIANN system satellite-based estimates of tropical rainfall, *Bulletin of the American Meteorological Society*, **81**, 2035–2046.

Su, F.G., Y. Hong, and D.P. Lettenmaier, 2008: Evaluation of TRMM Multi-satellite Precipitation Analysis (TMPA) and its utility in hydrologic prediction in La Plata Basin, *Journal of Hydrometeorology*, **9**(4), 622–640; DOI: 10.1175/2007JHM944.1.

Thornthwaite, C.W. and J.R. Mather, 1955: *"The Water Balance," Publications in Climatology VIII(1)*: 1–104, Drexel Institute of Climatology, Centerton, NJ, USA.

Turk, F.J., and S.D. Miller, 2005: Toward improving estimates of remotely-sensed precipitation with MODIS/AMSR-E blended data techniques. *IEEE Transactions on Geoscience and Remote Sensing*, **43**, 1059–1069.

USACE (US Army Corps of Engineers), 2000: HEC-HMS Technical Reference Manual, Hydrologic Engineering Center, Davic, CA, USA, http://www.hec.usace.army.mil/software/hec-hms/

USACE (US Army Corps of Engineers), 2002: HEC-HMS Application Guide, Hydrologic Engineering Center, Davis, CA; http://www.hec.usace.army.mil/software/hec-hms/

USDA (United States Department of Agriculture), 1986: *Urban Hydrology for Small Watersheds, Technical Release No. 55*, US Government Printing Office, Washington, DC.

USDA-NRCS (US Department of Agriculture, Natural Resources Conservation Service), 2005: National Soil Survey Handbook, title 430-VI. Online available: http://soils.usda.gov/technical/handbook/

Viessman, W. and G.L. Lewis, 1996: *Introduction to Hydrology*, Haper Colins College Publishers, New York.

Villarini, G. and W.F. Krajewski, 2007: Evaluation of the research version TMPA three-hourly 0.25 × 0.25° rainfall estimates over Oklahoma, *Geophysical Research Letters*, **34**, L05402, doi:10.1029/2006GL029147.

Vörösmarty, C., C.A. Federer, and A.L. Schloss, 1998. "Potential evaporation functions compared on US watersheds: possible implications for global-scale water balance and terrestrial ecosystem modeling," *Journal of Hydrology*, **207**, 147–169.

Voisin, N., A.W. Wood, and D.P. Lettenmaier, 2008: Evaluation of precipitation products for global hydrological prediction. *Journal of Hydrometeorology*, 9(3), 388–407.

World Disasters Report, 2003: International Federation of Red Cross and Red Crescent Societies, p239.

Yilmaz, K.K., R.F. Adler, Y. Hong, and H. Pierce (2008), Evaluation of a satellite-based near real-time global flood prediction system, *Eos Transaction AGU*, **89**(53), Fall Meet. Suppl., Abstract H43G-1108, SF, CA.

Real-Time Hydrology Operations at USDA for Monitoring Global Soil Moisture and Auditing National Crop Yield Estimates

Curt A. Reynolds

Abstract Global precipitation, temperature, soil moisture, vegetation health, and lake water heights data sets are several operational data sets continuously monitored by crop analysts from the Foreign Agricultural Service (FAS) of USDA to identify global weather and vegetation health anomalies that may affect national crop yield and production in foreign countries. Three relatively new satellite-derived precipitation data sets were recently introduced into the USDA/FAS crop monitoring system, along with two new soil moisture products that utilize passive microwave (PMW). Comparison results indicate no operational global precipitation data set is correct at all times for all geographic areas and those that combine station gauge (SG), polar-orbiting passive microwave and geo-stationary infrared (IR) data tend to perform better.

Keywords Precipitation · Evapotranspiration · Soil moisture · Passive microwave · Satellite radar altimetry · Crop yield estimates

Abbreviations

AFWA	Air Force Weather Agency
AGRMET	Agricultural Meteorology Model (AFWA)
AgRISTARS	Agriculture and Resources Inventory Surveys through Aerospace Remote Sensing
AMSR	Advanced Microwave Scanning Radiometer
AMSR-E	AMSR for EOS
AMSU-B	Advanced Microwave Sounding Unit B
ARS	Agricultural Research Service (USDA)
ASRC	Arctic Slope Regional Corporation (Crop Explorer)
AVHRR	Advanced Very High Resolution Radiometer
CADRE	Crop Condition Data Retrieval and Evaluation

C.A. Reynolds
USDA's Foreign Agricultural Service (FAS), Office of Global Analysis (OGA), International Production Assessment Division (IPAD), Washington, DC, 20250, USA
e-mail: curt.reynolds@fas.usda.gov

CMORPH	CPC MORPHed precipitation
CPC	Climate Prediction Center (NOAA)
DISC	Data and Information Services Center (NASA)
DMSP	Defense Meteorological Satellites Program
DSMW	Digital Soil Map of the World
DSSAT	Decision Support System for Agrotechnology Transfer
ECMWF	European Centre for Medium-Range Weather Forecast
EOS	Earth Observing System (NASA)
EnKF	Ensemble Kalman Filter
ERS	Economic Research Service (USDA)
ESSIC	Earth System Science Interdisciplinary Center (UMD)
FAS	Foreign Agricultural Service (USDA)
FAO	Food and Agriculture Organization (UN)
GDA Corp	Geospatial Data Analysis Corporation
GIS	Geographical Information System
GLAM	Global Agriculture Monitoring System (USDA/FAS and NASA)
GMS	Geostationary Meteorological Satellites
GOES	Geostationary Operational Environmental Satellite
GPM	Global Precipitation Mission
GRLM	Global Reservoir and Lake Monitor
GSFC	Goddard Space Flight Center (NASA)
GTS	Global Telecommunication System
HRSL	Hydrology and Remote Sensing Laboratory (USDA/ARS)
HSB	Hydrological Sciences Branch (NASA/GSFC)
IR	Infrared
IPAD	International Production Assessment Division (USDA/FAS)
JAWF	Joint Agricultural Weather Facility
JRC	Joint Research Center
LACIE	Large Area Crop Inventory Experiment
MARS	Monitoring Agriculture through Remote Sensing techniques
MODIS	Moderate Resolution Imaging Spectroradiometer (NASA/EOS)
NASS	National Agricultural Statistics Service (USDA)
NASA	National Aeronautics and Space Administration
NDVI	Normalized Difference Vegetation Index
NDWI	Normalized Difference Water Index
NEXRAD	Next-Generation Radar
NIR	Near Infrared
NOAA	National Oceanic and Atmospheric Administration
NPOESS	National Polar-orbiting Operational Environmental Satellite System
NWS	National Weather Service (NOAA)
OGA	Office of Global Analysis
PET	Potential evapotranspiration
PMW	Passive Microwave
PR	Precipitation Radar
PRISM	Parameter-elevation Regressions on Independent Slopes Model

PS&D	Production, Supply and Distribution (USDA/FAS)
RFC	River Forecast Centers (NOAA/NWS)
RTNEPH	Real Time Nephanalysis Cloud Model
SG	Station Gauge
SPE	Satellite Precipitation Estimates
SSM/I	Special Sensor Microwave/Imager
SWIR	Shortwave length Infrared
TMI	TRMM Microwave Imager
TMPA-RT	TRMM Multi-Satellite Precipitation Analysis-Real-Time TRMM Tropical Rainfall Measuring Mission
UMD	University of Maryland (UMD)
WPC	WeatherPredict Consulting (Surface Wetness)
USDA	United States Department of Agriculture
VI	Vegetation Indices
VIIRS	Visible Infrared Imager Radiometer Suite (NPOESS)
WAOB	World Agricultural Outlook Board
WASDE	World Agriculture Supply and Demand Estimates
WASP	Weighted Anomaly Standardized Precipitation
WMO	World Meteorological Organization
WPC	Weather Predict Consulting
WSR-88D	Weather Surveillance Radar, 1988, Doppler
3B42RT	TRMM Real-Time precipitation product (PMW and IR merged)
3B42(V6)	TRMM precipitation product merged with PMW, IR, and SG

1 USDA's Global Agriculture Economic Information System

The foundation of the US Department of Agriculture's (USDA) global agriculture economic information system is the monthly World Agricultural Supply and Demand Estimates (WASDE 2008) report distributed by the World Agricultural Outlook Board (WAOB 2008) and the Production, Supply and Distribution (PSD 2008) database archived by the USDA's Foreign Agricultural Service's (FAS). The WASDE report is essentially a timely and monthly audit of national crop statistics collected from around the world and it serves as an objective, reliable, and accurate benchmark of current global crop supply and demand for use by commodity markets, traders, producers, and government policy makers. USDA's monthly WASDE report and PSD historical archive were established to assist the private market place in price determination and adjustment, thus minimizing the risk of market manipulation and contributing towards an international pricing mechanism that accurately reflects real-world circumstances.

To enhance the accuracy and reliability of the monthly crop production forecasts and estimates published in WASDE, global weather data and satellite imagery are monitored by USDA's Joint Agricultural Weather Facility (JAWF) and USDA's

International Production Assessment Division (IPAD) of the Office of Global Analysis (OGA) within the Foreign Agricultural Service (FAS). The use of global weather data, satellite imagery, historical crop area/yield data from PSD, and numerous ancillary data, in most cases, can monitor and provide preliminary national crop production forecasts and estimates required by commodity markets at the beginning of each month. In addition, monitoring global weather data and satellite imagery helps to identify current and regional weather impacts on area and yield estimates in a timely manner and at appropriate national and regional scales.

Official crop statistics from other nations, where available, are critical in forming current crop estimates for the WASDE report, but in practice, not all countries have crop-estimating agencies capable of making reliable, timely, or objective production forecasts. Also, many major producing and trading countries do not publish crop reports until well after the crop has been harvested (Vogel and Bange 1999). In the interim, USDA must monitor precipitation, temperature, NDVI (Normalized Difference Vegetation Index) and other parameters over major crop producing regions that are economically important to the United States trade, especially for crops such as wheat, coarse grains, rice, oilseeds, and cotton. By comparing current weather and crop conditions with historical weather and crop yield data, the WASDE report provides American producers and commodity traders with equal access to global crop supply and demand information which helps to level the playing field and helps to create potential trade markets.

Crop analysts from FAS are responsible for interpreting all relevant weather, satellite imagery and ancillary data, and presenting these interpretations as evidence of support for proposed national crop area, yield or production changes made each month for the WASDE report. The proposed crop production changes are presented at the beginning of each month at crop production meetings held by USDA's Interagency Commodity Estimates Committees (ICEC 2008) and chaired by USDA's WAOB. The monthly ICEC meetings conclude with a final monthly "lockup" meeting and subsequent scheduled release of the WASDE report at 8:30 AM (Eastern Standard Time), on or slightly before the 12th of each month. ICEC was built upon a consensus or "interagency" approach that utilizes "all available information sources" (Vogel and Bange 1999), while FAS crop analysts utilize a "convergence of evidence" methodology for determining national crop area and yield estimates that minimizes risk of error and maximizes reliability.

Precipitation is the major agrometeorological variable that determines relative crop yield changes during the growing season, and, obviously, any improvements in satellite-derived precipitation data sets will help improve USDA's abilities to audit relative crop yield estimates and changes each month.

The first section below briefly describes the various precipitation data sets utilized by USDA's FAS. The next section is the operational soil moisture section and it describes how global precipitation estimates, potential evapotranspiration (PET), and soil water-holding capacity are utilized to run a global two-layer soil moisture model for monitoring potential crop stresses during the growing season. The Global Agriculture Monitoring (GLAM) section briefly describes how soil moisture and NDVI data are utilized and translated into monthly crop yield estimates. The final outlook section describes near-term global precipitation improvements required to improve the existing USDA/FAS global crop monitoring system.

2 Operational Precipitation Products Utilized by USDA/FAS

USDA/FAS relies on several different precipitation data sources to monitor global weather anomalies that affect national crop yield and production for several major agricultural commodities deemed economically important to the United States. The precipitation data sets utilized by USDA/FAS are summarized in Table 1 and include the following:

1. Ground Station (SG) data from the Global Telecommunication System (GTS) of the World Meteorological Organization (WMO)
2. Agricultural Meteorology (AGRMET) model from the Air Force Weather Agency (AFWA)
3. Tropical Rainfall Measuring Mission (TRMM) Multi-satellite Precipitation Analysis (TMPA-RT) real-time product from National Aeronautics and Space Administration (NASA)
4. CPC MORPHed precipitation (CMORPH) product from National Oceanic and Atmospheric Administration's (NOAA) Climate Prediction Center (CPC)
5. Next-Generation Radar (NEXRAD) from NOAA's National Weather Service (NWS)
6. Other Precipitation Data Sets (National, Regional and Commercial)

It should be noted that both the WMO and AFWA precipitation data sets also include daily minimum and maximum (min/max) temperatures and these three input parameters (i.e., precipitation and min/max temperatures) are the only daily input data used to run several operational crop models by FAS. Only precipitation and min/max temperatures are utilized as daily input parameters in the operational crop models in order to keep the global input data minimal and easily operational. These simple crop models utilized by USDA/FAS were established over 25-years ago and are still functional, as well as operational today.

The FAS Crop Assessment Data Retrieval and Evaluation (CADRE) system is the database that stores all the agro-meteorological data used by FAS, including the daily precipitation data sets listed above. CADRE's development began in 1979 (Tingley 1988), making it one of the first GIS (Geographic Information Systems) specifically designed for global agricultural monitoring purposes. CADRE is the operational outgrowth of the LACIE (Large Area Crop Inventory Experiment) and AgRISTARS (Agriculture and Resources Inventory Surveys through Aerospace Remote Sensing) programs which began in 1974 and 1980, respectively (Boatwright and Whitefield 1986). The main cooperating agencies for the LACIE and

Table 1 Summary of precipitation products utilized by USDA/FAS and displayed every 10-days in crop explorer

Product/ source[a]	Spatial resolution	Coverage	Infrared geostationary satellites (IR)	Passive microwave (PMW)	Active radar	Ground station gauge (SG)
GTS/WMO and NOAA/NWS (for USA)	Approx. 7500 stations report daily	Global	No	No	No	Yes
AGRMET/ AFWA	47-km at 60° latitude (true) and 25-km at the equator	Global 60° N-S	Yes	SSM/I	No	Yes, GTS/WMO, NOAA/NWS, and others
CMORPH/ NOAA-CPC	8-km at equator	Global 60° N-S	Yes	SSM/I,TMI, AMSR-E, AMSU-B	No	No
TMPA-RT (3B42RT)/ NASA-DISC	0.25° or approx. 28-km	Global/50° N-S	Yes	SSM/I,TMI, AMSR-E, AMSU-B	No	No[b]
NEXRAD/ NOAA-NWS	4-km	USA/lower 48	Yes[c]	No	Ground doppler	Yes, NOAA/NWS

[a]USDA/FAS' CADRE receives daily all precipitation products listed and CADRE aggregates the daily products into 10-day time periods for agricultural monitoring within Crop Explorer.
[b]Station gauges (SG) are added more than one month later to the 3B42RT product to produce an after real-time global precipitation product called 3B42 (V6).
[c]Satellite precipitation estimates (SPE) are incorporated in regions where there is limited or no radar coverage.

AgRISTARS programs were the National Oceanic and Atmospheric Administration (NOAA), National Aeronautics and Space Administration (NASA), and USDA.

CADRE stores other baseline data sets such as global soil water holding capacities (Reynolds, et al., 2000) and long-term climate normals (New, et al., 2002) so that weather anomalies and soil moisture model results can be monitored and displayed in Crop Explorer. Some of the value-added monitoring tools CADRE provides to Crop Explorer are the following maps and time-series charts:

1. Decadal precipitation (in mm) and temperature (in °C) departures from long-term normals for both WMO station data and AFWA grid cells.
2. Cumulative seasonal precipitation (in mm) with calculations beginning near the start of the crop's growing season.
3. Seasonal percent normal of precipitation (in %).
4. Number of Days since a Rain Day (in the past 30-days) and Maximum Consecutive Dry Days (in the past 30-days).
5. Daily snow cover and decadal departure from long-term average.
6. Daily snow depth (in cm) based on AGRMET output from AFWA.
7. Monthly Weighted Anomaly Standardized Precipitation (WASP).

8. Daily surface and sub-surface soil moisture (in mm).
9. Daily percent soil moisture storage when assuming a root depth of 1-m.

In summary, the following five daily precipitation data sets are downloaded daily into the CADRE database: GTS/WMO, AGRMET/AFWA, TMPA-RT/NASA, CMOPRH/NOAA-CPC and NEXRAD/NOAA-NWS. Daily precipitation data from these data sets are then processed into 10-day time sets and compared to 10-day climatology data sets stored in CADRE. Automated outputs from CADRE are displayed every 10-days in Crop Explorer in the form of weather maps and time-series graphs compared to climate long-term normals. The 10-day (or decadal) maps and graphs displayed in Crop Explorer are specifically designed to monitor and identify adverse weather conditions over the main agriculture regions within the world.

USDA/FAS crop analysts also have the ability to interactively query and extract all spatial and time-series data from CADRE for more detailed regional analysis, as required. These interactive CADRE data extractions and import tools have been developed to interface with ArcView and ArcMap GIS software. Both Crop Explorer and the CADRE spatial data extraction tools were developed by Arctic Slope Regional Corporation (ASRC 2008), who are the primary private contractors serving USDA/FAS crop analysts with spatial products.

In 2002, the FAS Crop Explorer web site was launched so that WMO and AFWA precipitation and temperature data sets, as well as decadal vegetation index (VI) data from several satellite sensors, could be viewed by the public and by commodity traders at 10-day intervals (Crop Explorer 2008). Before 2002, only USDA/FAS crop analysts viewed the WMO and AFWA global data sets for crop monitoring purposes, whereby analysts manually or interactively "pulled" limited amounts of data from CADRE rather than viewing larger volumes of data via automated "pushed" maps and time series charts produced every 10-days in Crop Explorer (Reynolds and Doorn 2001).

Soon after Crop Explorer was launched, it was decided that FAS should test other operational global precipitation data sets produced by other US government agencies such as NOAA and NASA. FAS crop analysts also wanted the ability to view several global precipitation data sets at once, reduce reliance on one particular global precipitation data set at any one time, and ensure they were receiving the best precipitation data available. For example, in some cases, WMO station data may adequately cover a country, while in other countries the WMO station network is sparse or the WMO stations do not consistently report. In these cases, AFWA data may have better spatial coverage or the crop analysts may have to rely on global VI data sets if the global satellite precipitation product had obvious errors. In addition, running five daily precipitation data streams in tandem also allowed FAS to operationally test each precipitation data set for reliability and accuracy for many different regions around the world.

Currently, the relatively new precipitation data sets behind Crop Explorer's firewall are CMORPH, TMPA-RT and NEXRAD products and currently these spatial precipitation products are not available to the public. However, there are plans to display within Crop Explorer the CMORPH, TMPA-RT, and NEXRAD precipitation data sets to the public by the end of 2009.

2.1 Ground Station Data From the World Meteorological Organization (WMO)

A global daily ground station data file with GTS/WMO station data is received by FAS from USDA's Joint Agricultural Weather Facility (JAWF) every morning and downloaded into CADRE every day. The original ground station data file received from JAWF is from NOAA, and NOAA merges daily GTS/WMO station data with additional station data reported within the United States. The merged global GTS/WMO and US station file from NOAA includes the following daily information:
1. Minimum and maximum temperature
2. Precipitation
3. Several data quality flags of unknown reliability

The GTS/WMO data has more than 6000 stations but many of the 6000 stations do not report to the GTS daily, with approximately 3800 GTS/WMO stations reporting each day. In addition, the global NOAA station file contains more than 7000 daily reporting stations after the reported stations from the United States are included.

2.2 AGRMET From the Air Force Weather Agency (AFWA)

The Agricultural Meteorology (AGRMET) model by AFWA was first developed in 1981 (Cochrane 1981). The AGRMET algorithms have evolved over several decades of work and these algorithms are constantly changing (AFWA 2002). The current global precipitation data set from AFWA is estimated by blending four different data sources together:
1. Special Sensor Microwave/Imager satellite (SSM/I) and rain rates are formulated from the brightness temperatures (Hollinger 1989).
2. Geostationary satellites: such as Geostationary Operational Environmental Satellite (GOES) over North and South America; METEOSAT over Europe, Africa and Middle East; and Geostationary Meteorological Satellites (GMS) over Asia and Australia.
3. Real Time Nephanalysis Cloud Model (RTNEPH) (Kiess and Cox 1988).
4. GTS/WMO ground station data and some additional ground station data from other countries.

AGRMET also provides FAS with the following global agrometeorology data sets which are downloaded daily into CADRE:
1. Minimum and maximum temperature
2. Precipitation
3. Snow depth

4. Solar and longwave radiation
5. Potential and actual evapotranspiration

FAS crop analysts find the global AGRMET/AFWA products useful for monitoring global precipitation and temperature anomalies; for determining the start of season based on an arrival of rains algorithm; for determining if start of season (or arrival of rains) was earlier or later than average; for determining how many days a region has not received rain, etc.

2.3 TMPA-RT From National Aeronautics and Space Administration (NASA)

The Tropical Rainfall Measuring Mission (TRMM) Multi-satellite Precipitation Analysis (TMPA) product from NASA is described by Huffman, et al., (2007), where the reader can obtain more detail information.

The TMPA dataset covers the latitude band between 50-degree North-South for the period from 1998 to present and TMPA-RT precipitation estimates are produced in four stages:

1. Passive microwave (PMW) precipitation estimates are calibrated and combined,
2. Infrared (IR) precipitation estimates are created using the calibrated microwave precipitation,
3. PMW and IR estimates are combined, and
4. Rain station gauge (SG) data are incorporated.

Several TMPA-RT products are available from NASA but FAS only imports the daily 3B42RT product into CADRE. It should be noted that the 3B42RT product does not incorporate SG measurements until more than one month after satellite data was acquired. SG data added more than one month later to the 3B42RT product is called the called 3B42 (V6) product, or the after real-time global precipitation product.

The technical documentation for 3B42RT product states it represents a new initiative and should be considered experimental. Formal validation studies are underway but are not yet available. NASA's technical documentation states that TMPA does reasonably well in detecting large daily events, but TMPA has lower skill in correctly specifying moderate and light event amounts on short time intervals.

FAS/OGA aggregates the daily data from 3B42RT into 10-day (or decadal) periods for agriculture monitoring. FAS crop analysts find NASA's 3B42RT product useful to compare with WMO and AFWA global data sets, but lack of SG data within the 3B42RT product is a great weakness. In addition, FAS staff has noticed that daily bias errors from the lack of SG data tend to accumulate over a growing period of 4-months or more and these accumulated errors become very large artifacts over several months of time.

2.4 CMORPH From National Oceanic and Atmospheric Administration (NOAA)

A high-resolution global precipitation analysis technique called "CMORPH" (CPC MORPHed precipitation) is described by Joyce, et al., (2004), where the reader can obtain more detail information.

CMORPH was developed at NOAA's Climate Prediction Center (CPC) for the real-time monitoring of global precipitation. CMORPH provides precipitation estimates on an 8 km latitude/longitude grid (at the equator) from 60°N-60°S with a temporal resolution of 30 min. Daily precipitation products are then produced by using the CPC MORPHing technique on a global basis and accumulating the 30-min segments into a 24-h time period.

The satellite data inputs from CMORPH include half-hourly geostationary satellite infrared (IR) temperature fields and polar orbiting passive microwave (PMW) brightness temperature retrievals with irregular-intervals. The morphing process involves using the relatively poor temporal resolution of passive microwave (PMW) precipitation estimate data and interpolating its movement between retrieval periods. The motivation for developing such precipitation products stem from the fact that passive microwave observations yield more direct information about precipitation than is available from infrared data, but PMW-derived precipitation estimates have poor spatial and temporal sampling characteristics due to their polar orbits. Conversely, while the IR data provide relatively poor estimates of precipitation, they provide extremely good spatial and temporal sampling. Therefore, CPC combines or morphs the data from these two disparate sensors to take advantage of the strengths that each has to offer (Joyce, et al., 2004).

Correspondingly, FAS/OGA aggregates daily CMORPH data into 10-day (or decadal) periods for agriculture monitoring. The CMORPH data is currently displayed behind Crop Explorer's firewall and it is not displayed in the original 8-km spatial resolution.

It should be noted that the CMOPRH product does not blend SG data into its estimates although studies have shown that algorithms which combine both satellite and SG data tend to provide more accurate precipitation estimates than those precipitation products that rely only on satellite sensors without SG data. The lack of SG data in the CMORPH product is a weakness and CMOPRH comparisons with the other rainfall products utilized by FAS tend to show positive biases when compared to daily SG data. In addition, for crop monitoring during the growing season of four months or more, these daily positive bias errors accumulate and make the CMOPRH product not very useful for crop monitoring purposes. However, the improved spatial resolution of 8-kilometer by CMORPH is greatly desired by FAS crop analysts and it is hoped the NOAA CPC developers will consider integrating SG data into the CMOPRH product to remove biases and improve daily rainfall estimates.

USDA Crop Management System 277

2.5 NEXRAD From National Weather Service (NWS)

NEXRAD or **Nexrad** (**Nex**t-Generation **Rad**ar) is a network of 158 high-resolution Doppler weather radars operated by the National Weather Service, an agency of the National Oceanic and Atmospheric Administration (NOAA) within the United States Department of Commerce. Its technical name is **WSR-88D**, which stands for **W**eather **S**urveillance **R**adar, 19**88**, **D**oppler. NEXRAD detects precipitation and atmospheric movement or wind.

The daily NEXRAD precipitation data are quality-controlled, multi-sensor (radar and rain gauge) precipitation estimates obtained from National Weather Service (NWS) River Forecast Centers (RFCs). The "observed" precipitation data is a byproduct of the National Weather Service (NWS) operations at the 12 RFCs, and is displayed as a gridded field with a spatial resolution of 4x4 kilometers over a 24-h time period (NWS 2008).

East of the Continental Divide, the RFCs derive the "observed" precipitation field using a multisensor approach. Hourly precipitation estimates from WSR-88D NEXRAD are compared to ground rainfall gauge reports, and a bias (correction factor) is calculated and applied to the radar field. The radar and gauge fields are combined into a "multisensor field", which is quality controlled on an hourly basis. In areas where there is limited or no radar coverage, satellite precipitation estimates (SPE) can be incorporated into this multisensor field. The SPE can also be biased against rain gauge reports (NWS 2008).

In mountainous areas west of the Continental Divide, a different method is used to derive the "observed" data. Gauge reports are plotted against long term climatologic precipitation (PRISM data), and derived amounts are interpolated between gauge locations (NWS 2008).

Studies have shown that algorithms which combine several sensor inputs such as radar, gauge, and satellite data yield more accurate precipitation estimates than those which rely on a single sensor such as radar-only, gauge-only, or satellite-only. Although the NEXRAD precipitation product is not perfect, this dataset covers the lower 48 states and it is one of the best sources of timely, high resolution precipitation information available (NWS 2008).

NEXRAD data was first introduced behind Crop Explorer's firewall in early 2008 and FAS crop analysts have noticed that it truly has superior data quality compared to the other global precipitation data sets that cover the United States. FAS crop analysts also greatly appreciate this state-of-the-art product.

2.6 Other Precipitation Data Sets (National, Regional and Commercial)

Global, regional, and national precipitation data is available from both public and private sources. However, at the moment, there are no indications that commercial systems provide more reliable global precipitation results than public sources, largely because satellite technology and coverage is the main limitation in producing better global precipitation data sets.

For example, MDA Federal produces a global precipitation data set of good quality through MDA EarthSat Weather. This product has been developed and improved for the past 20-years, and the developers also admit they have experienced the same technical limitations as experienced by precipitation products produced by different US government agencies as described in the previous sections. MDA Federal also provides crop monitoring services to subscribers with their Ag On-Demand Global Weather Interface product. (MDA Federal 2008). FAS crop analysts have been given limited complimentary service to MDA's precipitation product and initial impressions were good with respect to MDA's global precipitation product and quality.

Another satellite based crop monitoring project is the MARS Project (Monitoring Agriculture through Remote Sensing techniques) which is funded by the Joint Research Center (JRC) of the European Commission. The MARS project uses satellites and spatial crop monitoring products with global coverage but they do not report on crop conditions in North America, Southeast Asia and Australia. Besides not reporting on crop conditions worldwide, the MARS project is also slightly different than FAS because they do not release a global monthly report similar to USDA's WASDE that serves global commodity markets. However, MARS receives daily, 10-day and monthly meteorological information produced by Meteoconsult, and the global precipitation data utilized by the MARS project is derived from the ECMWF (European Centre for Medium-Range Weather Forecast) model at Reading, United Kingdom. In addition, MARS receives 10-day crop condition indicators and vegetation data derived from satellites by two European private companies called Alterra and VITO. All global weather, crop indicators, and vegetation data from satellites is available from the MARS interactive web site listed at MARS (2008). FAS crop analysts also utilize the MARS web site for weather and crop model data comparisons to weather maps displayed in Crop Explorer and crop models displayed behind Crop Explorer's firewall.

Many national meteorology departments have also developed web sites that display rainfall and temperature maps based on national station data, which may have greater station density than the global WMO station data currently used by FAS crop analysts. These national meteorological web sites vary in quality and functionality from country-to-country, while many countries still keep their meteorology data virtually hidden from the public. However, for countries with national meteorology web sites, FAS crop analysts often compare the national weather maps (with better ground station density or coverage) to global precipitation maps displayed in Crop Explorer and derived from GTS/WMO, AGRMET/AFWA, TMPA-RT/NASA, CMOPRH/NOAA-CPC and NEXRAD/NOAA-NWS.

3 Operational Soil Moisture Products Utilized by USDA/FAS

The LACIE and AgRISTARS programs were the first joint effort by the U.S. government to use satellite imagery and operational crop models to continuously monitor and assess crop production over selected areas of the world. The Cold War provided the motivation for the LACIE and AgRISTARS programs to monitor crop conditions via satellite imagery over the Former Soviet Union (FSU) and China during the 1970's and 1980's. Many of the spatial crop models from the AgRISTARS program were later expanded in CADRE during the late 1990's to cover most countries worldwide rather than only for the FSU countries and China (Reynolds 2001).

FAS inherited the operational soil moisture, crop stage, and relative-yield models developed by the AgRISTARS program and many of these same models are utilized by FAS worldwide today with some minor model modifications and major database expansion within CADRE. The CADRE database is the heart of the operation and CADRE was originally developed by the AgRISTARS program. The original AgRISTARS crop model codes and soil moisture algorithms utilized by CADRE today were written by an assortment of personnel from the USDA, NOAA, NASA and private contractors.

The two-layer soil moisture algorithm developed by the AgRISTARS program is the backbone algorithm that runs the crop calendar (growth stage) and crop stress (alarm) models. The original crop model code used by CADRE and developed by the AgRISTARS program were early versions of the DSSAT (Decision Support System for Agrotechnology Transfer) suite of crop models (Ritchie 1991, and Ritchie et al. 1998). However, crop varieties parameters were later generalized or re-classified into short, medium, or long-season varieties.

The crop calendar models utilized by DSSAT and CADRE are accretion models that model the crop growth incrementally, based on growing degree-days (or thermal units) for several different crop types and varieties. The growing degree-day algorithm uses daily minimum and maximum temperature measurements, as well as threshold temperatures defined by the particular crop type. The crop calendar models are initialized by average start of season data derived from national crop reports.

Crop-stress models developed by the AgRISTARS programs serve as agrometeorological data filters by alerting crop analysts of abnormal temperature or moisture stresses that may affect yields. Both soil moisture and crop calendar algorithms are used, as well as hazard algorithms. The hazard algorithms are based on temperature and soil moisture thresholds known to be outside the optimal range of growing conditions and which may cause crop damage at various crop stages. For example, optimal growing conditions for corn is critical during the reproductive phase and the soil moisture and temperature thresholds are most sensitive during this stage. Therefore, if the plant experiences extreme water deficits or temperature conditions during the reproductive phase, the alarm model alerts the analysts of the crop stress in the region.

The FAS-CADRE crop calendar and crop stress model results described above are not displayed to the public but are displayed behind Crop Explorer's firewall for use by FAS crop analysts. The operational two-layer soil moisture algorithm, which is available in Crop Explorer, is described in more detail in the next section.

3.1 Modified Palmer Two-Layer Soil Moisture Model

The two-layer soil moisture model is a bookkeeping method that accounts for the water gained or lost in the soil profile by recording the amount of water withdrawn by evapotranspiration and replenished by precipitation. The final aim of the soil moisture model is to estimate if soil moisture storage between dry spells was adequate for maximum plant growth. The two-layer soil moisture model used by FAS was first described by Palmer (1965).

The Palmer soil moisture model for two soil layers is calculated in daily time increments (mm/day of precipitation and evapotranspiration). The top-layer soil moisture is assumed to hold a maximum of one inch (or 25-mm) of available water, and the sub-layer soil moisture may hold 0–400 mm/m of water depending on the soil's water-holding capacity for the grid cell (i.e., based on soil texture and soil depth).

The soil moisture model assumes precipitation enters the two soil layers by first filling the top surface soil layer and then filling the lower soil layer. Moisture is extracted from the two soil layers by evapotranspiration, whereby water is first depleted from the top layer and then extracted from the sub-surface layer. When the water-holding capacity of both soil layers is reached, excess precipitation is lost from the model and treated as runoff or deep percolation.

The original Palmer (1965) two-layer soil moisture model was later modified over the years by FAS personnel and contractors by:

1. Allowing more gradual and realistic depletion of the top surface layer.
2. Allowing moisture to be depleted from the lower layer before the surface layer is completely dry as a better way to describe water extraction by plants.
3. Estimating potential evapotranspiration with the modified FAO Penman-Monteith equation described by Allen, et al., (1998) and not using the Thornthwaite (1948) equation proposed by Palmer. PET is therefore calculated from daily min/max temperature data sets and the location of each WMO station (i.e., latitude, longitude, and elevation) and AFWA grid cell.
4. Utilizing FAO's (1996) Digital Soil Map of the World (DSMW) to determine soil type, soil depth, and soil's water-holding capacity.
5. Assuming maximum root depth is one meter (or less depending on impermeable soil layers) for calculating total water-holding capacity. From these assumptions, water holding capacity for both layers normally range from 125–200 mm/meter of water depending on soil texture (ranging from sand to clay) and soil depth (one meter or less).

A summary of the global soil moisture products utilized by USDA/FAS are presented in Table 2.

3.2 Surface Wetness

The surface wetness product utilized by FAS is shown in Fig. 1, and it is available to subscribers from Weather Predict Consulting (WPC 2008). The surface

USDA Crop Management System

Table 2 Summary of soil moisture products utilized by USDA/FAS

Product	Data source (process agency)	Spatial resolution	Coverage	Input data	Infrared geostationary satellites	Passive microwave	Ground station with gauge
"Station" 2-layer soil moisture model (modified Palmer)	GTS/WMO and NOAA-NWS for USA (USDA/FAS)	16,000 total stations	Global	Daily rainfall and min/max temperatures for PET calculations	No	No	Yes, approx. 7500 stations report daily
"Grid cell" 2-layer soil moisture model (modified Palmer)	AGRMET/AFWA (USDA/FAS)	47-km at 60° latitude (true) and 25-km at the equator	Global 60° N-S	Daily rainfall and min/max temperatures for PET calculations	Yes	SSM/I	Yes, GTS/WMO, NOAA/NWS, and others
"Corrected" (AMSR-E) 2-layer soil moisture model (modified Palmer)	AGRMET/AFWA and MODIS/AMSR-E (USDA/ARS/ HRSL)	47-km at 60° latitude (true) and 25-km at the equator	Global 60°N-S	"Grid cell" Soil Moisture "corrected" every 3-days with MODIS/AMSR-E data	Yes	SSM/I and "corrected" with MODIS/AMSR-E data	Yes, GTS/WMO, NOAA/NWS, and others
Surface wetness	SSM/I (WeatherPredict Consulting-WPC)	1/3-degree or approx. 37-km	Global/80°N-S	Current week compared to 20-year climatology	No	SSM/I	No

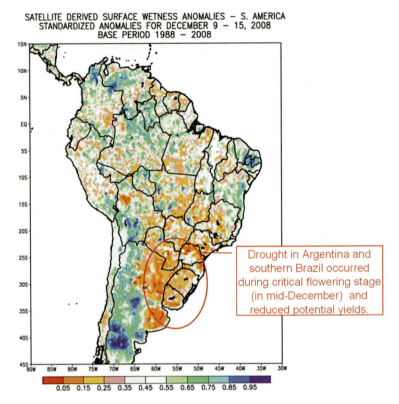

Fig. 1 The surface wetness anomaly product for South America is derived from the SMM/I sensor and produced by Weather Predict Consulting (WPC 2008)

wetness product is delivered to FAS every Monday morning. USDA/FAS crop analysts learned that real-time products delivered on Monday morning are more useful than delivering products every 10-days. Decadal, or 10-day products, is a common time frame used by agro-meteorologists but is not a useful delivery schedule for government agricultural economists at FAS or for commodity traders who need summary weather information early Monday morning when their work week begins.

The surface wetness product is derived from the Special Sensor Microwave Imager (SSM/I), which is a PMW sensor with global coverage, and is flown on polar orbiting satellites as part of the Defense Meteorological Satellite Program (DMSP). The frequencies observed by the SSM/I are sensitive to liquid water near the earth's surface. The surface water index quantifies the magnitude of the near surface wetness from precipitation, snow-melt, and irrigation. The signature of liquid water can originate from water: intercepted by the canopy, stored in the leaves, pooled on the surface, melting snow, and/or in the upper few centimeters of the soil.

The product represents an integration of all of these sources of liquid water, and is based on variation from a mean value with no smoothing performed (Basist, et al., 2001).

Surface wetness estimates are derived at 1/3-degree resolution (i.e., approximately 37-km), and have been has been calibrated and validated using independent high resolution in situ observations. A 20-year climatology (1988–2008) serves as the base period for monthly and weekly anomalies and the climatology is updated at the beginning of every year. Basist, et al., (2001) indicates the wetness product assumes the data have a gamma distribution and uses a pixel specific standardized cumulative probability (in %) to represent the anomalies. Basist, et al., (2001) also indicate the standardization procedure accounts for variation in surface features around a region (i.e. forest, lakes, farm land, mountains), time of year (i.e. wet versus dry season), and soil type (i.e. clay versus sandy soil).

FAS crop analysts find the surface wetness anomaly product useful for:

1. Comparing with global precipitation and vegetation indices data sets.
2. Determining if start of season rains arrived earlier or later than average.
3. Determining early signs of drought stress after the crop has been planted.
4. Determining if rains were excessive at time of harvest or if rains possibly caused damage when crop is in the field and has not yet been harvested.
5. Determining if rains were excessive during time of planting or harvesting and if heavy machinery was not able to enter the fields due to waterlogged soils.
6. Monitoring the entire crop season and observing when (i.e. during crop stage) cumulative surface wetness anomalies are above or below the seasonal average.

4 Global Agriculture Monitoring (GLAM) System

The National Aeronautics and Space Administration (NASA) and the USDA/FAS jointly fund a project called the Global Agriculture Monitoring (GLAM) project, which aims to:

1. Develop and improve FAS abilities for making operational quantitative estimates for crop area and yield based on satellite-derived data.
2. Improve global hydrologic data flows for FAS by integrating more satellite-derived products from NASA into the existing operational DSS at FAS and USDA's lockup process.

The GLAM project also provided a partnership with NASA's Hydrological Sciences Branch (HSB) at the Goddard Space Flight Center (GSFC); Hydrology and Remote Sensing Laboratory (HRSL) of USDA's Agricultural Research Service (ARS); and the University of Maryland's (UMD) Earth System Science Interdisciplinary Center (ESSIC). These new partnerships enabled FAS to:

1. Utilize AMRS-E or PMW data to improve the FAS two-layer soil moisture model by correcting for errors in the global precipitation data sets.
2. Utilize satellite radar altimetry for monitoring surface water levels for more than 70 global reservoirs and lakes.

These two new operational hydrologic data systems from PMW satellites and satellite radar altimetry are described below in separate sections.

4.1 Corrected Soil Moisture Model With Passive Microwave (PMW)

The GLAM project created another new hydrologic data stream for FAS by introducing NASA's EOS Advanced Microwave Scanning Radiometer (AMSR-E) data into the FAS two-layer soil moisture algorithm. This work was conducted in cooperation with NASA's HSB/GSFC at Greenbelt, Maryland and USDA's ARS/ HRSL at Beltsville, Maryland.

AMSR-E is a microwave radiometer launched in 2002 aboard the NASA-EOS Aqua satellite and measures vertically and horizontally polarized brightness temperatures at six frequencies: 6.92, 10.65, 18.7, 23.8, 36.5, and 89.0 GHz; and global coverage is possible every 2–3 days. At a fixed incidence angle of 54.8° and an altitude of 705 km, AMSR-E provides a conically scanning footprint pattern with a swath width of 1445 km. The mean footprint diameter ranges from 56 km at 6.92 GHz to 5 km at 89 GHz.

HRSL-USDA/ARS helped to improve FAS soil moisture estimates in CADRE by merging PMW satellite data from the AMSR-E sensor. The improved temporal resolution and spatial coverage of AMSR-E helped to provide a better characterization of regional-scale surface wetness and enable more accurate soil moisture monitoring in key agricultural areas. An operational data assimilation system and delivery system was developed that utilized Ensemble Kalman filtering (EnKF) and was calibrated using a suite of synthetic experiments and in situ validation sites. In this way, AMSR-E soil moisture retrievals are able to effectively compensate modeled soil moisture for the impact of poorly observed rainfall patterns (Bolten, et al., 2008a and 2008b).

The new and improved soil moisture system with AMSR-E data is currently operational and delivered to the FAS' CADRE in near real-time. For this system, the gridded AMSR-E Level-3 soil moisture product is used (Njoku 2004). The Level-3 product is a gridded data product using global cylindrical 25 km Equal-Area Scalable Earth Grid (EASE-Grid) cell spacing. Observations of soil moisture are calculated from Polarization Ratios of 10.7 and 18.7 GHz, plus three empirical coefficients to compute a vegetation/roughness parameter for each grid cell. Deviations from 18.7 GHZ polarization ratio baseline value for each grid cell are used to calculate daily soil moisture estimates for each grid cell (Njoku, et al., 2003). These soil moisture estimates represent a soil depth comparable to the first layer soil moisture used by the FAS/OGA/IPAD (Bolten, et al., 2009 and 2008).

Although there are several operational satellites providing multi-frequency brightness temperature observations, AMSR-E is the first satellite-based remote sensing instrument designed specifically for soil moisture retrieval. The launch of AMSR-E has improved the spatial resolution and frequency range of earlier satellite-based passive microwave instruments.

In addition, HSB-GSFC and HRSL-ARS are continuously assessing the difference between the different US government global precipitation data sets and updating FAS/OGA/IPAD with their latest findings (Tian and Peters-Lidard 2007, and Crow, et al., 2005). If one US Government precipitation data set should later be found to consistently out perform the AFWA precipitation data set, then arrangements would be made to introduce this data into the modified Palmer two-layer soil moisture model. However, to date, none of the US Government global precipitation data sets from NASA or NOAA have outperformed the AFWA precipitation to warrant such a change in FAS/OGA/IPAD's CADRE system.

4.2 Operational Surface Water Heights From Satellite Radar Altimetry

The GLAM project created another new hydrologic data stream for FAS by utilizing satellite radar altimetry to monitor changes in surface water levels for reservoirs and lakes located worldwide. The Global Reservoir and Lake Monitor (GRLM) system was developed in partnership with NASA's HSB/GSFC and the University of Maryland's ESSIC.

The GRLM was launched within Crop Eexplorer at the end of 2003, and the program focuses on the delivery of near-real time surface water elevation products within an operational framework for FAS and USDA's DSS. Phase 1 of GRLM recorded variations in surface water height for approximately 70 lakes worldwide using a combination of satellite radar altimetry data sets from the NASA/CNES TOPEX/Poseidon mission (1992–2002), NASA/CNES Jason-1 mission data (2002–current), and the US Naval Research Lab's GFO mission (2000–2008). Validation exercises show the products range in accuracy from a few centimeters to several tens of centimeters depending on target size and surface wave conditions (Birkett, et al., 2009).

The GRLM retrieves satellite radar altimetry data, converts it to relative surface water heights and updates the surface water heights on a weekly basis. Output is in the form of graphs and text files with web links to other imaging and information resources. The USDA/FAS utilize the products for irrigation potential considerations and as general indicators of drought and high-water conditions.

Phase 2 of GRLM will incorporate satellite radar altimetry products from the ESA ERS missions (1994–2002), the ESA ENVISAT mission (2002–current), and the NASA/Jason-2 mission (post 2009 and near real time). Figure 2 shows how ENVISAT's longer repeat time cycle has better spatial coverage and Fig. 3 illustrates the location of new potential lakes and reservoirs which can be monitored worldwide by ENVISAT.

Different Satellite Orbits and Repeat Cycles

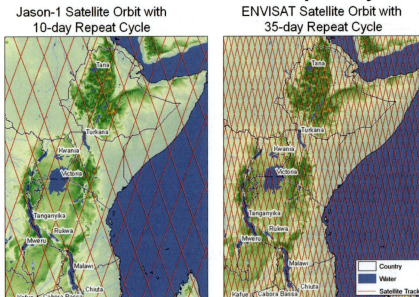

Fig. 2 More lakes can be monitored by ENVISAT than by Jason-1 due to ENVISAT having more orbits between longer (35-days) repeat cycles

Potential Improvement with ENVISAT

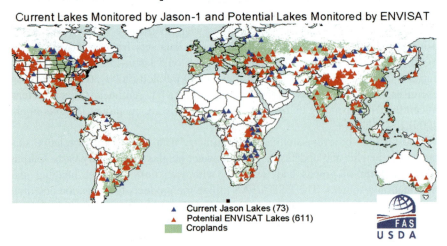

Fig. 3 More potential lakes are envisaged to be monitored during Phase 2 of the Global Reservoir and Lake Monitor (GRLM) project which will utilize ENVISAT radar altimeter data. The *blue triangles* indicate lakes monitored by Phase 1 (with Jason-1 data) and the *red triangles* indicate potential lakes to be monitored during Phase 2 (with ENVISAT data)

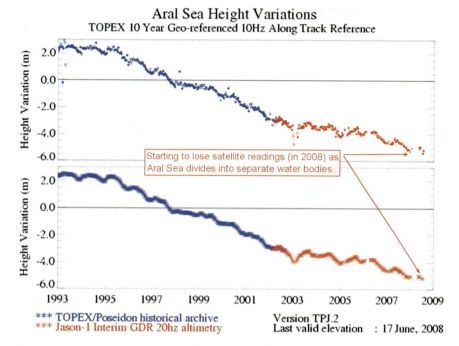

Fig. 4 Surface water levels of the Aral Sea have been decreasing from 1993 through 2008, and starting in 2008 the satellite radar altimeter started to lose the surface water as the Areal Sea dried up and separated into different water bodies

Examples of the surface water heights for the Aral Sea and Lake Victoria are shown in Figs. 4 and 5, respectively. The GRLM received international attention when Lake Victoria water levels dropped to historic lows and caused numerous environmental problems and economic losses to many businesses such as maritime trade, fisheries, tourism, etc. The Lake Victoria "drought" turned out to be a man-made drought because water was withdrawn from Lake Victoria's only outlet at Owens Falls Dam during 2000–2006 at rates greater than the Agreed Curve (Sutcliffe and Petersen 2007). The excessive water withdrawals from 2000 to 2006 dropped the lake's water level to the lowest levels since 1923. Lake Victoria's manmade "drought" caused lots of human suffering along shorelines of Lake Victoria where the population density is very high and the entire region is one the poorest in the world.

4.3 Operational Yield-Regression and Analog-Year Analysis

Operational quantitative crop yield estimates have been developed from NASA's Moderate Resolution Imaging Spectroradiometer (MODIS) data and NDVI products with 250-m spatial resolution. The improved spatial resolution of MODIS

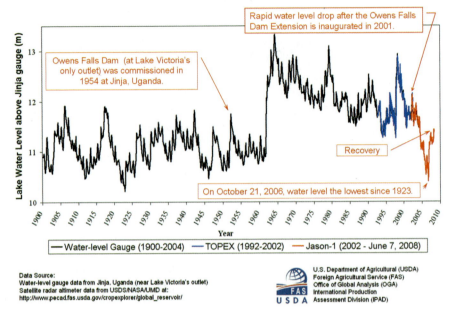

Fig. 5 Satellite radar altimeter data for Lake Victoria (in east Africa) was merged with elevation gauge data to create a historical water level record greater than 100-years. Excessive lake withdrawals during 2000–2006 at Jinja, Uganda (the lake's only outlet) caused the lake's water level to drop to its lowest level since 1923

data and products (with 250-m) helps to improve FAS abilities to monitor vegetation changes (i.e., crop yield) in the key agricultural areas in a more timely fashion, than previously possible with NOAA-AVHRR (8-km spatial resolution) and SPOT-VEGETATION (1-km spatial resolution) time-series satellite data.

NASA's MODIS sensor is onboard two platforms of the Earth Observing System (EOS), which was designed in part to monitor subtle vegetation responses to stress, vegetation production and land cover with regional-to-global coverage. Although MODIS is a NASA experimental mission, the instrument's capabilities will be extended by the launch of the Visible Infrared Imager Radiometer Suite (VIIRS), which will be part of the National Polar Orbiting Environmental Satellite System (NPOESS) and NPOESS Preparatory Project (NPP). Thus, the MODIS methods and system developed through GLAM will be easily transitioned into the fully operational VIIRS sensor onboard the NPP and NPOESS missions.

The GLAM project developed a new operational quantitative crop yield system for FAS that uses a MODIS (250-m) time-series database from 2000 to present for the red, near infrared (NIR) and shortwave length infrared (SWIR) bandwidths. The MODIS (250-m) time-series database enables FAS crop analysts to track the evolution of the growing season and make inter-annual comparisons of seasonal vegetation dynamics (i.e., green-up, peak greenness, and green-down) between individual years as well as relative to reference long-term mean conditions. The FAS methodology tracks vegetation "greenness" by utilizing VI (vegetation indices) such as NDVI (Normalized Difference Vegetation Index) and NDWI (Normalized Difference Water Index).

These NDVI time series curves are then used to operationally estimate quantitative yields (i.e. tons/hectare) from mid-season to end of season at national and sub-national levels. The rapid yield estimating system was initially designed to operate with MODIS-NDVI/NDWI time-series databases but later can be transferred to CADRE to perform similar yield-regression analysis for several agrometeorological variables stored within CADRE (such as percent soil moisture, cumulative seasonal rainfall, cumulative crop evapotranspiration, etc.). The operational yield estimates for the MODIS-NDVI system are made a few days after the first of each month so that the NDVI-derived yield maps can be used by FAS crop analysts for their monthly pre-lockup meetings with ICEC.

The heart of the rapid yield estimating MODIS-NDVI system is the historical sub-national yield database (from 2000 to current) which was constructed by the Geospatial Data Analysis Corporation (Hulina and Varlyguin 2008). The historical sub-national yield database from 2000 to present is used to perform a series of NDVI-yield regressions whereby the NDVI regressions comprise of several different seasonal crop metric parameters. The NDVI seasonal metric parameters are first derived by first smoothing the NDVI time-series data to obtain the critical seasonal metrics. The critical seasonal metrics include start of the season, end of the season, end of green-up, start of senescence, peak of the growing season, mid-senescence, length of season, rate of green-up, and rate of senescence. These seasonal metrics and several different areas under the NDVI time-series curves are used to estimate yields from NDVI-yield regressions.

In addition, an analog-year analysis is performed for MODIS-NDVI (250-meter) time series data from 2000 to present because even the best NDVI-regression model will under estimate yields during bumper crop years and over-estimate yields during drought years. Therefore, an analog-year analysis is performed to indentify which historical year best simulates the current NDVI time-series curve or "regime". The technology trend yield is then calculated and added to the yield of the analog-year to provide a current yield estimate. The analog-year analysis is expected to be better than yield-regression analysis, especially during extreme weather years (with high deviations from the mean) that produce bumper or drought-reduced crops.

In short, the GLAM project recognizes that monitoring global vegetation changes in tandem with global precipitation is essential for maintaining the "convergence of evidence" methodology used by FAS to determine regional weather anomalies worldwide and corresponding impacts on national crop yields. In addition, FAS crop analysts cannot rely exclusively on a single remote sensing data source due to limitations in temporal and spatial coverage, timeliness, accuracy, data quality, and life expectancy of satellite sensors.

NDVI also has it limitations in cloudy regions which implies agrometeorological crop models are probably more suitable for cloudy regions in the tropics and NDVI time-series analysis more suitable in temperate regions with less clouds. For example, the high-producing soybean region of Mato Grosso, Brazil (which borders the Amazon forest in the south), receives excessive rainfall and MODIS-NDVI time-series curves indicate great NDVI variability during the mid-season due to increased cloud-cover near the forest. In addition, the highest number of rain days in Mato Grosso (or south of the Amazon Forest) occurs during the mid-season which is when the NDVI signal is most important for estimating relative crop yields. Therefore, NDVI time-series analysis is not ideal for high rainfall regions because numerous clouds during mid-season lowers NDVI data quality during critical growing periods, making agrometeorological crop models more suitable for analysis in high rainfall regions in the tropics. In addition, vegetation response to excess/deficit precipitation lags behind rainfall events by 10–30 days, which makes NDVI yield analysis 10–30 days behind crop models or soil moisture models.

In contrast, the spatial resolution of 250-m is better for estimating relative crop yields from vegetation indices (VI) products rather than global rainfall products because the best spatial resolution for global precipitation products is 8-km resolution from CMORPH. However, unfortunately CMOPRH does not remove satellite data biases by merging global SG data with IR satellite imagery which prevents it from being used more intensively by crop monitoring systems.

5 Future Outlook

For daily global precipitation data, USDA/FAS essentially is dependent on three U.S. government agencies (i.e., AFWA, NOAA, and NASA) for developing and improving their daily global precipitation products. All three global precipitation products vary in the methods used to merge data from SG, PMW, and IR sources.

The AFWA global precipitation product has been serving FAS the longest, or more than 25-years, and it currently is the only global precipitation product that merges SG, PMW, and IR at near real-time. Studies have also shown that merging SG, PMW, and IR data helps to improve reliability and accuracy (Gruber and Levizzani 2008).

Summary future developments for global precipitation products from US Government agencies need to work on better data assimilation methods for improving spatial resolution, better retrieval algorithms for improved precipitation estimates, and introduction of new observation data or platforms, whenever possible.

Concerning better spatial resolution, AFWA global precipitation data set plans to improve their spatial resolution from 47- to 24-km in the very near future (Eylander 2008). This improvement in spatial resolution should greatly improve the two-layer soil moisture model maintained by USDA/FAS. In addition, the best spatial resolution for global precipitation products is CMOPRH with 8-km resolution, but unfortunately CMOPRH does not remove biases by merging global SG data with satellite imagery. It is hoped that CMORPH developers at NOAA will consider integrating SG data into their algorithms.

Concerning better retrieval algorithms, bias systematic errors must be removed from SPE products by integrating SG data because SPE products tend to generate false artifacts for agricultural monitoring during 4-month or longer growing seasons, as FAS has found with the CMORPH product. Also, better retrieval algorithms need to be improved near water bodies and in mountainous regions where most of IR satellite-derived precipitation products tend to have larger errors.

Concerning new observation platforms, satellite precipitation radar (PR) has great potential and ideally future generation precipitation products will have global PR coverage. Fortunately, NASA's Global Precipitation Mission (GPM 2008) has plans to address this issue by launching in 2013 their core precipitation monitoring satellite with two radars onboard.

In summary, FAS crop analysts have for more than 25-years utilized AWFA and WMO data sets, and during the past several years FAS began utilizing relatively new global precipitation data products from NASA and NOAA's CPC and NWS. These three new precipitation data sources provide additional data to USDA's "all sources data approach" for auditing national crop estimates each month. However, data quality for these global satellite-derived precipitation products need to be improved, which in turn should directly improve USDA's DSS systems for auditing and estimating crop yields in foreign countries. It also is envisaged that USDA will always need to monitor several global daily precipitation data streams in tandem, just as several global VI data streams are currently monitored in tandem with AVHRR, MODIS, and SPOT-VEGETATION sensors onboard three different satellite platforms.

References

AFWA (2002) Data Format handbook for AGRMET. (AFWA Agricultural Meteorology Modeling System) Air Force Weather Agency (AFWA) November 6 2002

ASRC (2008) Arctic Slope Regional Corporation, http://www.asrc.com/splash.asp , Accessed on December 22 2008

Allen RG, Pereira LS, Raes D, and Smith M (1998) Crop evapotranspiration Guidelines for computing crop water requirements. FAO Irrigation and Drainage, Paper 56, pp 27–65

Basist A, Williams C Jr, Ross TF, Menne MJ, Grody N, Ferraro R, and Shen S (2001) Using the Special Sensor Microwave Imager to Monitor Surface Wetness. Journal of Hydrometeorology, Vol 2, No 3, pp 297–308, June

Birkett C, Reynolds C, Beckley B, and Doorn B (2009) The global reservoir and lake monitor near-realtime monitoring of water surface levels using satellite radar altimeters. In Press with Springer Books

Boatwright GO and Whitefield VS (1986) Early Warning and Crop Condition Assessment Research. IEEE Transactions on Geosciences and Remote Sensing, Vol GE-24, No 1, pp 56–64 January

Bolten J, Crow WT, Zhan X, Reynolds C, and Jackson TJ (2009) Evaluating the Utility of Remotely-Sensed Soil Moisture Retrievals for Operational AgriculturalDrought Monitoring. IEEE Journal of Selected Topics in Applied Earth Observations and Remote Sensing. In Press with IEEE J-STARS

Bolten J, Crow WT, Zhan X, Reynolds C, and Jackson TJ (2008) Assimilation of a Satellite-Based Soil Moisture Product in a Two-Layer Water Balance Model for a Global Crop Production

Decision Support System. In press, in Data Assimilation for Atmospheric, Oceanic, and Hydrologic Applications, SK Park (Ed.), Springer-Verlag, New York

Cochrane MA (1981) Soil Moisture and Agromet Models. Technical Report USAF Air Weather Service (MAC), USAFETAC Scott AFB Illinois, TN-81/001, March 1981

Crop Explorer (2008) US Department of Agriculture (USDA), Foreign Agricultural Service (FAS), Office of Global Analysis (OGA), http://www.pecad.fas.usda.gov/cropexplorer/,Accessed on December 22 2008

Crow W, Koster RD, Reichle RH, and Sharif HO (2005) Relevance of Time-Varying and Time-Invariant Retrieval Error Sources on the Utility of Spaceborne Soil Moisture Products. Geophysical Research Letters, Vol 32, p L24405, December 2005

Eylander J (2008) Verbal Communication in USDA/FAS/OGA/IPAD Meeting in Washington DC, Chief Data Assimilation for AFWA, Offutt Air Force Base (AFB)

FAO (1996) The Digitized Soil Map of the World Including Derived Soil Properties, CD-ROM, Food and Agriculture Organization, Rome

GPM (2008) Global Precipitation Mission (GPM), NASA Goddard Space Flight Center, http://gpm.gsfc.nasa.gov/, Accessed on December 22 2008

Gruber A and Levizzani V (2008) Assessment of Global Precipitation Products. A project of the World Climate Research Programme Global Energy and Water Cycle Experiment (GEWEX) Radiation Panel, WCRP-128, WMO/TD, No 1430, May 2008

Hollinger J (1989) DMSP Special Sensor Microwave/Imager Calibration/Validation Report, Vol I, Naval Research Laboratory, Washington, DC

Huffman GJ, Adler RF, Bolvin DT, Gu G, Nelkin EJ, Bowman KP, Stocker EF, and Wolff DB (2007) The TRMM Multi-satellite Precipitation Analysis: Quasi-Global Multi-Year Combined-Sensor Precipitation Estimates at Fine Scale. Journal of Hydrometeorology, Vol 8, No 1, pp 38–55, February 2007

Hulina SM and Varlyguin DL (2008) Geospatial Data Analysis (GDA) Corporation. http://www.gdacorp.com/,Accessed on December 22 2008

ICEC (2008) Interagency Commodity Estimates Committees. US Department of Agriculture (USDA). WAOB. http://www.usda.gov/oce/commodity/committee.htm,Accessed on December 22 2008

Joyce RJ, Janowiak JE, Arkin PA, and Xie P (2004) CMORPH: A Method that Produces Global Precipitation Estimates from Passive Microwave and Infrared Data at High Spatial and Temporal Resolution. Journal of Hydrometeorology, Vol 5, pp 487–503

Kiess RB and Cox WM (1988) The AFWGWC Automated Real-Time Cloud Analysis Model. Technical Report AFGWTC/TN-88/001, (AD-B121615), Air Force Weather Agency, Offutt AFB, Nebraska

MARS (2008) Monitoring Agriculture Through Remote Sensing Techniques. Joint Research Center (JRC) of the European Commission, http://www.marsop.info/,Accessed on December 22 2008

MDA Federal Inc (2008) MDA EarthSat Weather. http://weather.earthsat.com/announcements/new-portal,and MDA CROPCAST (Ag Weather), http://www.mdafederal.com/mda-earthsat-weather/cropcast-ag-services,Accessed on December 22 2008

NWS (2008) Advanced Hydrologic Prediction Service. National Oceanic and Atmospheric Administration's (NOAA), National Weather Service (NWS). http://water.weather.gov/about.php, Accessed on December 22 2008

New M, Lister D, Hulme M, and Makin I (2002) A High-Resolution Data Set of Surface Climate Over Global Land Areas. Climate Research, Vol 21, pp 1–25

Njoku EG (2004) AMSR-E/Aqua Daily L3 Surface Soil Moisture, Interpretive Parameters & QC EASE-Grids, National Snow and Ice Data Center Digital Media, Boulder, CO, March to June

Njoku EG, Jackson TJ, Lakshmi V, Chan TK, and Nghiem SV (2003) Soil Moisture Retrieval from AMSR-E. IEEE Transactions on Geosciences and Remote Sensing , Vol 41, No 2, pp 215–229

PSD Online (2008) Production Supply and Demand (PSD) Online. US Department of Agriculture (USDA), Foreign Agricultural Service (FAS), Office of Global Analysis (OGA), http://www.fas.usda.gov/psdonline/,Accessed on December 22 2008

Palmer WC (1965) Meteorological Drought US Weather Bureau. Research Paper 45

Reynolds CA (2001) Input Data Sources, Climate Normals, Crop Models, and Data Extraction Routines Utilized by PECAD. Third International Conference on Geospatial Information in Agriculture and Forestry, Denver, CO, November 5–7, 2001

Reynolds CA and Doorn B (2001) Automation of USDA's Global Agrometeorological Databases. Editors RP Motha and MVK Sivakumar, Software for Agroclimatic Data Management, Proceedings of an Expert Group Meeting, October 16–20, 2000, Washington DC, USA; US Department of Agriculture and World Meteorological Organization. Staff Report WAOB-2001-2 and AGM-4 WMO/TD No. 1075, October, 2001, pp. 33–41

Reynolds CA, Jackson TJ, and WJ Rawls (2000) Estimating Soil Water-Holding Capacities by Linking the FAO Soil Map of the World with Global Pedon Databases and Continuous Pedotransfer Functions. Water Resources Research, Vol 36, No 12, pp 3653–3662, December

Ritchie, JT (1991) Wheat Phasic Development. In Modeling Plant and Soil Systems, SH Mickelson (Ed) J Hanks, and JT Ritchie (co-eds), American Society of Agronomy (ASA), Crop Science Society of America (CSSA), Soil Science Society of America (SSSA), Madison, WI, Number 31 Agronomy Series, Chapter 3, pp 31–54

Ritchie JT, Singh U, Godwin DC, and Bowen WT (1998) Cereal Growth, Development, and Yield. GY Tsuiji, et al., (eds), Understanding Options in for Agricultural Production, Kluwer Academic Publishers, Dordrecht, pp 79–98

Sutcliffe JV and Petersen G (2007) Lake Victoria: Derivation of a Corrected Natural Water Level Series. Hydrological Sciences Journal, Vol 52, No 6, pp 1316–1321 December, 2007, IAHS Press

Thornthwaite CW (1948) An Approach Toward a Rational Classification of Climate. Geographical Review, Vol 38, pp 55–94

Tian Y and Peters-Lidard CD (2007) Systematic Anomalies Over Inland Bodies in Satellite Based Precipitation Estimates. Geophysical Research Letters, Vol. 34, p L14403

Tingley W (1988) Crop Condition Data Retrieval and Evaluation (CADRE) DBMS Dictionary. Lockheed Engineering and Sciences Company, Inc. Unpublished

Vogel FA and Bange GA (1999) Understanding USDA Crop Forecasts. National Agricultural Statistics Service (NASS) and World Agricultural Outlook Board (WAOB), Office of the Chief Economist (OCE), US Department of Agriculture. Miscellaneous Publication No. 1554. http://www.usda.gov/nass/nassinfo/pub1554.pdf, Accessed on December 22 2008

WAOB (2008) World Agricultural Outlook Board. US Department of Agriculture (USDA), Office of the Chief Economist (OCE), http://www.usda.gov/oce/commodity/index.htm,Accessed on December 22 2008

WASDE (2008) World Agricultural Supply and Demand Estimates. US Department of Agriculture (USDA), Office of the Chief Economist (OCE). World Agricultural Outlook Board (WAOB), http://www.usda.gov/oce/commodity/wasde/index.htm , Accessed on December 22 2008

WPC (2008). Weather Predict Consulting Inc. http://www.weatherpredict.com/products/agriculture.html, Accessed on December 22 2008

Real-Time Decision Support Systems: The Famine Early Warning System Network

Chris Funk and James P. Verdin

Abstract A multi-institutional partnership, the US Agency for International Development's Famine Early Warning System Network (FEWS NET) provides routine monitoring of climatic, agricultural, market, and socioeconomic conditions in over 20 countries. FEWS NET supports and informs disaster relief decisions that impact millions of people and involve billions of dollars. In this chapter, we focus on some of FEWS NET's hydrologic monitoring tools, with a specific emphasis on combining "low frequency" and "high frequency" assessment tools. Low frequency assessment tools, tied to water and food balance estimates, enable us to evaluate and map long-term tendencies in food security. High frequency assessments are supported by agrohydrologic models driven by satellite rainfall estimates, such as the Water Requirement Satisfaction Index (WRSI). Focusing on eastern Africa, we suggest that both these high and low frequency approaches are necessary to capture the interaction of slow variations in vulnerability and the relatively rapid onset of climatic shocks.

Keywords Early warning · Drought · Food security · Climate change · Crop modeling · Hydrology

1 Introduction

The rhythms of plant emergence, vegetative increase, reproduction, and grain filling still dominate and organize the activities of half the world. Cycles of good, bad, and intermediate harvests continue to help shape the fate of nations. Cycles of recurrent bad harvests punctuated by a few seasons with good harvest continue to aggravate the fate of developing countries. In many developing nations, coping with hydrologic extremes is equivalent in cost and potential outcome to war (Kates 2000). The

---

C. Funk (✉)
Research Geographer, USGS EROS/UCSB Geography, USA, Santa Barbara, CA, 93106
e-mail: funk.cc@gmail.com

M. Gebremichael, F. Hossain (eds.), *Satellite Rainfall Applications for Surface Hydrology*, DOI 10.1007/978-90-481-2915-7_17,
© Springer Science+Business Media B.V. 2010

impacts of drought are not limited to the poorest nations. Even though only 2% of the Republic of South Africa's GDP is based on agriculture, season rainfall totals are tightly coupled to economic growth, with a correlation of 0.7 (Jury 2002). In the United States, severe drought years, such as 2002, may result in billion dollar losses. Global per capita water supplies will likely drop by a third over the next 20 years (WWD 2003), and 2 to 7 billion people may face chronic water shortages by 2050. Food crises (Natsios and Doley 2009) will continue to emerge as the world's population grows faster than crop yields (Funk and Brown 2009); per capita cereal production peaked in 1986 and will likely decline by 14% over the next 20 years. In Kenya, it's estimated that arable land is declining by 2% per year due to population growth and human settlements in key agricultural areas. This figure is very likely to increase with declining rainfall trends and associated land degradation (personal communication). At present, 1 billion people in 50 nations face chronic food shortages, with 20% or more of that population undernourished (FAO 2007). Food security early warning systems seek to mitigate shocks to these vulnerable populations. This chapter briefly discusses the work of one such system: the US Agency for International Development's Famine Early Warning Systems Network (FEWS NET).

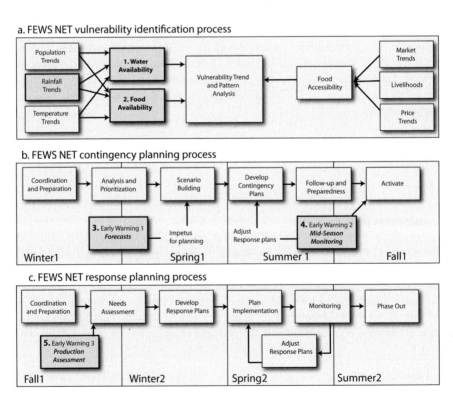

Fig. 1 FEWS NET contingency planning and response schema. Preseason, midseason, and postseason opportunities of hydrologic early warning

1.1 The Three Components of the FEWS NET Planning Process

Most food, especially in the developing world, is produced and consumed on a fairly local scale. Local food deficits related to agricultural and pastoral drought can have devastating impacts. Drought, however, is a "slow onset disaster" and, as such, is amenable to early warning applications tied to hydrologic monitoring and modeling. "Droughts," however, must be understood as a water deficit defined against a given human need. Thus, effective hydrologic early warning must evaluate changes in both demand and supply. Supply and demand will change at seasonal and decadal time scales, and effective monitoring requires modeling at both these temporal horizons.

The FEWS NET process can be conceptually divided into three components (Fig. 1). In the first process, "vulnerability identification," at-risk populations are mapped and trends in food insecurity are analyzed (Fig. 1a). This process is informed both by water and food availability studies and more detailed food economy studies focused on markets, prices, and livelihoods. The vulnerability identification stage guides long-term decision making and planning by aid agencies.

The second FEWS NET process involves the development of food security contingency plans (Fig. 1b). These contingency plans, supported by food security outlooks and forecasts, enable disaster response planners to initiate strategic planning. Seasonal rainfall forecasts and Water Requirement Satisfaction Index (WRSI) imagery play an important role in supporting agrohydrologic modeling and monitoring. The third and final FEWS NET planning process (Fig. 1c) supports and informs the design and implementation of timely and appropriate disaster relief packages. USGS FEWS NET scientists primarily support these three activities by studying trends in rainfall, food, and water availability by providing seasonal rainfall forecasts, midseason crop water assessments, and postseason crop production assessments based on Normalized Difference Vegetation Index imagery (Funk and Budde 2009). This chapter discusses our contributions to the Vulnerability Identification and Contingency Planning (Shaded boxes 1–4 in Fig. 1a and b).

1.2 Focus on Eastern African Food Insecurity in 2009

As of February 2009, 17 million eastern Africans face extremely high levels of food insecurity. These individuals live primarily in the water insecure eastern parts of these countries. These food insecurity crises have arisen through a combination of both non-climatic and climatic underlying factors, such as increasing population pressure, hyperinflation, trans-boundary human and livestock diseases, conflicts and civil insecurity, climatic constraints on water availability, anomalous climate conditions in the Indian and Pacific Oceans, and a recurrence of drought over the past several years. The "real-time" applications discussed and presented in this chapter are therefore germane to a current and grave food security crisis. After a brief discussion of the background of FEWS NET (Section 2), we describe approaches for modeling agro-hydrologic risk (Section 3) use these tools to analyze Kenyan agricultural hydrologic conditions (Section 4), and summarize our approach (Section 5).

2 Background

2.1 A Brief History of FEWS NET

In 1984–1985, catastrophic droughts hit Ethiopia and Sudan, leading to more than a million deaths. These large-scale famines shocked the world. Famine is a slow onset disaster. The tragic lack of timely information and intervention led to widespread human suffering. Responding to concerned citizens, the US Congress called on USAID to create the Famine Early Warning System (FEWS) in 1985.

FEWS has been implemented in roughly 5-year phases since its inception. The prime contract for implementation in each phase is awarded by USAID to a private sector firm through a competitive procurement process. Support in the form of remote sensing, modeling, forecasting, geographic information systems (GIS), data archive, training, and product dissemination is provided by US Government science agencies: The US Geological Survey (USGS), National Aeronautics and Space Administration (NASA), the National Oceanic and Atmospheric Administration (NOAA), and the US Department of Agriculture (USDA) were engaged as scientific implementing partners through interagency agreements with USAID. Since the late 1980s, FEWS has steadily evolved from being a Washington-based activity with a few expatriates in the field to one that is primarily African-based, with African professionals composing the majority of the staff. The latest phase of the activity places an emphasis on networking among individuals and institutions (governmental, inter-governmental, and non-governmental) across disciplines at the local, national, regional, and continental levels, hence the new name: FEWS NET.

USGS participation has evolved in step with the overall shift to African-based analyses. Regional scientists have been recruited for West Africa, the Greater Horn of Africa (GHA), and southern Africa. These experienced scientists are African nationals with expertise in drought monitoring, remote sensing, and GIS. They work closely with food security analysts to interpret the nature of drought and flood threats to livelihood systems (especially subsistence agriculture) and articulate their findings in bulletins and reports disseminated to the international community. The field scientists devote significant time to technical capacity building through formal and informal training on remote sensing, GIS, hydrology, agroclimatology, and other topics. They work with the following African regional institutions: Agronomy-Hydrology-Meteorology Regional Center in Niamey, Niger; IGAD Climate Predictions and Applications Centre (ICPAC) Intergovernmental Authority on Development in Nairobi, Kenya; the Regional Center for Mapping of Resources for Development (RCMRD) in Nairobi, Kenya; and the Southern Africa Development Community's Regional Remote Sensing Unit in Harare, Zimbabwe. They play a central role in research to improve techniques, algorithms, and methods of geospatial hydroclimatology. They are well positioned to provide scientific insights and local data that complement the work of US-based colleagues. They also have invaluable links to African institutions of higher education.

In 2002, USAID reorganized and moved FEWS NET out of the Bureau for Africa and into the Bureau for Democracy, Conflict, and Humanitarian Assistance. The scope of activity was expanded beyond Africa to include Afghanistan, Haiti, and four countries of Central America. The global price shocks of 2007 and 2008 have spread food security concerns across a broad swath of developing nations, and the geographic scope of FEWS NET activities is expanding as well, in synch with these spreading concerns. The twenty first century will require the effective remote monitoring of agriculture and pastoral conditions. Without a doubt, satellite rainfall estimates will play a critical role in achieving this goal.

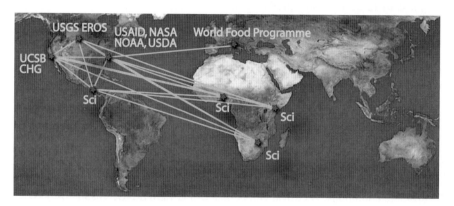

Fig. 2 The FEWS NET science network

2.2 The FEWS NET Early Warning System

The FEWS NET early warning system combines information from multiple sources into coherent food security outlooks, alerts, and briefs for decision makers. These products support decision making by the USAID Office of Food for Peace, the USAID Office of US Foreign Disaster Assistance, and the United Nation's World Food Programme (WFP) that is critical to protecting lives and livelihoods. The national governments of food insecure countries often use this information as well. Early warning can help mitigate the political and humanitarian impacts of food shortages by triggering food, health, and market-related interventions. Satellite observations can contribute substantially to both the contingency planning and disaster response planning phases of FEWS NET (Fig. 1), supporting decisions that save lives and livelihoods, and lessen the impacts of climate extremes – droughts and floods During the contingency planning phase, relatively uncertain information, such as climate forecasts (Funk, et al., 2006b; Brown, et al., 2007) and climate indicators (Box A in Fig. 1), can help guide scenario building and food security outlooks. This typically occurs before or during the early phase of the crop growing season. In the middle of the season (Box B in Fig. 1), satellite rainfall fields are used to monitor crop growing conditions. These simple water balance models use grids of rainfall and potential evapotranspiration (Verdin and Klaver 2002; Senay and Verdin 2003) to estimate the sufficiency of soil moisture for crop growth. At the close of the crop growing season (Box C in Fig. 1), satellite-observed vegetation is used to estimate crop production and/or yield (Funk and Budde 2009). In this report, we focus on early-to-mid-season analysis of conditions in Zimbabwe and Kenya/Somalia. While improved monitoring tools cannot make up for inadequate agricultural inputs (seeds and fertilizer) or rainfall, they can help guide the early identification of agricultural drought, which can lead to more timely and effective response to dangerous food insecurity.

The FEWS decision support system DSS process can be seen as an interactive filtering process by which enormous amounts of data are transformed into fair, objective, reproducible, and defensible analyses. For physical observations, FEWS NET relies primarily on satellite rainfall retrievals provided by the Climate Prediction Center (CPC) and the Tropical Rainfall Monitoring Mission (TRMM) a NASA product, augmented by in situ observations from the Global Telecommunications System (GTS). Other important inputs include satellite-observed Normalized Difference Vegetation Index (NDVI), snow extent, prevailing global climate conditions, and local soil and topography. Such information is used by experienced early warning analysts from USGS, NOAA, NASA, USDA, University of California, Santa Barbara UCSB, and Africa (Fig. 2) to monitor agro-hydrologic conditions. A critical component of the FEWS NET DSS is its network of in-country food security analysts. In Africa, Central America, and Afghanistan, these experts track market, vulnerability, livelihood, and agricultural conditions. These extensive analyses are compiled by a team of experts in Washington, DC (currently led by Chemonics International), who also maintain the primary FEWS NET Web portal (http://www.fews.net). Interactions between the physical and social components are vital. For example, in an area where people depend on export cash crop employment (e.g., coffee) rather than subsistence agriculture, global price shocks may be much more harmful than local drought. Availability of agricultural inputs, such as the distribution of seeds, can moderate or amplify the effects of growing season moisture deficits. Effective early warning combines a successful blend of earth observations, hydrologic modeling, food economics, weather and climate modeling, and much more. The remainder of this chapter, however, will focus on applications of satellite remote sensing to agrohydrologic early warning.

2.3 A Synopsis of USGS FEWS NET Early Warning Research

Early Warning Systems can help mitigate the political and humanitarian impacts of food shortages by supporting food, health, and market-related interventions. Satellite observations can contribute substantially to both the contingency planning and disaster response planning phases of FEWS NET (Fig. 1), supporting decisions that save lives and lessen the impacts of drought. A broad suite of early warning products (Rowland, et al., 2005) can be viewed at http://earlywarning.usgs.gov.

These products are primarily driven by satellite rainfall estimates (RFE) provided by the NOAA CPC (Xie and Arkin 1997) or the NASA TRMM multisatellite precipitation analysis (TMPA, Huffman, et al., 2007). Early work by the USGS science team involved using remotely sensed rainfall estimates to monitor the onset of rains (Verdin and Senay 2002) and generate WRSI maps (Verdin and Klaver 2002; Senay and Verdin 2003). These simple water balance models use grids of rainfall and potential evapotranspiration to estimate whether sufficient soil moisture is available for crop growth. A stand-alone version of the Geospatial WRSI (Magadzire 2009) is available from the Climate Hazard Group at the University of California, Santa Barbara (UCSB).[1] The USGS team has also developed early warning tools based on NDVI (Funk and Budde 2009).

Beginning in the late 1990s (Verdin, et al., 1999), the USGS FEWS NET team has also evaluated the impact of El Niño and Indian Ocean climate variations (Funk, et al., 2002, 2006a; Brown, et al., 2007; Funk 2009), occasionally producing ad hoc forecasts as needed to support early warning.

2.4 A Synopsis of FEWS NET-Related Climate Change and Food Security Research

One focus of our FEWS NET research has been the evaluation of climate change and vulnerability trends in food insecure eastern and southern Africa. This work began with the creation of historical rainfall time series for Africa (Funk, et al., 2003b; Funk and Michaelsen 2004). In 2003, FEWS NET evaluated the predictive potential of early growing season rainfall in Ethiopia and provided USAID with food balance projections (Funk, et al., 2003a). That analysis revealed two disturbing tendencies. First, agriculturally critical regions of Ethiopia had experienced substantial precipitation declines. Second, population growth and food balance analyses suggested that Ethiopia faces chronic and increasing food deficits.

FEWS NET followed up on this study with a careful study of thousands of eastern African rainfall gauge observations. The analysis suggested that a warming Indian Ocean was likely to produce increasing dryness in extremely vulnerable areas of eastern and southern Africa. These results were presented in an extensive FEWS NET report (Funk, et al., 2005). The work was also published by the United Kingdom's Royal Society (Verdin, et al., 2005) and presented in 2005 at its meeting on Climate Change and Agriculture. Lord May, the President of the Royal Society, referred to this work in an open letter to the G8 Ministers, asking them to "recognize the impacts of increasing drought conditions in Ethiopia ... that may already be occurring due to climate change, and to agree to further action to combat greenhouse gas emissions."[2] Satellite observations of vegetation greenness also reveal these declines (Funk and Brown 2005).

---

[1] http://chg.geog.ucsb.edu/wb/geowrsi.php

[2] http://www.royalsociety.ac.uk/news.asp?id=3833

Over the past several years, FEWS NET has continued multidisciplinary research on this topic. Reporting in the *Proceedings of the National Academy of Sciences* (Funk, et al., 2008) suggests that the dangerous warming in the Indian Ocean is likely to be at least partially caused by anthropogenic greenhouse gas emissions. Thus, further rainfall declines across parts of eastern and southern Africa appear likely. These drought projections run counter to the recent 4th Intergovernmental Panel for Climate Change (IPCC) assessment. The authors have suggested in *Science* that climate change assessments, based on inaccurate global climate precipitation fields, probably understate the agricultural risks of the warming Pacific and Indian Oceans (Brown and Funk 2008). The interaction of growing populations and limited potential water and cultivated areas increases food and water insecurity, amplifying the impacts of drought. A more recent paper, for the new journal *Food Security*, focuses on global risks implied by these tendencies (Funk and Brown 2009).

3 Techniques for Evaluating Hydrologic Risk

3.1 Low Frequency and High Frequency Models for Food Security Risk Monitoring

In general terms, we can represent the risk of food insecurity (r) as a function of shocks (s) and vulnerabilities (v).

$$r = F(s,v) \tag{1}$$

In this equation, shocks represent any serious disruption of food availability or access. Shocks may be related to global price increases, fertilizer shortages, political instability, or outbreaks of epizootic diseases such as Rift Valley Fever. For many semiarid areas dependent on rainfed agriculture, however, soil moisture deficits are commonly a potential shock. Shocks alone, however, do not create risks. The underlying vulnerability of livelihoods determines the impact of a given shock, such as agricultural drought. Complex economies, integrated into world markets, have the means to transport food (virtual water), making up for local rainfall deficits. In many parts of Africa, Asia, and Central and South America, where most people still subsist by farming, local rainfall deficits often translate into local food shortages.

In examining food security risks, it is important to consider both low frequency (years-to-decades) and high frequency (weeks-to-seasons) changes in shocks and risks. Theoretically, we can write a somewhat more complicated equation for risk.

$$r = F(s_{\text{low}} + s_{\text{high}}, v_{\text{low}} + v_{\text{high}}) \tag{2}$$

In this revised formula, hydrologic shocks might arise as a function of both weather and slowly varying changes in growing conditions. This latter category might include deleterious tendencies of declining rainfall and increasing temperatures, or degrading soil conditions. Shifts in agricultural practices (crop selection, fertilizer use, water retention, and harvesting practices) will also modify a shock. In a similar fashion, globalization, urbanization, biofuel usage, economic development and growth, and the burden of diseases such as HIV/AIDS and malaria act to slowly change baseline vulnerability patterns. We discuss techniques for evaluating the patterns in the next two Sections 4.1 and 4.2.

3.2 Evaluating Low Frequency Changes in Food Security Risks with Food and Water Balance Models

While they can often miss the complexity of individual food or water insecurity crises, at low frequencies, simple water and food balance calculations can usefully represent the slow evolution of risk, especially in less economically developed societies. It often holds, both in space and in time, that food and water vulnerability are strongly coupled to per capita supply. This is especially true in landlocked, poor, semiarid countries with nominal food and water transport infrastructures. Most food is used near where it is produced, and most rainfall is used near where it falls. Understanding this fact allows us to relate low frequency spatial and temporal variations in vulnerability (v_{low}) to basic per capita food and water balances.

$$v_{low} \propto supply \cdot person^{-1} \qquad (3)$$

In this equation, supply may typically be cereal grain production, total caloric production, or available water. While these balance equations clearly miss a great deal of the local variations between societies and governments, they do help define significant variations in the geography of food and water insecurity. Insecurity often arises from limited food and water availability, and balance equations provide a first order approximation of vulnerability.

Figure 3 shows an example drawn from an updated version of a 2003 FEWS NET analysis. This report provided USAID with historical and projected estimates of a "theoretical number of people without food" based on an assumed per capita cereal requirement. Historical trends in this food balance (Fig. 3.a) indicated increasing levels of food insecurity. Projections based on flat production trends and a population growth of 1.8 million people per year (Fig. 3.b) suggested that the theoretical number of people in Ethiopia without food would increase by some 1.5 million per year. In fact, since 2003, the number of people in Ethiopia has increased from 7 to 12 million, an increase of about 1 million per year.

Spatial per capita water availability measures can also provide useful guidance. In 2005 (Funk, et al., 2005), runoff built on the water harvest potential mapping work of Senay and Verdin (2004) was used to evaluate per capita water availability for Ethiopia. This work used the SCS Curve Number method to estimate annual runoff. The derivation of the curve numbers can be found in Artan, et al., (2001).

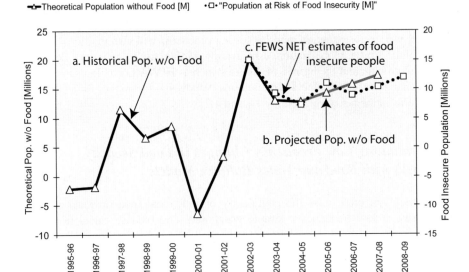

Fig. 3 Theoretical food balance results from our 2003 FEWS NET report (1995–2008, left axis), superimposed with actual FEWS NET food insecurity estimates (2002–2008, *right axis*). Historical population without food estimates (*a*) were based on observed crop production and population data. Projected population without food estimates (*b*) assumed constant crop production and a growing population. The *dashed line* (*c*) shows actual FEWS NET estimates of acutely food insecure people. These FEWS NET estimates are based on extensive in-country analysis, and are one important basis for international food aid requests

Daily RFE2 data were used to derive annual mean runoff values for 10 km grid cells. This mean runoff was divided by gridded population (Dobson, et al., 2000) to estimate spatial patterns of household water availability (Fig. 4). This map is presented with a reference unit volume of 1000 m^3 of water, after considering evaporation and seepage losses from reservoirs. The 1000 m^3 is suggested based on the amount of water that can be used to grow enough grain and biomass to support an average farm family in Africa. Taking into account system inefficiencies, regions with two or fewer units may be labeled as highly vulnerable. Areas with 2–4 units may be considered vulnerable. In general, Ethiopia may be roughly partitioned into three sections: water insecure areas with low rainfall (Fig. 4a), relatively wet areas with high population densities (Fig. 4b), and relatively wet areas with water surpluses (Fig. 4c).

Spatially, there is a very strong correspondence between areas of low rainfall and water availability (Fig. 4a) and areas in eastern Ethiopia currently experiencing chronic food insecurity (red areas in Fig. 5). These food insecure conditions have arisen through a combination of increasing population pressure (Fig. 3), climatic constraints on water availability (Fig. 4), and recurrent drought. The next section evaluates this latter tendency using a combination of downscaled 2.5° long-term

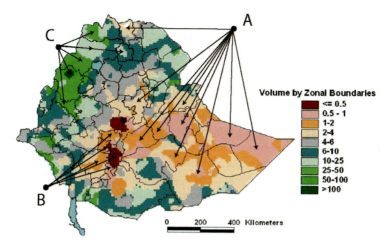

Fig. 4 Volume of potentially available annual surface water per family in 1,000 m^3 units (assumes 7 persons per family)

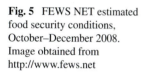
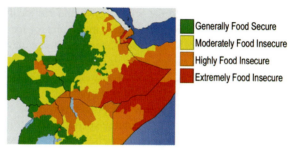

Fig. 5 FEWS NET estimated food security conditions, October–December 2008. Image obtained from http://www.fews.net

(1979–2005) Global Precipitation Climatology Project (GPCP, Adler, et al., 2003) monthly rainfall fields and RFE2 precipitation.

3.3 Combining Long-Term and Real-Time Satellite Rainfall Records

While extremely useful for crop modeling and early warning applications, high resolution satellite products, such as the RFE2 (Xie and Arkin 1997) and the TMPA (Huffman, et al., 2007), have relatively short periods of record. To overcome this limitation, we have developed two analogs to the satellite estimates: the 1960–1996 Collaborative Historical African Rainfall Model (CHARM) time series (Funk, et al., 2003b) and a gauge-enhanced downscaled version of the GPCP (Funk, et al., 2008). The CHARM time series used a reanalysis-driven model of orographic rainfall (Funk and Michaelsen 2004). Unfortunately, the reanalysis model data can produce spurious trends in the resulting CHARM data. For this reason, our new work focuses on the enhanced GPCP product. We describe this product here, evaluate its accuracy in Kenya, and use the combination of enhanced GPCP and RFE2 data to examine recent rainfall trends and anomalies in Kenya, where the station support for both products is quite high, and current food insecurity is very substantial, with more than 10 million people at risk.

The GPCP enhancement procedure began with the creation of a set of high quality, monthly 0.1° resolution long-term mean fields. These orographically enhanced mean fields were produced by combining three sources of information: (i) 0.1° long-term average monthly satellite rainfall estimate (RFE2, Xie and Arkin 1997) grids $\bar{p}$, (ii) 0.1° grids of elevation e and slope s, and (iii) observations ($\bar{o}$) of long term mean rainfall measured at a large number of stations. The use of satellite rainfall averages as a basis for deriving improved gridded climatologies, as far as we know, is new. This innovation grows naturally out of the fact that there are strong local regressions between station normals and monthly mean satellite precipitation ($\bar{p}$). Because variations in infrared and microwave emissions covary in space with rainfall, these estimates represent well large scale precipitation gradients. Local variations within these large scale climate gradients are often induced by topography, and strongly related to the product of $\bar{p}$ and the local elevation e and slope s. The term $\bar{p}s$ describes the multiplicative interaction of local slopes and satellite rainfall estimates. The term $\bar{p}e$ describes the interaction of elevation and mean satellite rainfall. The observed station normals ($\bar{o}$) can be reasonably fit by local regressions of the form $\bar{o} \approx b_o + b_1\bar{p} + b_2\bar{p}e + b_3\bar{p}s$.

Because these models use long term monthly mean rainfall $\bar{p}$ and the interaction of these rainfall mean fields with topography ($\bar{p}e, \bar{p}s$), they benefit from the ability of satellite rainfall estimates to capture spatial gradients in rainfall. These models were fit as described in Funk and Michaelsen (2004), except that a series of moving spatial windows with a 7° radius (~770 km) were used to develop localized regression models, based on distance-weighted subsets of 6965 FAOCLIM2.0 precipitation. The period represented by these climate normals varies by station but typically corresponds to the 1950–1980 era. These moving window regressions produced 12 monthly 0.1° grids of average rainfall. Block kriging was then used to interpolate the 6,965 at-station differences (residuals) between the FAOCLIM2.0 climate normals and regression estimate grids. The regression estimates and kriged anomalies were combined yielding 12 monthly FEWS NET climatology fields (FCLIM). The at-station accuracy of the FCLIM monthly long-term mean fields was evaluated numerically by comparing the regression estimates at each of the 6965 points to the observed value for each month. The error statistics were promising, with a coefficient of determination of 0.9, a mean bias error of 0.06 mm month^{-1}, and mean absolute error of 18 mm month^{-1}. As a reference, the mean monthly rainfall in sub-Saharan Africa is 80 mm month^{-1}, and typically ranges between 0 and 200 mm month^{-1}.

In the second step of the GPCP enhancement procedure, the monthly, 0.1°, African (20°W-55°E, 40°S-40°N) FCLIM fields were used to downscale the 2.5° 1979–2005 GPCP dataset. Monthly GPCP data were translated into fractions of their long-term means, downscaled to 0.1° degree fields via cubic convolution interpolation, and multisplied against the corresponding 0.1° FCLIM grids. This produced monthly, 1979–2005, 0.1° downscaled GPCP fields. The second stage of the GPCP enhancement used a modified inverse distance weighting procedure to blend a moderately dense, quality controlled set of rain gauge observations with the downscaled GPCP fields. Some of these gauges would have been included in the 2.5° GPCP estimates. We will refer to the blended gauge-GPCP-FCLIM dataset as the "enhanced GPCP."

Figures 6 and 7-top panel show March-April-May validation results for the enhanced GPCP dataset. The validation is based on 22 years (1979–1998) of a large number (73) of high-quality daily gauge observations located the western edge of Kenya between 34.15°E and 35.55°E and 1°S and 1°N. While the study site has an area equal to 45% of a GPCP grid cell, the downscaled enhanced GPCP means correspond fairly well at 0.1° resolution (Fig. 6), and the spatial R^2 of these fields is about 0.65. Temporally, the enhanced GPCP and validation data track very well (Fig. 7), with a seasonal R^2 of 0.87. The monthly 0.1° mean absolute error of the data is 14 mm month^{-1}, and the mean bias is 0 mm month^{-1}. This compares favorably with error statistics from the first set of rainfall estimates used by FEWS NET (the RFE1, Herman et al. 1997). Previous analysis for this area found monthly 0.1° mean absolute errors of 20 mm month^{-1}, and mean bias values of 15 mm month^{-1}, Funk and Verdin 2003).

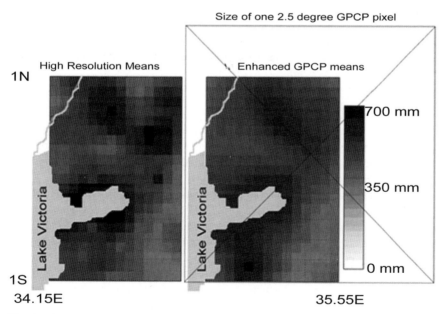

Fig. 6 Monthly March–April–May mean 1979–1998 high density gauge and enhanced GPCP rainfall estimates over the Kenya test site

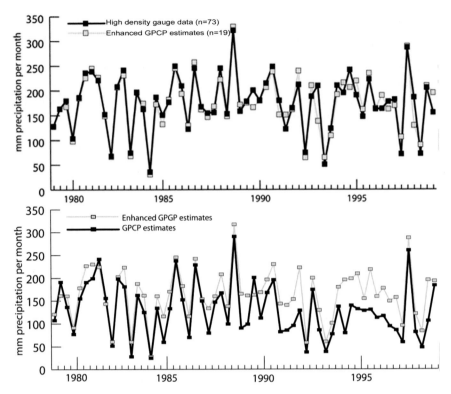

Fig. 7 Regionally averaged 1979–1998 March–May rainfall over the western Kenya test site. The first three *boxes* represent 3 months from 1979 (March–May). Each consecutive set of three *boxes* represents one of the following years. The *top panel* shows high density gauge observations and enhanced GPCP time series. The *bottom panel* shows the GPCP data and enhanced GPCP time series

Comparison between the enhance GPCP and raw GPCP data (Fig. 7, bottom panel) show substantial discrepancies between the two data sets: the GPCP tends to be substantially lower than the enhanced GPCP data, especially after 1986. This shift in performance is likely due to the degradation of the publically available station data sets over the past 20 years.

Further validation can be achieved by comparing the enhanced GPCP and RFE2 data. These results, evaluated across provinces in Kenya, are shown in Table 1 for the two main growing seasons. The long rains are centered on March–May. The short rains are centered on October–December. In general, the correlations are high (over 0.8), especially during the short rains. A very small province (Nairobi) has a low correlation (0.49) during March–May. This is likely due to a difference in spatial scale and underlying station support. The low correlation for the coastal province's March–May time series may be attributable to the low station density here in the RFE2 and the known difficulty with rainfall retrievals near the coast.

Table 1 Correlations between 2001 and 2005 enhanced GPCP and RFE2

Province	Correlation March–May	Correlation October–December	Correlation with long rains yields
Eastern	0.81	0.97	0.66
North Eastern	0.83	0.86	NA
Coast	0.61	0.97	0.38
Nairobi	0.49	0.98	0.12
Central	0.85	0.85	0.60
Rift Valley	0.91	0.97	0.58
Nyanza	0.93	0.99	0.87
Western	0.98	0.94	0.95

3.4 Monitoring High Frequency Shocks Using Water Requirement Satisfaction Index Maps

The primary agrohydrologic monitoring tool used by USGS FEWS NET is a gridded version of the WRSI.[3] Originally developed by the FAO (1977, 1979, 1986), the WRSI is a measure of how much moisture is available to a crop relative to the crop's phenologically changing demands. The USGS FEWS NET team (Verdin and Klaver 2002; Senay and Verdin 2003) has created a spatially explicit version of the WRSI, driven by gridded estimates of satellite rainfall (Xie and Arkin 1997; Huffman, et al., 2007) and potential evapotranspiration (PET) derived using the Penman-Monteith equation (Shuttleworth 1992; Verdin and Klaver 2002; Senay, et al., 2008) which uses numerical weather prediction model data. In addition to rainfall and PET, the WRSI also uses grids of soil parameters and length of the crop growing season (Senay and Verdin 2003). This last parameter is determined by examining the ratio of rainfall and PET and may vary from 60 days for very fast maturing crops in arid zones to 180 days in moist high-altitude locations. In addition to these grids of data, the WRSI requires crop-specific water demand coefficients (K_c) as a function of the current phenology of the crop.

Before looking at the specifics of the WRSI calculation, it is worth a quick review of crop phenology. To represent this, we show time-series data from an early study (Tucker 1979) of vegetation index observations of a cornfield in the United States (Fig. 8). As the plants mature, plant height, percent cover, vegetation index values, and the crop coefficient increase linearly out to about 80 days. At this time, the first tassels appear, and the plants go from the vegetative to reproductive stage. The mass of cereal grains increases during the reproductive stage, so this transition is important. Soil water deficits during this critical grain filling period are the most

---

[3] This section builds strongly on the FEWS NET readme (http://earlywarning.usgs.gov/adds/readme.php?symbol=ws), written by Gabriel Senay, as well as the GeoWRSI technical manual, written by Tamuka Magadzire.

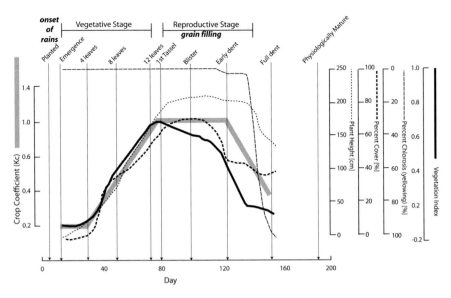

Fig. 8 Crop phenology for a maize plot, modified from Tucker (1979)

damaging. Conversely, late season soil water deficits, after the grain biomass accumulation is complete, may actually lead to higher yields by protecting the grains from loss due to disease, insects, and mold.

Because of the different water needs of the plant at different phenological stages, timing is critical to the successful calculation of the WRSI, which measures the relative crop water availability from the Start of the Season (SOS) to the End of the Season (EOS). This time period corresponds to the typical phenological curve shown in Fig. 8. Standard FEWS NET WRSI modeling is done using ~10 day (dekadal) accumulations. Each month's rainfall is divided into the sum of the first 10 days, the middle 10 days, and the remaining 8–11 days. The SOS date is then determined by finding the first dekad with more than 25 mm of rain, followed by two dekads with a total rainfall of at least 20 mm. This threshold is linked to the necessary moisture availability triggering the crop's emergence. The EOS date is a function of the length of growing period, LGP (EOS=SOS+LGP).

For a given grid cell, calculation of the WRSI initializes several months before the SOS date with a standard water balance calculation. Once at SOS dekad, the WRSI calculation begins. At this, and each following dekad d, up to the EOS, the WRSI estimates the running ratio of actual plant evapotranspiration (AET_c) to the full plant water requirement (WR).

$$\text{WRSI} = 100 \frac{\sum_{t=\text{SOS}}^{d} \text{AET}_c}{\sum_{t=\text{SOS}}^{d} \text{WR}} \qquad (4)$$

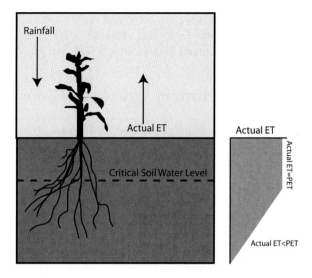

Fig. 9 WRSI soil water balance

The WR value is a function of the PET and the phenologically dependent crop coefficient (K_c), and the WRSI index is accumulated from the SOS to a given dekad (d).

$$\text{WR} = \text{PET} \cdot K_c \tag{5}$$

The K_c parameter peaks during the reproductive stage of the crop (Fig. 8). The WR is a measure of how much water the crop would need under ideal growing conditions. Full satisfaction of WR constitutes growing conditions without water stress, that is, WRSI values of 100. When WRSI falls below 50, a crop is considered to have failed. This threshold of 50 is based on empirical analysis (FAO 1986, Senay and Verdin 2003).

AET_c is determined by a modified water balance calculation, with the AET_c value representing the water withdrawn from the soil water reservoir (Fig. 9) at each time step. Depending on the soil water level, root depth, and WR, AET_c may be equivalent, or less, than WR. Please refer to Senay and Verdin (2003) for details. Each time AET_c is less than WR, the WRSI value lowers, indicating increasing water stress. It is standard practice to produce "extended WRSI" predictions. These extended WRSI maps continue integrating the WRSI value forward in time from dekad d using long-term average rainfall and PET. This provides an approximation of the final crop water status of the crop. These projections will become increasingly accurate as the EOS date approaches and are typically quite stable by the middle to the end of the reproductive stage. Since this date is typically several months before the crops are harvested, the WRSI provides a valuable early warning tool.

Operational WRSI runs are hosted at the USGS early warning portal.[4] A stand-alone version of the GeoWRSI (Magadzire 2009) has been created for PCs and is available at the Climate Hazard Group Web site: http://chg.geog.ucsb.edu.

4 Analysis of Kenyan Agricultural Hydrologic Conditions

4.1 WRSI Anomalies for the 2007 and 2008 Long and Short Rains

Using 2nd generation satellite rainfall estimates (RFE2) from NOAA CPC, Penman-Montieth PET (Shuttleworth 1992, Senay, et al., 2008) fields from the USGS,[4] and the stand-alone GeoWRSI tool obtained from the Climate Hazard Group Web site,

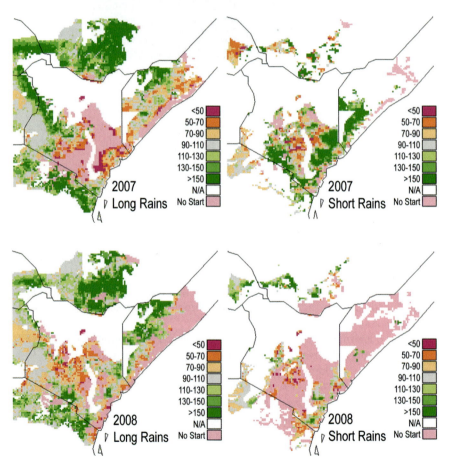

Fig. 10 GeoWRSI end-of-season maize percent anomalies for the long rains (March–September) and short rains (October–February)

---

[4]http://earlywarning.usgs.gov

we have calculated the 2007 and 2008 maize WRSI anomalies for the long (March–September) and short (October–December) rainfall seasons (Fig. 10). These figures show the end of season WRSI, expressed as percent deviations from the long-term mean (2001–2007). In general, the arid northern parts of Kenya depend on pastoral livelihoods. These areas are masked in the WRSI runs and shown in white in Fig. 10. Across the southern two-thirds of the country, the western parts rely more upon the long rains, and the eastern parts depend more upon the less reliable short rainy season. In general, the rainfall performance for the 2007 long, 2008 long, and 2008 short seasons was very poor across the entire eastern half of the country. Many areas never received sufficient moisture to even initialize the WRSI model with an "onset of rains" signal. This could indicate that the 25 mm SOS-threshold, originally developed for the Sahel during the 1970s, might not be appropriate in eastern Kenya. More research into this component of the model seems warranted.

The 2007 short rainy season provided some relief near the coast, but not further inland. Substantial agrohydrologic shortages have contributed significantly to the current food insecurity (Fig. 5). Using 2001–2006 long rains maize FEWS NET yield data pooled across the eastern provinces, we can establish a reasonable relationship to the log of seasonal March–May rainfall ($R^2=0.63$). This simple relationship, in turn, can be used to make estimates of very low long rain yields across the eastern provinces (Fig. 11). Because the main rainy season ends several months before the actual harvest, satellite rainfall can be a good early warning trigger. In February 2009, maize prices in Kenya are almost twice the 2003–2008 average. Without assistance, the food security situation there is likely to degrade substantially.

4.2 The 2007 and 2008 Seasons in Historical Context

How uncommon is the multiseason combination of crop water deficits presented in Fig. 10? To address this question, we extracted long (March-April-May) and short

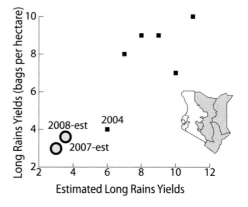

Fig. 11 Actual yields and estimated long rain maize yields for the eastern provinces of Kenya (the area *shaded* in the map above). Actual 2001–2006 yields (*y*-axis) were obtained from FEWS NET collaborators in Kenya. Yield estimates (*x*-axis) based on the log of March–May rainfall ($R^2=0.63$). No yield estimates were available for 2007 and 2008 – the values shown are estimated from rainfall

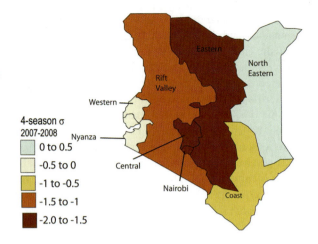

Fig. 12 Combined rainfall performance for last four seasons (2007 MAM, 2007 OND, 2008 MAM, 2008 OND), measured as standard deviations over the 1979–2008 era

(October–December) province-scale rainfall time series. The well-correlated RFE2 data (Table 1) were bias corrected using the period of overlap (2001–2005) and the 2006–2008 seasons to produce a complete 1979–2008 record. For each season, and for each province, the ratio of the 3-year (2002–2005) enhanced GPCP and RFE2 average was estimated. The 2006–2008 RFE2 values were multiplied by this scalar, and added to the end of the enhanced GPCP time series.

The rainfall data were next transformed into ranks, which minimized the impact of a few extremely wet short rainy seasons associated with El Niño years. Time series of 4-season averages were then calculated and expressed as standard deviations from the average. These sigma (σ) values range from about −2 to +2, with values above ±1 denoting exceptional 4-season groupings. Figure 12 shows the sigma values for the combined 2007 long, 2007 short, 2008 long, and 2008 short seasons. In the middle of Kenya (the Eastern, Central and Nairobi provinces), four-season rainfall performance has been extremely poor, compared to 1979–2008 records, with sigma values of less than −1.5. The Rift Valley province, by far Kenya's most productive crop growing region, is not far behind, with a sigma of −1.4. In the arid pastoral North Eastern province and in the tropical Western province, four-season rainfall performance has been near normal. The Coast Province received modestly below normal rainfall in each of the four 2007 and 2008 seasons, resulting in a 4-season sigma of −0.8.

4.3 1979–2008 Trends in Kenyan Rainfall and WRSI

We can use the enhanced GPCP and RFE2 rainfall grids to examine trends in rainfall and WRSI. These results are presented in Figs. 13 and 14. In order to run the GeoWRSI over the 1979–2001 era, dekadal rainfall estimates were derived by equally dividing each months total into three dekadal estimates. Correlation analysis

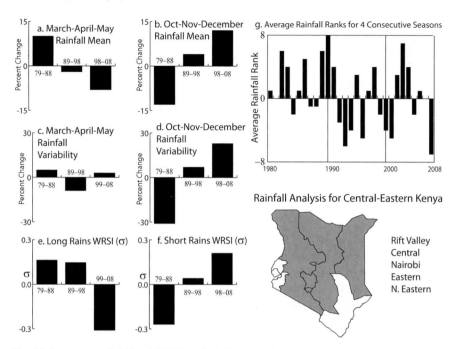

Fig. 13 Long-term rainfall and WRSI analysis for Central-Eastern Kenya. The area analyzed is *shaded* in the map of Kenya

of the seasonal provincial rainfall time series indicated strong homogeneity (1979–2008 correlations of > 0.8) among the Rift Valley, Nairobi, Central, Eastern, and North Eastern provinces. Hence, these regions have been pooled (Fig. 13). Coastal Kenya displayed different interannual variations, so it is presented alone (Fig. 14). The humid Western and Nyanza provinces displayed little decadal variation, so results for these provinces are not displayed here.

Both the central-eastern and coastal areas exhibit substantial shifts in seasonality, with long rains decreasing (panel a) and short rains increasing (panel b) by 20–30%. This shift has been previously noted by the regional FEWS NET scientist (Galu 2008), who has also suggested that the intraseasonal variability of the rainfall has increased in recent years, leading to less reliable crop performance. We test this hypothesis by estimating the 3-month standard deviation for each long and short rain season. The standard deviation estimated from the monthly 1979 rainfall for March, April, and May represents the variability for that season. These values, broken out by region, decade, and season, are shown in panels c and d in Figs. 13 and 14. For the March-April-May season, no increase in variability is apparent. For the October–December rains, on the other hand, there does appear to be a large (>30%) increase in the intraseasonal rainfall variability, from about 38 mm month^{-1} in 1979–1988 to about 50 mm month^{-1} in the 10 years between 1999 and 2008. The combination of panels b and d suggests that while October-November-December

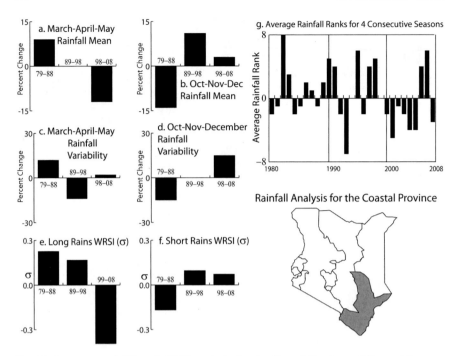

Fig. 14 Long-term rainfall and WRSI analysis for the Coastal Province of Kenya. The area analyzed is *shaded* in the map of Kenya

rainfall has been increasing, on average this rainfall tends to arrive poorly distributed throughout the season (again, as suggested by Gideon Galu, 2008). We can examine the impacts of intraseasonal rainfall variations by running the WRSI model over the 1979–2008 era, expressing the end-of-season WRSI values as standard deviations (σ), and estimating decadal averages. As expected, long rain WRSI values (panel e in Figs. 13 and 14) appear to have dropped substantially across both Central-Eastern and Coastal Kenya. In Central-Eastern Kenya, short rain WRSI (Fig. 13f) has increased, in line with recent rainfall increases (panel Fig. 13d). The case in coastal Kenya, however, appears quite different. While both the short and WRSI seem to have increased by a small amount, the increase in variability appears much more substantial.

Figures 13 and 14 also show time series displaying successive 2-year combinations of long and short rainy season. The first bar on the left in panel g represents the combined performance of the 1979 long and short rains together with the 2008 long and short rains. The last bar on the right represents the most recent 4 seasons: the 2007 long and short rains and 2008 long and short rains. The intervening dry seasons are not included. The data have been ranked to minimize the effect of a few extremely wet El Niño October–December seasons. For each season, ranks for the past 30 years have been calculated from lowest to highest and offset by 15. A value of –15 indicates the worst season on record, 0 a median season, and 15 the best on record. For the areas analyzed, these individual ranks were then averaged, producing values between −8 (very low 4-season rainfall) and 8 (very good 4-season rainfall). For central and eastern Kenya (i.e., most of the country), the tendency toward poorer rainfall is apparent. In the early 2000s, rainfall performance was quite good, but the combined 2007–2008 long-short rains appear to be the worst over the period analyzed. Coastal Kenya also exhibits a downward tendency, again driven by the decreasing long rains. Except for a few positive years, linked to wet 2006 long and short rains, the average rainfall performance for Coastal Kenya has been substantially below normal.

5 Summary and Discussion

In Africa, 90% of farmers are smallholders, reliant on small plots of land, limited technological inputs, and rainfed agriculture (Rockstrom 2000). These farmers and their societies are tightly coupled to the environment and climate. This makes them vulnerable to hydrologic extremes. Satellite rainfall estimates, especially when linked to agrohydrologic models, such as the WRSI, can provide valuable early indication of weather-induced shocks. The WRSI filters the rainfall data in space and time. The particular impact of midseason rainfall receipts will vary by the soil characteristics, the length of growing period, the crop type, antecedent rainfall and PET, and the phenological stage of the plant. The most damaging crop water deficits arise during the reproductive stage of the crop (Fig. 8), when the cereal plants switch from growing leaves to growing grains. Late planting (Funk and Budde 2009) or midseason water deficits (Senay and Verdin 2003) can dramatically reduce yields. The WRSI allows these disruptions to be identified months before the actual harvest date, providing early warning and time to develop disaster response strategies (Fig. 1).

Food security responses by USAID and partner agencies are saving thousands of lives. A good example would be the 2002–2003 food crisis in Ethiopia. Rainfall performance was very poor (Funk, et al., 2003a, 2005), perhaps analogous to conditions accompanying the devastating 1984–1985 famine. This dryness, combined with low planted area due to low cereal prices, produced a large spike in food insecurity (Fig. 3). This food crisis provided a benchmark test for the international food security organizations, and effective response prevented widespread hunger, disease, and social disruption. These responses were enhanced substantially by real-time satellite rainfall applications.

In addition to effective early warning, agrohydrologic modeling can also inform long-term food security decision making through water and food budget analysis. This perspective helps explain, in part, the increasingly chronic food insecurity in eastern Africa. The Ethiopian 2002–2003 food crisis in Ethiopia was associated with about 15 million food insecure individuals. Recent food insecurity levels appear to be trending toward this amount at a rate of about 1 million people per year. Growing population and stagnant yields help create this problem (Fig. 3), as has the low water availability across the more arid parts of eastern Africa (Fig. 4).

Focusing on Kenya, we have shown that the WRSI model, driven by satellite rainfall fields, can effectively monitor anomalous hydrologic conditions (Fig. 10). Across most of Kenya, hydrologic growing conditions for the 2007 and 2008 long rains and the 2008 short rains were very poor, indicating failure or near-failure of maize crops, as suggested by our empirical estimation of yields (Fig. 11). Performance of the 2008 short rains was mixed but poor in the center of the country.

The combination of these 4 seasons appears unusually bad, indicating that a rare and intense multiyear drought has impacted most of the country (Fig. 12). Examination of pooled enhanced GPCP/RFE2 data support the assertion (Galu 2008) that a shift in seasonality may be occurring. Consistent with our previous research (Funk, et al., 2005; Verdin, et al., 2005; Funk, et al., 2008), March–May rainfall appears to be decreasing by almost 10% a decade (Figs. 13 and 14), producing a -0.5σ reduction in WRSI over the 1979–2008 era over both coastal and central-eastern Kenya.

The October–December short rains, on the other hand, appear to be increasing. There appears to have been a substantial increase in intraseasonal variability in the October–December short rains across central-eastern and coastal Kenya, and the March–May long rains in coastal Kenya. Increasing intraseasonal variability tends to reduce crop performance due to the occurrence of midseason dry spells. WRSI analysis suggests that this increasing variability may be reducing the beneficial impact of rainfall increases in coastal Kenya (Fig. 13f), consistent with reports coming from Kenya (Galu 2008).

We suggest that satellite observations can contribute to both short- and long-term monitoring of food security in Africa. Furthermore, we believe that both these perspectives are necessary. As the number of urban poor rises rapidly and global food prices soar due to increased consumption by biofuels and livestock, there has been a broad increase in three classic coping mechanisms (Natsios and Doley 2009): food hoarding, migration, and increased banditry. This expanding food stress disrupts societies and creates political unrest; over the next decade we are likely to see "food coups" emerge as modern counterparts to the famines of the past. We have shown that agricultural development can help reduce these impacts (Funk, et al., 2008; Brown and Funk 2008; Funk and Brown 2009). Without addressing the key issues of resource scarcity, short-term food aid responses may in fact act to create future risk, moving African societies into imbalance and helping to create greater need. More analysis of the shift in seasonality, discussed briefly here, could help guide future agricultural development strategy.

References

Adler RF, Huffman GJ, Chang A, Ferraro R, Xie P-P, Janowiak J, Rudolf B, Schneider U, Curtis S, Bolvin D, Gruber A, Susskind J, Arkin P, Nelkin E (2003) The version-2 global precipitation climatology project (GPCP) monthly precipitation analysis (1979–Present). Journal of Hydrometeorology 4:1147–1167

Artan G, Verdin J, Asante K (2001) A wide-area flood risk monitoring model. Proc of the Fifth International Workshop on Applications of Remote Sensing in Hydrology. Montpelier, France

Brown M, Funk C (2008) Food security under climate change. Science 319:580–581

Brown ME, Funk CC, Galu G, Choularton R (2007) Earlier famine warning possible using remote sensing and models. EOS, Transactions of the American Geophysical Union 88(39):381–382

Dobson JE, Bright EA, Coleman PR, Durfee RC, Worley BA (2000) Landscan: global population for estimating population at risk. Photogrammetric Engineering and Remote Sensing 66(7):849–857

FAO (1977) Crop water requirements. FAO Irrigation and Drainage Paper No. 24, by Doorenbos J and W.O. Pruitt. FAO, Rome, Italy

FAO (1979) Agrometeorological crop monitoring and forecasting. FAO Plant Production and Protection Paper No. 17, by M. Frère and G.F. Popov. FAO, Rome, Italy

FAO (1986) Early agrometeorological crop yield forecasting. FAO Plant Production and Protection Paper No. 73, by M. Frère and G.F. Popov. FAO, Rome, Italy

FAO (2007) The State of Food and Agriculture. United Nations Food and Agriculture Organization, Rome

Funk C (2009) New satellite observations and rainfall forecasts help provide earlier warning of drought in Africa. The Earth Observer http://earlywarning.usgs.gov/adds/pubs/Funk_EarthObserver_Jan_Feb09.pdf. Accessed 17 March 2009

Funk C, Asfaw A, Steffen P, Senay G, Rowland J, Verdin J (2003) Estimating *Meher* Crop Production Using Rainfall in the 'Long Cycle' Region of Ethiopia. FEWS NET Special Report.http://earlywarning.usgs.gov/adds/pubs/EthProductionOutlook.pdf. Accessed 17 March 2009

Funk C, Brown M (2005) A maximum-to-minimum technique for making projections of NDVI in semi-arid Africa for food security early warning. Rem Sens Environment 101:249–256. http://earlywarning.usgs.gov/adds/pubs/ndvi_projections.pdf. Accessed 17 March 2009

Funk C, Brown M (2009) Emerging threats to globalfood security. Food Security 1(3):271

Funk C, Budde M (2009) Phenologically-tuned MODIS NDVI-based production anomaly estimates for Zimbabwe. Remote Sensing of Environment 113(1):115–125

Funk C, Dettinger MD, Brown ME, Michaelsen JC, Verdin JP, Barlow M, Hoell A (2008) Warming of the Indian Ocean threatens eastern and southern Africa, but could be mitigated by agricultural development. Proceedings of the National Academy of Sciences 105:11081–11086

Funk C, Magadzire T, Husak G, Verdin J, Michaelsen J, Rowland J (2002) Forecasts of 2002/2003 Southern Africa Maize Growing Conditions Based on October 2002 Sea Surface Temperature and Climate Fields. FEWS NET Special Report

Funk C, Michaelsen J (2004) A simplified diagnostic model of orographic rainfall for enhancing satellite-based rainfall estimates in data poor regions. Journal of Applied Meteorology 43:1366–1378

Funk C, Michaelsen J, Verdin J, Artan G, Husak G, Senay G, Gadain H, Magadzire T (2003) The collaborative historical African rainfall model: description and evaluation. International Journal of Climatology 23:47–66

Funk C, Schmitt C, LeComte D (2006) El Niño and Indian Ocean Dipole conditions likely into early 2007, with drought and flooding implications for Southern and Eastern Africa. FEWS NET Special Report

Funk C, Senay G, Asfaw A, Verdin J, Rowland J, Michaelsen J, Eilerts G, Korecha D, Choularton R (2005) Recent Drought Tendencies in Ethiopia and equatorial-subtropical eastern Africa. FEWS NET Special Report. http://chg.geog.ucsb.edu/pub/pubs/RecentDroughtTendenciesInEthiopia.pdf. Accessed 17 March 2009

Funk C, Verdin J (2003) Comparing satellite rainfall estimates and reanalysis precipitation fields with station data for western Kenya. Proceedings of the International Workshop on Crop Monitoring for Food Security in Africa, European Joint Research Centre/UN Food and Agriculture Organization, Nairobi, Kenya, January 28–30

Funk C, Verdin J, Husak G (2006) Integrating observation and statistical forecasts over sub-Saharan Africa to support Famine Early Warning. American Meteorological Society Meeting, Nov. 2006, Extended Abstract

Galu G (2008) Recent changes in seasonal rainfall patterns in the Greater Horn of Africa. FEWS NET Internal Report

Herman A, Kumar VB, Arkin PA, Kousky JV (1997) Objectively determined 10-day African rainfall estimates created for famine early warning. International Journal of Remote Sensing 18:2147–2159

Huffman G, Adler RF, Bolvin DT, Gu G, Nelkin EJ, Bowman KP, Hong Y, Stocker EF, Wolff DB (2007) The TRMM Multisatellite Precipitation Analysis (TMPA): quasi-global, multi-year, combined-sensor precipitation estimates at fine scales. Journal of Hydrometeorology 8(1):38–55

Jury MR (2002) Economic impacts of climate variability in South Africa and development of resource prediction models. Journal of Applied Meteorology 41:46–55

Kates RW (2000) Cautionary tales: adaptation and the global poor. Climatic Change 45:5–17

Magadzire T (2009) The Geospatial Water Requirement Satisfaction Index Tool, TechnicalManual. USGS Open-File Report. In Review

Natsios AS, Doley KW (2009) The coming food coups. The Washington Quarterly 32(1):7–25

Rockstrom J (2000) Water resources management in smallholder farms in Eastern and Southern Africa: an overview. Physics and Chemistry of the Earth (B) 25:275–283

Rowland J, Verdin J, Adoum A, Senay G (2005) Drought monitoring techniques for famine early warning systems in Africa. Chapter 19 in Monitoring and Predicting Agricultural Droughts: A Global Study, Boken VK, Cracknell AP, Heathcote RL(Eds.), Oxford University Press, New York

Senay G, Verdin J (2003) Characterization of yield reduction in Ethiopia using a GIS-based crop water balance model. Canadian Journal of Remote Sensing 29(6):687–692

Senay G, Verdin J (2004) Developing index maps of water-harvest potential in Africa. Applied Engineering in Agriculture, American Society of Agricultural Engineers 20(6):789–799

Senay G, Verdin J, Lietzow R, Melesse A (2008) Global daily reference evapotranspiration modeling and validation. Journal of the American Water Resources Association 44(4):969–979

Shuttleworth J (1992) Evaporation. In: Maidment D (Ed.) Handbook of Hydrology. McGraw-Hill, New York

Tucker CJ (1979) Red and photographic infrared linear combinations for monitoring vegetation. Remote Sensing of Environment 8:127–150

Verdin J, Funk C, Klaver J, Roberts D (1999) Exploring the correlation between Southern African NDVI and ENSO sea surface temperatures: results for the 1998 growing season. International Journal of Remote Sensing 20(10):2117–2124

Verdin J, Funk C, Senay G, Choularton R (2005) Climate science and famine early warning. Philosophical Transactions of the Royal Society B 360:2155–2168 http://earlywarning.usgs.gov/adds/pubs/Climate%20Science%20and%20Famine%20EW.pdf. Accessed 17 March 2009

Verdin J, Klaver R (2002) Grid cell based crop water accounting for the famine early warning system. Hydrological Processes 16:1617–1630

Verdin J, Senay G (2002). Evaluating the performance of a crop water balance model in estimating regional crop production. Proceedings of the Pecora 15 Symposium, Denver CO

WWD (2003) UN World Water Development Report, Water for People, Water for Life. UNESCO. http://www.unesco.org/water/wwap/wwdr/table_contents.shtml. Accessed 17 March 2009

Xie P, Arkin PA (1997) A 17-year monthly analysis based on gauge observations, satellite estimates, and numerical model outputs. Bulletin of the American Meteorological Society 78(11):2539–2558

Index

A

Advanced Infrared Sounder (AIRS), 20
Advanced Microwave Scanning Radiometer (AMSR), 5, 6, 7, 9, 10, 19, 25, 26, 27, 28, 33, 71, 86, 87, 94, 108, 112, 113, 114, 198, 262, 272, 281, 284, 285
Advanced Microwave Sounding Unit (AMSU), 6, 25, 41, 71, 86, 198
Advanced Microwave Sounding Unit B (AMSU-B), 6, 7, 25, 26, 27, 33, 41, 71, 86, 182, 198, 272
Advanced TOVS (ATOVS), 20
Advanced Very High Resolution Radiometer (AVHRR), 288, 291
Agricultural Meteorology Model (AFWA) (AGRMET), 271, 272, 273, 274, 275, 278, 281
Agricultural Research Service (USDA) (ARS), 281, 283, 284
Agriculture, 95, 127, 249, 258, 269–271, 273, 275, 276, 278, 283–290, 296, 298, 299, 300, 301, 302, 317
Agriculture and Resources Inventory Surveys through Aerospace (AgRISTARS), 271, 272, 278, 279
Air Force Weather Agency (AFWA), 271, 272, 274–275, 280, 285, 290
Albedo, 39
Algorithm, 4, 5, 6, 7, 9, 10, 11–13, 14, 20, 25, 26, 27, 29, 33, 40, 41, 42, 43–45, 46, 49–65, 69–81, 93, 107, 108, 109, 112, 113, 114, 115, 119, 120, 121, 129, 137, 154, 161, 173, 175, 194, 195, 198, 199, 201, 202, 203, 207, 211, 218, 219, 225, 230, 233, 235, 247, 274, 275, 276, 277, 279, 284, 290, 291, 298
AMSR-E–GPROF, 5
AMSR for EOS (AMSR-E), 108, 284
AMSU–NESDIS, 5

Antecedent precipitation index (API), 252, 253, 254, 260, 261, 262
Aqua, 6, 25, 86, 87, 94, 99, 108, 113, 284
Aral Sea, 287
Arctic Slope Regional Corporation (Crop Explorer) (ASRC), 273
Arkansas, 96, 98
Artificial neural networks, 40, 50, 71, 86, 171, 207, 233
Australia, 13, 34, 78, 79, 92, 114, 120, 129, 133, 135, 137, 138, 172, 258, 274, 278

B

3B43, 11, 12, 19
Bacchiglione, 217, 218, 219, 221, 222, 223, 224, 225
Baiu, 117
Basins, 14, 50, 52, 76, 96, 98, 99, 154, 162, 172, 216, 217, 218, 219, 220, 221, 222, 223, 224, 225, 232, 233, 235, 246, 248, 250, 251, 254, 255, 256, 257, 258, 260, 261, 262
Beressa watershed, 206, 207
Bicubic interpolation, 88
Bilinear interpolation, 29
Brier score, 130, 237
Brightness temperature, 7, 24, 39, 41, 42, 43, 50, 51, 54, 87, 107, 108, 109, 110, 111, 216, 233, 274, 276, 284, 285
3B40RT, 12
3B41RT, 12, 153, 156, 157, 158, 159, 160, 162, 163, 164, 217
3B42RT, 12, 15, 19, 36, 76, 77, 78, 120, 121, 137, 155, 156, 157, 158, 159, 160, 171, 173, 198, 199, 201, 272, 275

C

Calibration, 5, 7, 8, 9, 10, 11, 12, 13, 14, 15–18, 20, 24, 27, 33, 40, 41, 42, 43, 44, 45, 55, 71, 72, 73, 74, 75, 76, 80, 81, 147, 150, 151, 154–159, 162, 163, 165, 198, 202, 216, 218, 219, 222, 230, 232, 248
Canopy, 96, 282
Climate Assessment and Monitoring System (CAMS), 8
Climate Prediction Center (NOAA) (CPC), 7, 8, 10, 20, 34, 35, 65, 72, 86, 93, 107, 108, 113, 171, 174, 175, 177, 182, 183, 184, 185, 186, 188, 189, 198, 207, 271, 276, 300
Climatology, 5, 8, 9, 10, 12, 13, 14, 15–18, 20, 70, 71, 72, 74, 76, 98, 106, 117, 154, 195, 237, 239, 249, 252, 253, 255, 273, 277, 281, 283, 298, 305, 306
Cloud patch, 50, 51, 52, 53, 54, 55
Colombia, 194, 195, 196, 197, 200
Complex terrain, 15, 94, 205–214, 215–225
Continental United States (CONUS), 13, 40, 42, 54, 93, 95
Convective, 13, 26, 32, 33, 40, 43, 45, 52, 55, 56, 71, 75, 76, 78, 91, 92, 116, 155, 171, 172, 173, 176, 179, 180, 181, 190, 194, 213
Correlation length, 149, 150, 151–152, 154, 158, 161, 163
Correlogram, 154, 160
CPC Morphing algorithm (CMORPH), 20, 23–36, 65, 71, 76, 77, 78, 80, 86, 91, 106, 107, 109, 111, 120, 121, 128, 129, 133–140, 155, 156, 157, 158, 159, 160, 171, 172, 173, 178, 179, 180, 181, 182, 186, 193–203, 207, 208, 209, 210, 211, 212, 213, 214, 271, 272, 273, 276, 290, 291
    precipitation, 271, 276
Crop Condition Data Retrieval and Evaluation (CADRE), 271, 272, 273, 274, 275, 279, 284, 285, 289
Crop yield, 267–291, 296
Curve number, 250, 251, 303

D

Data and Information Services Center (NASA) (DISC), 272
Decision Support System for Agrotechnology Transfer (DSSAT), 279
Defense Meteorological Satellite Program (DMSP), 4, 6, 25, 71, 72, 81, 86, 87, 102, 108, 112, 113, 282
Digital Soil Map of the World (DSMW), 280
Discriminant analysis, 40, 42
Drift, 4, 87
Drought, 91, 146, 283, 285, 287, 289, 296, 297, 298, 299, 300, 301, 302, 304, 318

E

Earth Observing System (NASA) (EOS), 6, 25, 86, 198, 284, 288
Earth System Science Interdisciplinary Center (UMD) (ESSIC), 283, 285
East African Rift Valley, 195
Economic Research Service (USDA) (ERS), 285
Ensemble, 94, 147, 161, 162, 216, 217, 219, 220, 222, 223, 225, 230, 231–233, 234–239, 240, 284
Ensemble Kalman filter (EnKF), 284
ENVISAT, 285, 286
Equitable threat score (ETS), 93, 94, 102, 134, 135, 136, 138
Error
    metric, 129, 134, 140, 147, 148, 149, 150–154, 155–159, 161, 162–165, 166, 224
    propagation, 147, 215–225
Ethiopia, 194, 195, 196, 197, 200, 201, 202, 205–214, 298, 301, 303, 304, 317
European Centre for Medium-Range Weather Forecast (ECMWF), 278
European Operational Meteorological (MetOp) satellite, 4, 6, 19, 86, 87
Evaporation, 70, 98, 99, 182, 185, 213, 235, 255, 304
Evapotranspiration, 270, 275, 280, 289, 300, 301, 309, 310

F

False alarm, 93, 94, 131, 132, 134, 136, 137, 138, 140, 148, 149, 150, 151, 157, 159, 163, 172, 199, 200, 235, 239, 240
Famine, 295–318
Famine Early Warning System (FEWS), 295–318
Flash flood, 40, 106, 127, 152
Food and Agriculture Organization (UN) (FAO), 249, 258, 280, 296, 309, 311
Food security, 95, 296, 297, 298, 299, 300, 301–305, 313, 317, 318
Foreign Agricultural Service (USDA) (FAS), 269, 270, 271, 272, 273, 275, 278–283, 284, 285, 290

Index 323

Fractions skill score (FSS), 130, 131, 132, 133, 135, 136, 137, 138, 140, 141
Frequency bias score (FBS), 130, 131, 138, 139, 199, 200, 201

G

Gauge data, 5, 8, 11, 32, 34, 35, 72, 76, 78, 79, 172, 178, 197–198, 200, 201, 202, 208, 254, 256, 288
Gaussian filter, 74
Geographical information system (GIS), 64, 271, 273, 298
Geospatial Data Abstraction Library (GDAL), 61
Geospatial Data Analysis Corporation (GDA Corp), 289
Geostationary Meteorological Satellites (GMS), 24, 86, 274
Geosynchronous (GEO), 5, 6, 28, 50, 54, 65, 85, 86, 87–88, 89, 107, 113, 114, 194
Geosynchronous Operational Environmental Satellite (GOES), 5, 7, 24, 26, 40, 41, 42, 43, 44, 45, 46, 52, 65, 70, 72, 86, 112, 113, 114, 274
Global Agriculture Monitoring System (USDA/FAS and NASA) (GLAM), 270, 283–290
Global flood monitoring, 250, 261
Global Forecast System (GFS), 173, 262
Global Precipitation Climatology Project (GPCP), 4, 5, 7, 10, 11, 70, 71, 106, 107, 154, 305, 306, 307, 308, 309, 314, 318
Global Precipitation Measurement (GPM), 20, 106, 107, 115, 216, 246, 247, 262, 291
Global Precipitation Mission (GPM), 81, 92, 93, 95, 97, 98, 100, 102, 103, 115, 116, 216, 246, 291
Global Reservoir and Lake Monitor (GRLM), 285, 286, 287
Global Runoff Data Center (GRDC), 255, 256, 257, 260
Global Satellite Map of Precipitation (GSMap), 20, 24, 36, 105–122
Global Telecommunication System (GTS), 195, 196, 197, 198, 200, 201, 271, 272, 273, 274, 278, 281, 300
Goddard Profiling Algorithm (GPROF), 5, 6, 7, 14, 15, 19, 27
Goddard Space Flight Center (NASA) (GSFC), 113, 283, 284, 285
GOES Multispectral Rainfall Algorithm (GMSRA), 70

Ground validation (GV), 91–95, 148, 149, 150, 154, 155, 156, 165, 205

H

Hanssen and Kuipers discriminant (HK), 131, 132, 136, 140
HEC-HMS, 250, 260
Heidke Skill Score (HSS), 42, 181, 199, 200, 201
High Resolution Precipitation Products (HRPP), 85–91, 92, 93, 94, 95, 102, 107, 120, 127–142, 155, 165, 247
Hurricane
  Ernesto, 56–58, 65
  Katrina, 49, 58–60, 65, 246
Hydro-Estimator, 45, 46, 80, 171, 173, 176, 178, 179, 180, 181, 182
Hydrological Sciences Branch (NASA/GSFC) (HSB), 283, 284, 285
Hydrologic Engineering Center (HEC), 250, 260
Hydrologic model (HyMOD), 216, 217, 218, 219, 222, 225, 229, 230, 231, 234–235, 239, 247, 249, 250, 297, 300, 317
Hydrologic Modeling System (HMS), 250, 260
Hydrology, 9, 46, 72, 85–103, 146, 147, 150, 165, 216, 248, 267–291, 298
Hydrology and Remote Sensing Laboratory (USDA/ARS) (HRSL), 281, 283, 284, 285

I

Ice water path (IWP), 7, 26
Infrared (IR), 5, 6, 7, 8, 9, 10, 11, 12, 13, 14, 19, 20, 23, 24–25, 26, 28, 29, 33, 39, 40, 41, 42, 43, 50, 51, 52, 54, 57, 65, 69–81, 86, 87, 88, 90, 106, 107, 108, 109, 110, 111, 112, 113, 115, 116, 162, 171, 172, 193, 207, 230, 233, 272, 275, 276, 281, 288, 306
Intermittency, 145, 148
International Geoshpere Biosphere Programme (IGBP), 62, 63
International Precipitation Working Group (IPWG), 13, 78, 79, 80, 92, 114, 120, 172, 173–178
International Production Assessment Division (USDA/FAS) (IPAD), 270, 284, 285
Inter Tropical Convergence Zone (ITCZ), 117, 195, 196, 197, 206
Italy, 153, 154, 162, 163, 164, 165, 215–225

J

Japan, 16, 24, 106, 107, 112, 114, 115, 116, 117, 119, 120
Joint Agricultural Weather Facility (JAWF), 269, 274
Joint Research Center (JRC), 278

K

Kalman filter, 24, 33–34, 105–122, 284
Kansas, 173, 178
Kenya, 296, 298, 300, 306, 307, 308, 312, 313, 314–317, 318
Kuril Islands, 117

L

Lagrangian, 20, 65, 148
Lake Victoria, 287, 288
Land Information System (LIS), 95, 97, 103, 261
Landslide monitoring, 146
Land surface model (LSM), 95–102, 103, 231, 261
Large Area Crop Inventory Experiment (LACIE), 271, 278, 279
Latency, 12, 29, 30, 39–47
Leaf River Basin, 233, 235
Look up tables (LUT), 25, 43, 87, 88, 89, 114, 151
Low earth orbit (LEO), 4, 23, 40, 50, 54, 85, 86, 87–88, 89, 92, 93, 106, 107, 108, 116

M

Mesoscale model, 7
Meteorological satellites (Meteosat), 24, 26, 65, 81, 86, 112, 113, 114, 274
MHS–NESDIS, 5
Microwave, 4, 5, 6, 7, 8–11, 12, 13, 14, 19, 20, 23, 25–26, 28, 30, 33, 40, 41, 43, 50, 65, 69–81, 85, 86, 87, 91, 93, 99, 101, 102, 103, 106, 107, 108, 109, 110, 111, 112, 113, 114, 115, 118, 119, 120, 121, 128, 171, 173, 193, 198, 207, 211, 216, 218, 230, 246, 272, 274, 275, 276, 281, 282, 284–285, 306
Microwave Humidity Sounders (MHS), 6, 7, 9, 10, 19, 86, 87, 93
Modeling, 56, 65, 87, 95, 105, 146, 147, 148–150, 151, 154, 155, 158, 159, 161, 165, 166, 217, 230, 245–262, 297, 298, 300, 305, 310, 317
Moderate Resolution Imaging Spectroradiometer (MODIS), 62, 249, 251, 281, 287, 288, 289, 290, 291

Monitoring Agriculture through Remote Sensing techniques (MARS), 278
Monte Carlo, 239
Mosaic, 35, 36, 71, 95, 96, 97, 98, 99, 100, 101, 102, 103
Motion vector, 28, 65, 80, 107, 108, 109, 171, 198
Multi-event contingency table (MECT), 131, 133, 136, 137, 140, 141
Multi-Function Transport Satellite (MTSAT), 24, 86, 112, 113, 114

N

Nash-Sutcliffe efficiency, 135
National Aeronautics and Space Administration (NASA), 20, 72, 198, 271, 272, 275–276, 283, 298
National Agricultural Statistics Service (NASS) (USDA), 268
National Centers for Environmental Prediction (NCEP), 29, 32, 34, 35, 173, 178, 182, 234
National Environmental Satellite Data and Information Service (NESDIS), 5, 7, 9, 26, 28, 43, 45
National Oceanic and Atmospheric Administration (NOAA), 6, 7, 8, 19, 25, 26, 28, 34, 45, 46, 49, 86, 87, 93, 102, 112, 113, 194, 198, 271, 272, 276, 277, 298
National Polar-orbiting Operational Environmental Satellite System (NPOESS), 288
National Weather Service (NOAA) (NWS), 7, 34, 49, 232, 233, 234, 240, 271, 272, 273, 277, 278, 281, 291
Natural Resources Conservation Service (NRCS), 250–253, 254, 256, 260
Navy Operational Global Atmospheric Prediction System (NOGAPS), 77, 78, 79, 88, 90, 91, 92, 94, 173, 178
Navy Research Laboratory (NRL), 77, 78, 79, 80, 85–103, 106, 107, 128, 171, 178, 179, 180, 181
NCEP GFS, 178, 182
Near Infrared (NIR), 288
Neighborhood verification, 127–142
Neighborhood verification score, 135, 137
Next-Generation Radar (NEXRAD), 155, 156, 157, 158, 233, 271, 272, 273, 277, 278
Noah, 95, 96, 97, 98, 99, 100, 101, 102, 103
Normalized Difference Vegetation Index (NDVI), 270, 287, 289, 290, 297, 300, 301

Normalized Difference Water Index (NDWI), 289
North America Land Data Assimilation System (NLDAS), 97
North American Monsoon Experiment (NAME), 173, 181–188, 189, 213
Northern hemisphere, 17, 75, 77, 105, 255
NRL-Blended, 171, 178, 179, 180, 181
Numerical weather prediction (NWP), 77, 86, 88, 90–91, 92, 93, 94, 127, 172, 190, 309

O

Ocean, 7, 8, 9, 10, 13, 17, 18, 19, 25, 26, 27, 69, 77, 91, 92, 113, 117, 119, 197, 230, 301, 302
Oceanic rainfall, 33
Office of Global Analysis (OGA), 270, 275, 276, 284, 285
Oklahoma, 147, 155, 156, 157, 158, 159, 172, 173, 178, 213
Optimization, 157, 222, 235
Orographic rain, 194, 213, 305

P

Parameter-elevation Regressions on Independent Slopes Model (PRISM), 277
Passive Microwave-Infrared (PMIR), 69–81
Passive microwave (PMW), 7, 23, 24, 25–26, 27, 28, 29, 30, 33, 34, 35, 69–81, 85, 86, 87, 89, 91, 92, 94, 102, 107, 113, 115, 128, 171, 172, 173, 182, 183, 188, 193, 198, 216, 218, 230, 272, 275, 276, 281, 282, 284–285, 290
Pilot Evaluation of the High Resolution Precipitation Products (PEHRPP), 92, 107, 120
Posina, 218, 219, 220, 222, 223, 224, 225
Potential evapotranspiration (PET), 270, 280, 281, 300, 301, 309, 311, 312, 317
Precipitation Estimation from Remotely Sensed Information using Artificial Neural Networks (PERSIANN), 24, 35, 49–65, 71, 86, 106, 107, 128, 155, 156, 157, 158, 159, 160, 171, 172, 173, 178, 179, 180, 181, 182, 183, 186, 207, 208, 209, 210, 211, 212, 213, 214, 231, 232, 233, 234, 240
Precipitation Processing System (PPS), 7, 11, 12
Precipitation radar (PR), 7, 41, 81, 87, 115, 116, 199, 291

Probability of detection (POD), 93, 94, 102, 131, 140, 149, 150, 151, 156, 157, 158, 161, 162, 163, 165, 172, 199, 200, 201, 235, 238, 239, 240
Probability matching, 8, 10, 55, 199
Production, Supply and Distribution (USDA/FAS) (PS&D/PSD), 269, 270

Q

QMORPH, 29, 30

R

Radar, 7, 14, 28, 30, 31, 34, 36, 39, 41, 45, 46, 54, 56, 57, 58, 59, 60, 65, 69, 73, 79, 81, 85, 86, 87, 95, 107, 114, 115, 116, 119, 120, 121, 128, 129, 130, 133, 134, 135, 137, 146, 155, 169, 170, 171, 172, 176, 178, 179, 182, 185, 199, 217, 218, 219, 220, 222, 223, 225, 230, 231, 232, 233, 234, 235, 239, 247, 271, 272, 277, 284, 285–287, 288, 291
Rainfall estimation (RFE), 27, 50, 51, 54, 65, 70, 193, 194, 198, 201, 230, 248, 261, 262, 301
Real Time Nephanalysis Cloud Model (RTNEPH), 274
Red River, 96, 98
Remote sensing, 50, 61, 113, 146, 165, 218, 219, 247, 251, 254, 261, 271, 278, 283, 285, 289, 298, 300
Retrieval, 14, 26, 39–47, 50, 65, 70, 71, 80, 102, 103, 112, 114, 121, 128, 148, 149, 150, 151, 152, 154, 157, 158, 160, 161, 162, 163, 166, 194, 195, 198, 201, 207, 216, 217, 219, 225, 229–240, 271, 276, 284, 285, 290, 291, 300, 308
RFE1, 201, 307
RFE2, 201, 304, 305, 306, 308, 309, 312, 314, 318
River forecast centers (NOAA/NWS) (RFC), 231, 234, 277
River network, 248, 249, 254

S

Sacramento, 234
Satellite precipitation estimates (SPE), 34, 95, 99, 128, 129, 188, 205, 208, 272, 277, 291
Self-Calibrating Multivariate Precipitation Retrieval (SCaMPR), 39–47
Self-Organizing Feature Map (SOFM), 53, 54, 55
Shortwave length Infrared (SWIR), 288

Shuttle Radar Topography Mission (SRTM), 247, 248, 249
Skewness, 152
Soil moisture, 95, 96, 97, 99, 102, 103, 146, 231, 234, 246, 250, 252, 262, 267–291, 300, 301, 302
Soil water content (SWC), 97–102, 103, 234
Somali jet, 206
Southern Great Plains, 178
Special Sensor Microwave/Imager (SSM/I), 5, 6, 7, 8, 10, 19, 25, 26, 40, 41, 43, 71, 72, 74, 81, 86, 108, 112, 113, 114, 198, 272, 274, 281, 282
Spinning Enhanced Visible/InfraRed Imager (SEVIRI), 43, 44, 65
SSM/I–GPROF, 5
Station gauge (SG), 272, 275
Stream flow, 127, 146, 225, 234, 251

T
TAMSAT, 202
Television Infrared Observation Satellite (TIROS), 20
Temporal correlation, 148, 149, 154, 175
Thermal infrared (TIR), 193, 194, 198, 199, 201
TIROS Operational Vertical Sounder (TOVS), 20
TMI–GPROF, 5
TOGA-COARE, 75
Total column precipitable water (TPW), 88
Triangulated irregular network Integrated Basin Simulator, (tRIBS), 217, 222
TRMM combined instrument (TCI), 5, 8, 10, 11, 12, 14, 199
TRMM On-line Visualizations and Analysis System (TOVAS), 13
TRMM Microwave Imager (TMI), 6, 25, 41, 71, 87, 108, 115, 198
TRMM Multi-Satellite Precipitation Analysis-Real-Time (TMPA-RT), 18, 106, 128, 129, 137, 138, 139, 140, 271, 272, 273, 275–276, 278
TRMM Multi-Satellite Precipitation Analysis (TMPA), 3–20, 24, 36, 86, 106, 107, 133–140, 171, 172, 178, 179, 180, 181, 194, 203, 248, 255, 256, 258, 259, 260, 261, 262, 271, 272, 273, 275–276, 278, 301, 305
TRMM precipitation product merged with PMW, IR, and SG (3B42(V6)), 15, 16, 153, 162, 164

TRMM precipitation radar (PR), 5, 7, 8, 41, 43, 73, 87, 88, 94, 99, 115, 116, 199, 251, 291
TRMM Real-Time precipitation product (PMW and IR merged) (3B42RT), 246
Tropical Rainfall Measuring Mission (TRMM), 3–20, 25, 27, 41, 43, 71, 73, 86, 87, 88, 92, 94, 106, 108, 112, 113, 115, 116, 129, 171, 193–203, 217, 218, 245–262, 271, 275, 300, 301
Two dimensional histograms, 16, 18
Two dimensional satellite rainfall error model (SREM2D), 146, 147–165, 166, 217, 219, 220, 222, 223, 231
Typhoon Nabi, 117, 118

U
Uncertainty, 13, 14, 15, 110, 146, 148, 161, 165, 205, 216, 217, 225, 230, 231, 235, 240, 248, 249, 251, 261
UNESCO, 50
UNESCO's International Hydrological Program (IHP), 61
University of Maryland (UMD), 114, 283, 285
Upscaling, 129, 130, 132, 133, 134, 136, 137, 138, 139, 141
US Agency for International Development (USAID), 258–259, 262, 296, 298, 299, 301, 303, 317
US Army Corps of Engineers (USACE), 250, 254, 260
US Department of Agriculture (USDA), 251, 267–291, 298, 300

V
Validation, 13, 40, 72, 78, 79, 80, 91–95, 99, 102, 103, 114, 119–121, 128, 130, 148, 165, 169, 171, 172, 173–181, 188, 194, 195, 196, 197, 198, 199, 200, 203, 205, 275, 284, 285, 307, 308
Vegetation indices (VI), 270, 273, 283, 289, 290, 291, 297, 300, 309
Version 5 (MM5), 7, 26
Visible Infrared Imager Radiometer Suite (NPOESS/VIIRS), 288
Visible (VIS), 39, 40, 41, 43, 45, 50, 65, 70, 86, 87, 116, 288

W
Warm rain, 7, 14, 75, 194, 201
Warm season, 15, 34, 171, 172, 173, 175, 176, 178, 179, 180, 181–188, 190
Water balance model (WBM), 255, 300, 301, 303–305

Water And Development Information for Arid Lands-A global Network (G-WADI), 50, 62, 65
Water Requirement Satisfaction Index (WRSI), 297, 301, 309–313, 314–317, 318
WeatherPredict Consulting (Surface Wetness) (WPC), 281
Weather Predict Consulting (WPC), 280, 282
Weather Surveillance Radar, 1988, Doppler (WSR-88D), 233, 277
Weighted Anomaly Standardized Precipitation (WASP), 272

World Agricultural Outlook Board (WAOB), 269, 270
World Agriculture Supply and Demand Estimates (WASDE), 269, 270, 278
World Food Programme (WFP), 299
World Meteorological Organization (WMO), 271, 272, 273, 274, 275, 278, 280, 281, 291

Z

Zenith angle, 7, 10, 44, 46
Zimbabwe, 200, 201, 298, 300